Peter Marland is a former RN Weapon Engineer Officer with sea jobs in HM Ships *[...]* and *Euryalus*, and postings ashore in research and in procurement. He subsequently *[...]* Operational Analyst, and has contributed a number of articles to *Warship* on postwar *[...]* weapons and electronics. He is a Chartered Engineer, with post-graduate qualifications in Project Management and in teaching.

Stephen McLaughlin retired in 2017 after working for 35 years as a librarian at the San Francisco Public Library. In addition to contributing regularly to *Warship*, he is the author of *Russian and Soviet Battleships* (US Naval Institute Press, 2003) and is co-editor of an annotated version of the controversial *Naval Staff Appreciation of Jutland* (Seaforth Publishing, 2016).

Kathrin Milanovich has been researching the history of the Imperial Japanese Navy and has contributed a number of recent articles to *Warship*.

Jean Moulin did his national service in the Marine Nationale, and worked as an IT technician until his retirement in 2000. He has written widely on the French Navy, and is the author of several monographs on warships of the interwar period. Jean is the co-author, with John Jordan, of *French Cruisers 1922–1956* and *French Destroyers 1922–1956* (Seaforth 2013 and 2015). His book on the 'stealth' frigates of the *La Fayette* class was published by Lela Presse in 2018.

Dirk Nottelmann is a marine engineer by profession, and is currently working for the German shipping administration. He has contributed for several years to *Warship International* and various German magazines, as well as being author of *Die Brandenburg-Klasse* (Mittler, 2002) and co-author of *Halbmond und Kaiseradler* (Mittler, 1999).

Ian Sturton, who retired recently, is a regular contributor of articles and illustrations to naval publications, including *Warship*, *Warship International* and *Jane's Fighting Ships*. He edited *Conway's Battleships: The Definitive Visual Reference to the World's All-Big-Gun Ships* (Revised and Expanded Edition, 2008).

Sergei Trubitsyn is a history teacher by profession and currently works in a family business. He is particularly interested in the war at sea during the 20th century and the service history of the ships of the same period. The author of twenty books and many articles, his work has been published in the Russian magazines *Gangut* and *Arsenal Collection*, and in the Polish magazine *Okrety Wojenne*.

Sergei Evgenevich Vinogradov trained as a civil engineer, graduating from the Moscow Engineering and Construction Institute in 1982. He has published many works on the capital ships of the Imperial Russian Navy, including books on the predreadnought battleship *Slava*, the dreadnought *Imperatritsa Mariia*, and never-built designs for Russian battleships with 16in guns. He currently works at the Central Museum of the Armed Forces in Moscow.

Michael Whitby is Senior Naval Historian at the Directorate of History and Heritage, National Defence Headquarters, Ottawa. He has published widely on Second World War and Cold War naval history, including co-authoring the official histories of the Royal Canadian Navy in the Second World War, and editing *Commanding Canadians: The Second World War Diaries of Commander AFC Layard, RN* (UBC Press, 2007).

WARSHIP 2021

WARSHIP 2021

Editor: **John Jordan**

Assistant Editor: **Stephen Dent**

OSPREY
PUBLISHING

Title pages: The German light cruiser *Regensburg* after rearmament with 15cm guns at the Imperial Dockyard, Kiel. The new configuration of the artillery on the forecastle is particularly evident in this picture, as are the two cut-outs amidships for the four 50cm torpedo tubes. *Regensburg*, together with her sister *Graudenz*, features prominently in Dirk Nottelmann's article on the German *Kleiner Kreuzer* published on pages 44–60. (Author's collection)

OSPREY PUBLISHING
Bloomsbury Publishing Plc
Kemp House, Chawley Park, Cumnor Hill, Oxford OX2 9PH, UK
29 Earlsfort Terrace, Dublin 2, Ireland
1385 Broadway, 5th Floor, New York, NY 10018, USA
E-mail: info@ospreypublishing.com
www.ospreypublishing.com

OSPREY is a trademark of Osprey Publishing Ltd

First published in Great Britain in 2021

A catalogue record for this book is available from the British Library.

ISBN: HB 9781472847799; eBook 9781472847782; ePDF 9781472847775; XML 9781472847768

21 22 23 24 25   10 9 8 7 6 5 4 3 2 1

Typeset by Stephen Dent
Printed and bound in India by Replika Press Private Ltd.

FSC MIX
Paper from responsible sources
FSC® C016779
www.fsc.org

Osprey Publishing supports the Woodland Trust, the UK's leading woodland conservation charity.

To find out more about our authors and books visit www.ospreypublishing.com. Here you will find extracts, author interviews, details of forthcoming events and the option to sign up for our newsletter.

CONTENTS

EDITORIAL

Most of our feature articles deal with warships that were not only designed for a particular navy, but were completed and served for perhaps 25–35 years before being discarded, thereby providing ample opportunity for an evaluation of their qualities and of their suitability for adaptation to changed tactical or strategic imperatives. However, some designs never proceeded beyond the drawing board, while others failed to materialise because of circumstances: the outbreak of war and a consequent shift in priorities, or even invasion and its impact on military-industrial infrastructure. We are then restricted to the original plans, and at best photographs of hulls partially assembled on the slipway. Evaluation of a design for a ship or submarine that remained uncompleted is necessarily speculative and incomplete.

Despite this, 'paper' designs, or 'might-have-beens', have an enduring fascination for naval enthusiasts, and the internet has a number of forums dedicated to them. And for this year's annual we have opted to lead with an article on one such design: that of the Soviet 'super-battleships' of the *Sovetskii Soiuz* class, by regular contributor Stephen McLaughlin. The design process for these massive ships was enormously complex, and the result was a ship 'designed by committee', with every additional requirement being met by increasing size and weight to the extent that, had *Sovetskii Soiuz* been completed, she would have approached the Japanese *Yamato* in her overall dimensions and displacement. Considering the embryonic state of Soviet industrial infrastructure during the 1930s, this was a hugely ambitious project, which in the end came to nought – though not before four hulls had been laid down.

The West got its first glimpses of the *Sovetskii Soiuz* design in the late 1980s thanks to Gorbachev's policy of *glasnost*. However, the story told in these early articles was incomplete, and the drawings and model photographs published were not those of the final design. More detailed technical accounts, accompanied by plans of the many variants drawn up during the protracted design process, have recently been published in Russia, and Stephen has used these to shed a new light on these giant but never-completed battleships.

The Imperial Japanese Navy is represented this year by two unusual designs: the submarines of the *I 15* class and the seaplane carriers *Chitose* and *Chiyoda*. The *I 15* class was the culmination of Japanese interwar development of the large, fast 'fleet' submarine designed to operate at long range against the American main body during its transit across the Pacific, with the aim of reducing its numerical strength to a level at which it could be defeated by Japan's own battle fleet. Equipped with a catapult and collapsible floatplane, these submarines could operate independently or in packs. However, as Kathrin Milanovich's article makes clear, the Pacific War failed to develop in the way that the IJN anticipated, and these large, unhandy boats failed to make a contribution commensurate with the enormous investment of resources involved, achieving only a handful of spectacular but isolated successes. By contrast, the IJN's *Chitose* and *Chiyoda* suffered from being designed for multiple potential roles, only one of which could be performed at any given time. Hans Lengerer outlines the complex requirements and design process of these ships, which served first as seaplane carriers, then as mother ships for midget submarines (*Chiyoda* only) and finally, following a lengthy reconstruction, as light fleet carriers.

Coverage of the period 1930–45 is completed by two contrasting articles. Michele Cosentino follows up his feature in last year's annual on the Italian Navy's interwar carrier projects with an article detailing the redesign and reconstruction of the liner *Roma* as the aircraft carrier *Aquila*. The article takes advantage of material only recently unearthed from the Italian archives, and includes many plans and photographs which have not previously been published. Michael Whitby, on the other hand, addresses issues that are primarily tactical and strategic with an account of the employment of Royal Navy 'Fleet' destroyers to form the backbone of fast support groups at the height of the battle against the U-boats in the North Atlantic in the spring of 1943. The key quality of the fleet destroyers was their high speed, which enabled them to move quickly to support whichever convoy faced an imminent threat; however, contrary to what has been stated elsewhere, these newly-completed ships were not always equipped with the latest centimetric radars or HF/DF.

Elsewhere in the annual, Dirk Nottelmann continues his ground-breaking series of articles on the German *Kleiner Kreuzer*, this time covering the turbine-powered ships that accompanied the High Sea Fleet and the Scouting Groups during the First World War, together with the cruisers that were still on the stocks or fitting out when the war ended in November 1918. There is a particular focus on the adoption of the side belt, turbine development, and the rearmament of the older German light cruisers with the 15cm gun. To complete our coverage of the pre-WWI period, Philippe Caresse continues his series on the French battleships of the *Flotte d'échantillons* with a study of the battleship *Carnot*, arguably the least successful of the series. This year also sees the publication of a major new article by Ian Sturton on the Royal Yacht *Victoria and Albert (III)*, the design of which suffered from constant interventions by prominent members of the British royal family, resulting in weight miscalculations that led to the ship all but capsizing when floated out of Pembroke Dock, and the end of the otherwise unblemished career of the Director of Naval Construction, Sir William White.

The Soviet 'Flotilla Leader' *Leningrad* in her last years in combat service, photographed from the English Wharf in the Neva River in 1956. These ships will be the subject of a detailed study by Przemysław Budzbon and Jan Radziemski to be published in *Warship* 2022. (Przemysław Budzbon collection)

Conrad Waters has taken a well-earned break from his series on modern warship developments this year, leaving the field to the Editor and Jean Moulin, who have collaborated on an article on the French frigates of the *La Fayette* class that essentially ushered in the era of 'stealth' technology. Constructed of steel and glass-reinforced plastic (GRP), these ships featured completely smooth outer surfaces, with the hull and superstructures angled in such a way as to minimise the electronic signature. They inspired a new generation of frigates and destroyers, but proved difficult to modernise, in part due to funding issues but also because of the need to retain the integrity of the 'stealth' design. Finally, the Editor follows his short drawing feature on the French postwar 'fleet escorts' of the T 47 type with a similar feature on their successors of the T 53 class, which were laid down during the mid-1950s and were intended to accompany France's new carriers, *Clemenceau* and *Foch*.

Next year's annual will include a major study of the Soviet Flotilla Leaders of the *Leningrad* class by Przemysław Budzbon and Jan Radziemski, an article by Stephen McLaughlin on Soviet battleship design 1939–41 (Projects 23*bis*, 23NU and 24), an account of Operation 'Tunnel' and the loss of HMS *Charybdis* by Michael Whitby, and a feature by Kathrin Milanovich on the design of the IJN fleet carriers *Soryu* and *Hiryu*. Dirk Nottelmann will return with an article on the German cruiser gunboats of the late 19th century, and Peter Marland will continue his series on postwar developments in the Royal Navy with a study of radar.

John Jordan
March 2021

Clive Taylor (1947–2020)

We are sorry to report the death of Clive Taylor, who contributed the photographs of Royal Navy warships and the drafts of the accompanying captions for *Warship 2020*.

Clive began taking photographs of warships as a hobby in 1965, and after his marriage to Sue in 1970 they used the byline C & S Taylor for the specialist warship photographic agency they ran jointly until their retirement in 1995. During this period Clive and Sue were regular visitors to the Round Tower and the Walls at Portsmouth at weekends and during holidays, photographing many of the RN and foreign warships that entered or left harbour. The regular Round Tower photographers of the day formed something of a clique that tended to keep information about upcoming movements to themselves. As a young man who had only recently moved down from London and purchased his first SLR camera, I found Clive and Sue refreshingly open; Clive's enthusiasm was infectious, and he had a wealth of amusing stories.

Clive used an unusual medium-format camera: a British-made KL Biggs GP in an aerial body with a 180mm Zeiss Sonnar lens and a Linhof 6cm x 9cm roll film back. The camera had a fixed focal range, and was generally set up on a tripod on the wall close to the Round Tower, which for Clive provided the ideal angle for a vessel of frigate/destroyer size. He would sometimes charter a Cessna light aircraft for aerial photography; on other occasions, when offered the use of a helicopter by the Royal Navy, he would strap himself to the frame of the open door.

Clive and Sue were to become the foremost warship photographers of the day, contributing photos to international naval magazines and to prominent reference source books such as *Jane's Fighting Ships* and *Combat Fleets*. They also supplied photographs to several international intelligence agencies, including those of the USA, Germany and Japan.

The recently-published *Cold War Fleet* (Osprey Publishing, 2019), a compilation of the C & S Taylor photographs of Royal Navy warships taken between 1966 and 1991, will be a fitting legacy.

STALIN'S SUPER-BATTLESHIPS: THE *SOVETSKII SOIUZ* CLASS

The West got its first glimpses of the *Sovetskii Soiuz* design in the late 1980s thanks to Gorbachev's policy of *glasnost*. But the story told in these early articles was incomplete, and the drawings and model photographs published were not those of the final design. **Stephen McLaughlin** takes advantage of recent Russian publications to describe and illustrate the design of these giant but never-completed battleships.

Iosef Vissarionovich Stalin, chairman of the Communist Party and de facto head of the Soviet government, wanted a battle fleet. Why he wanted it is an open question, but by the mid-1930s the international situation certainly looked threatening. The economies of the capitalist nations were still mired in depression; there were ongoing clashes with Japan in the Far East, and Hitler's virulently anti-Communist Nazi party was firmly in power in Germany. To a dedicated Communist – and Stalin was indeed a dedicated Communist – all of this signalled the long-anticipated 'crisis of capitalism'. In his 'Report to the XVII Congress' of the Communist Party, delivered on 26 January 1934, Stalin predicted that this crisis would mean war, either between capitalist nations – in which case the Soviet Union had to be prepared to intervene in support of the proletarian revolutions that such wars might engender – or directly against the Soviet Union.[1] In either case a strong navy would be vital, especially if intervention were necessary in areas that the Red Army could not reach overland. Perhaps we need look no further than this for his motive in initiating a massive naval construction programme.

As early as 11 July 1931 Stalin had declared to his inner circle: 'It is necessary to start the construction of a great navy with small ships. It cannot be ruled out that in five years we will build battleships'.[2] But over the next few years the anti-battleship 'Young School' was allowed to dominate naval policy, and the survivors of the tsarist navy, the chief supporters of battleship construction, were viciously purged. Once Stalin believed that the USSR's economy and industry had reached a point where they could sustain a programme of battleship construction – almost exactly five years after his 1931 prediction – it was the turn of the Young School to be eliminated.

The entire machinery of the Soviet state would eventually be drawn into the battleship programme. At the top of that vast bureaucracy was the Council of People's Commissars, chaired by Stalin's long-time crony Vyacheslav Molotov and composed of the commissars (heads) of the various commissariats (ministries) – all Stalin's picked men. Another important body was the Council of Labour and Defence (from April 1937 simply the Defence Committee), also chaired by Molotov; it was essentially a subset of the Council of Commissars, with many of the same men serving in both. Stalin was a night owl, so the meetings began in the evening and lasted into the early morning hours; after an issue had been discussed it was common practice for Molotov to turn to Stalin and ask, 'How do we decide?'[3] All major decisions thus came from Stalin. Through these organs Stalin would approve ship characteristics and resolve technical disputes. One naval constructor noted:

> All of us … were greatly impressed by the detailed and deep examination of the complex tactical and engineering issues that took place at such a high-level meeting, and in particular the active and knowledgeable participation … of I V Stalin.[4]

The two principal institutions involved in designing the *Sovetskii Soiuz* class, the Navy and the shipbuilding industry, would both undergo administrative changes in the latter half of the 1930s. The Navy was initially part of the Red Army before being elevated to its own commissariat on 31 December 1937, which gave it direct representation on the Council of People's Commissars. Shipbuilding and ship design were concentrated in the Commissariat of Heavy Industry until December 1936, when the newly formed Commissariat of the Defence Industry took over that responsibility. In January 1939 this unwieldy organisation was broken up, and a Commissariat of the Shipbuilding Industry was created. In order to avoid confusion, throughout this article reference will be made simply to 'the Navy' and 'the shipbuilding industry'.

Designing ships was a back-and-forth process. The Navy would explore potential warship designs through its Scientific-Research Institute for Warship Construction (*Nauchno-isledovatelskii institut voennogo korablestroeniia*, NIVK), which included a small cadre of naval constructors. They produced what amounted to feasibility studies to determine what was broadly possible. The result of NIVK's work would be a set of Tactical-Technical Requirements (*Taktiko-tekhnicheskii zadanie*, or TTZ) that would be sent to the shipbuilding industry, where they would be given a design (*proekt*) number and assigned to a construction bureau or, in the case of major warships, to two bureaux. Each would produce a sketch design (*eskiznyi proekt*), and the Navy would select the one it considered superior. Inevitably, the Navy would demand modifications to the chosen sketch design, and the winning design bureau would set to work on a technical design (*tekhnicheskii proekt*), which was equivalent to a contract design in the US Navy or a detailed design in the Royal Navy. In the case of the *Sovetskii Soiuz*, there were several successive technical designs as the Navy and the chosen design bureau sought to reconcile expectations with the realities of weights and hydrodynamics.

Each of these major steps in the design process had to be approved by the Government, usually by the Defence Committee. In effect, instead of being a direct negotiation between the Navy and the shipbuilding industry, the process became one of advocacy, with each institution arguing for its point of view before the highest officials in the nation – an analogy would find the merits of different design choices being judged by the British cabinet.

Designing the *Sovetskii Soiuz* Class

In the autumn of 1935 the Naval Academy (the Soviet naval war college) was ordered by the head of the Navy, V M Orlov, to study 'large armoured artillery ships' – the term 'battleship' was avoided, but that would soon change. The impulse behind this certainly came from Stalin, for Orlov would never have dared to launch such an initiative without his approval. The Academy's report, dated 8 September 1935, concluded that the Soviet Union required two types:

– Battleship 'A': Large ships for the Pacific and Northern theatres, capable of engaging any foreign ships in service or likely to be built in the near future; and
– Battleship 'B': Smaller ships for 'enclosed seas' – that is, the Baltic and Black Sea – whose primary purpose would be the destruction of Washington Treaty cruisers and German *Panzerschiffe*.

Work on determining the initial characteristics for the two types moved forward on two fronts, at the Navy's NIVK and the shipbuilding industry's Central Construction Bureau for Special Shipbuilding No 1 (*Tsentralnyi konstruktorskii biuro spetsialnogo sudostroeniia No 1*, or TsKBS-1). The result was a series of 'pre-sketch' (*predeskiznyi*, that is, preliminary) designs for a range of battleships. Most of these studies were completely unrealistic, but over the course of several months the Navy's more extravagant hopes were brought down to earth.

After reviewing all of these preliminary designs Orlov

ordered that development be concentrated on a 55,000-ton ship with nine 406mm (16in) guns and a 450mm (17.5in) armour belt for Battleship A, and a 35,000-ton ship with the same main battery but a 350mm (14in) belt for Battleship B. The corresponding TTZ were issued to NIVK, TsKBS-1, and Construction Bureau No 4 (*Konstruktorskoe biuro* 4, or KB-4) – based at the Ordzhonikidze (Baltic) Works – on 21 February 1936.

However, international events soon forced a major change in priorities. On 25 March 1936 the Second London Naval Treaty was signed by France, Great Britain, and the United States. It confirmed the 35,000-ton displacement limit established by the Washington Treaty, but reduced the maximum gun calibre to 14in (356mm). Although the Soviet Union was not a signatory, at this time it was pursuing a policy of 'collective security' in an attempt to curb German and Japanese aggression, so in May 1936 negotiations began with Great Britain for a bilateral naval agreement that would bring the USSR into the treaty system. As a result, Battleship A was downgraded to a 35,000-ton ship, while Battleship B became a 26,000-ton ship armed with 305mm (12in) guns. Some work continued on 55,000-ton designs, still regarded as necessary for the Pacific theatre to counter the powerful Japanese fleet.

In June 1936 TsKBS-1 and KB-4 submitted their 35,000-ton designs. KB-4's strongly resembled HMS *Nelson*, with all the main-battery turrets forward of the superstructure, while TsKBS-1's proposal featured two turrets forward and one aft. The Navy preferred the latter arrangement for tactical reasons, and it would thereafter be used in all the design work. However, this phase made it very clear that the inexperienced Soviet designers would need a great deal of assistance if real progress were to be made: many features of these 35,000-ton designs were vague, almost cartoonish.

Italian Input

The most promising source of such help was Fascist Italy. Italian technical assistance – including Italian constructors working in TsKBS-1 – had played an important role in the design of the *Kirov* (Project 26) class cruisers and the construction of other warships, so when the Italian firm of Ansaldo offered to draw up battleship designs in March 1936, the Soviets eagerly accepted. The head of TsKBS-1, V L Bzhezinskii, was sent to Italy to work out the details. On 10 June the head of the shipbuilding industry, R A Muklevich, telegraphed instructions and encouragement to Bzhezinskii: 'Try to get the *Littorio* design. What is needed is a battleship of 35,000 tons.[5] This is the main task. The next design should be this: displacement 26,000 tons …'.[6] Muklevich also wanted Italian designs for large cruisers and an 'armoured scout'.

The fruits of Ansaldo's labours arrived in Moscow in July 1936. The large battleship design, designated UP.41, was for a ship with a standard displacement of 42,000 tons that bore a strong resemblance to the *Littorio* class. This was no accident, as it was in fact a design worked out for the Italian Navy by Umberto Pugliese's department in 1934–35 as a potential follow-on to that class.[7] That possibility had been set aside in favour of building a second pair of *Littorios*, so UP.41 was available for trading to the Soviets; apparently the only modification made to it was the replacement of the triple 152mm secondary turrets by triple 180mm turrets as in the *Kirov* class cruisers – ironically, a calibre the Soviets had never intended to use in their battleships.

The Ansaldo materials arrived at an opportune moment, for the Navy had begun to doubt that the desired characteristics in speed, protection, and firepower could be achieved in a 35,000-ton ship. UP.41, produced by the highly-regarded Italian designers, probably confirmed this view. Although the Soviets never considered building a battleship to the Italian design, it did have a considerable influence on the next stage of the design work, as can be seen if it is compared to the TTZ

Table 1: Battleship A – Preliminary Designs, 1936

Column	A	B	C	D
	UP.41	TTZ[1]	TsKBS-1	KB-4
Date	14 Jul 1936	3 Aug 1936	Oct 1936	Oct 1936
Displacement:				
standard	42,000 tonnes	41,500 tonnes	44,900 tonnes	45,930 tonnes
trials	45,470 tonnes	N/S[2]	46,700 tonnes	51,030 tonnes
full load	—[3]	N/S	50,000 tonnes	—
Length:				
overall	252m	N/S	251m	255m
waterline	245m	N/S	245m	—
Beam:				
maximum	35.5m	N/S	33.6m	33.5m
waterline		N/S	33.1m	31.5m
Draft (maximum)	9.4m (normal)	9.5m (damaged)	9.8m	9.5m
Block coefficient	—	N/S	0.595	0.635
GM (standard)	—	N/S	3.0m	2.2m
Armament:				
main guns	9 x 406mm (3 x III)	9 x 406mm (3 x III)	9 x 406mm (3 x III)	9 x 406mm (3 x III)
secondary guns	12 x 180mm (4 x III)	12 x 152mm (6 x II)	12 x 152mm (6 x II)	12 x 152mm (6 x II)
HA guns	24 x 100mm (12 x II)	12 x 100mm (6 x II)	12 x 100mm (6 x II)	12 x 100mm (6 x II)
light AA	48 x 45mm (12 x IV)	40 x 37mm (10 x IV)	40 x 37mm (10 x IV)	40 x 37mm (10 x IV)
Catapults	1	2	2	2
Aircraft	4	4	4	4
Protection:				
main belt	370mm at 6°	380mm	380mm at 5°	380mm at 5°
upper belt	150mm at 6°	250mm	200mm at 5°	220mm at 5°
belt forward	—	200mm	125–90mm	200mm
forecastle deck	55mm	30mm	30mm	30mm
upper deck	10mm	50mm	50mm	50mm
main deck	25mm, 100mm[4]	135mm machinery 180mm magazines	135mm machinery 180mm magazines	135mm machinery 180mm magazines
turrets	400mm faces	425mm faces	425mm faces	420mm faces
barbettes	350mm	425mm	425mm	420mm
Underwater protection:				
system	Ansaldo	N/S	Ansaldo	Pugliese or Ansaldo
depth	9.8m (amidships)	not less than 7.5m	7.3–7.5m	—
Machinery	4-shaft turbines 4 x 45,000shp	N/S	3-shaft turbines 3 x 60,000shp	3-shaft turbines 3 x 66,700shp
Speed	32 knots	30 knots	30 knots	30 knots
Range	6,300nm/20kts	6–8,000nm/14kts	7,000nm/14kts	7,000nm/14 kts
Complement	1,600	1,373	—	1,360

Notes:

[1] TTZ = *Taktiko-tekhnicheskii zadanie* (Tactical-Technical Requirements).

[2] N/S = Not Specified

[3] — = Data not available

[4] UP.41 deck armour in four layers; the middle deck was 100mm.

Sources: Vasil'ev, 21, 52; Garzke & Dulin, *Battleships: Allied Battleships*, 310.

worked out by NIVK and approved by the government in August 1936 (see Table 1, cols A & B). According to the major historian of the *Sovetskii Soiuz* class, the displacement of 41,500 tons was 'based on Italian experience and [NIVK's] own previous studies .... The authors of the TTZ were well aware that it was almost impossible to establish visually such a small deviation from the treaty limit'.[8] So, like the Italians and the Germans, the Soviets hoped to pass off battleships of more than 40,000 tons as 35,000-ton ships.

The Early Soviet Designs

At this point the TTZ were handed over to the shipbuilding industry, which designated the work as Project 23 and assigned TsKBS-1 and KB-4 to work out sketch designs for both battleships A and B. The constructors faced a difficult task: experimental work on underwater protection had barely been started, and all the guns and mountings, as well as fire control equipment and much

else, were in the earliest phases of development, so many weights could only be estimated. As a result, both sketch designs were very incomplete when they were examined by the Council of People's Commissars, with Stalin in attendance, on 2 November 1936.

Both designs for Battleship A exceeded the specified displacement by a considerable margin; moreover, in an attempt to minimise the excess tonnage, the two design groups had shaved some armour thicknesses (Table 1, cols C & D). The designs shared a number of features, including the general hull form (inclined sides, bulges, a long forecastle deck), which had been developed by NIVK in early June. Another common element was the use of a three-shaft machinery plant, based on the belief that a four-shaft plant would be heavier and make it difficult to provide a full-fledged side protection system in the after part of the citadel. And both designs had very densely packed citadels, squeezing some of the 100mm AA guns out onto the quarterdeck.

But there were also significant differences: TsKBS-1's

Table 2: **Project 23 Technical Designs, 1937–1938**

Column	A	B	C	D	E
	TTZ	Variant I	Revised TTZ	Variant III	Variant IIIu
Date	26 Nov 1936	June 1937	July/Aug 1937	Nov 1937	Feb 1938
Displacement:					
standard	46–47,000 tonnes	48,415 tonnes	55–57,000 tonnes	57,850 tonnes	58,500 tonnes
full load	N/S	—	N/S	63,900 tonnes	64,460 tonnes
Length:					
overall	N/S	—	N/S	271m	271m
waterline	N/S	237m	N/S	260m	260m
Beam:					
maximum	N/S	36.5m	N/S	38.5m	38.9m
waterline	N/S	—	N/S	36m	36.4m
Draft (maximum)	10.0m	10.3m normal	10.25m	10.35m	10.4m
Armament:					
main guns	9 x 406mm (3 x III)	9 x 406mm (3 x III)	9 x 406mm (3 x III)	9 x 406mm (3 x III)	9 x 406mm (3 x III)
secondary guns	12 x 152mm (6 x II)	12 x 152mm (4 x III)	12 x 152mm (6 x II)	12 x 152mm (6 x II)	12 x 152mm (6 x II)
HA guns	12 x 100mm (6 x II)	12 x 100mm (6 x II)	12 x 100mm (6 x II)	12 x 100mm (6 x II)	8 x 100mm (4 x II)
light AA	40 x 37mm (10 x IV)	40 x 37mm (10 x IV)	40 x 37mm (10 x IV)	40 x 37mm (10 x IV)	32 x 37mm (8 x IV)
Catapults	2	2	2	2	2
Aircraft	4	4	4	4	4
Protection:					
main belt	380mm	380mm	380mm	380mm	380mm
upper belt	220mm	200mm	220mm	220mm	220mm
belt forward	220mm	200mm	220mm	220mm	220mm
forecastle deck	30mm	25mm	20mm	20mm	25mm
upper deck	50mm	50mm	50mm	50mm	140mm
middle deck	180mm	135mm/180mm	180mm	180mm	60mm
lower deck fwd	135mm	200mm	220mm	220mm	220mm
Depth u/w protection	7.5m	not less than 7.0m	7.0–7.5m	7.1–8.1m	7.0–8.2m
Machinery:					
natural draught	N/S	—	N/S	3 x 67,000shp	3 x 67,000shp
forced draught	N/S	3 x 75,000shp	N/S	3 x 77,000shp	3 x 77,000shp
Speed:					
natural draught	—	—	29 knots	28.7 knots	28.5 knots
forced draught	30 knots	30 knots	30 knots	29.5 knots	29.5 knots
Range	6–8,000nm	—	6–8,000nm/14kts	6,150–6,750nm/ 14 knots	6,480nm/14kts

Source: Vasil'ev, 51

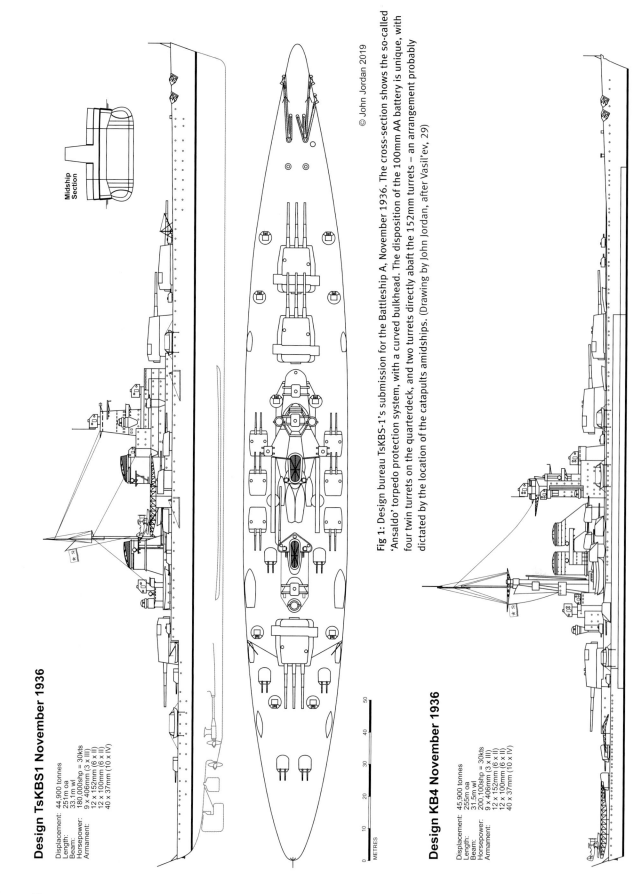

Design TsKBS1 November 1936

Displacement: 44,900 tonnes
Length: 251m oa
Beam: 33.1m wl
Horsepower: 180,000shp = 30kts
Armament: 9 x 406mm (3 x III)
               12 x 152mm (6 x II)
               12 x 100mm (6 x II)
               40 x 37mm (10 x IV)

Midship
Section

© John Jordan 2019

Fig 1: Design bureau TsKBS-1's submission for the Battleship A, November 1936. The cross-section shows the so-called 'Ansaldo' torpedo protection system, with a curved bulkhead. The disposition of the 100mm AA battery is unique, with four twin turrets on the quarterdeck, and two turrets directly abaft the 152mm turrets – an arrangement probably dictated by the location of the catapults amidships. (Drawing by John Jordan, after Vasil'ev, 29)

0   10   20   30   40   50
METRES

Design KB4 November 1936

Displacement: 45,900 tonnes
Length: 255m oa
Beam: 31.5m wl
Horsepower: 200,100shp = 30kts
Armament: 9 x 406mm (3 x III)
               12 x 152mm (6 x II)
               12 x 100mm (6 x II)
               40 x 37mm (10 x IV)

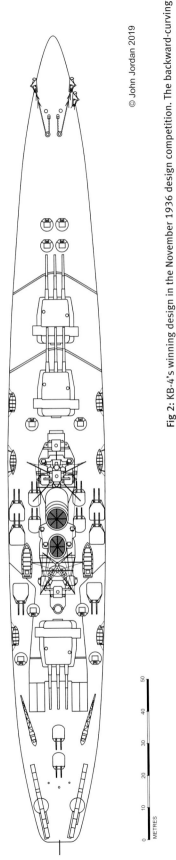

© John Jordan 2019

Fig 2: KB-4's winning design in the November 1936 design competition. The backward-curving forward funnel was due to the placement of the forward boiler room directly below the forward superstructure. Several elements of the design, including location of the catapults and a pair of 100mm turrets on the quarterdeck, as well as the hangar arrangements, would be a feature of subsequent KB-4 designs. (Drawing by John Jordan, after Vasil'ev, 22)

0 10 20 30 40 50
METRES

submission (Fig 1) had a finer hull form and used Wagner boilers, while the KB-4 version (Figs 2, 3, 4) used three-drum boilers and had a fuller hull form that required more horsepower to achieve the specified speed. A distinctive feature of KB-4's design was the curved forward funnel, similar to that of several Japanese battleships; this was a consequence of the extreme compression of the vitals, which led to the forward boiler room being placed directly under the conning tower, so that its uptakes had to be trunked back. The magazines for the secondary and AA batteries were jammed into narrow compartments outboard of the two forward boiler rooms and an equally narrow compartment sandwiched between the turbines of the wing shafts. The long, unencumbered forecastle made it possible to maintain fine lines despite the full-depth side protection system abreast the forward magazines. KB-4 offered two different versions of underwater protection: the Pugliese system and the 'Ansaldo' system, a multi-bulkhead type that featured a concave main bulkhead that had been used in UP.41. TsKBS-4's design also used the Ansaldo system.

On the whole, both the Navy and shipbuilding industry preferred the KB-4 design, which had been worked out in a number of variants and was considered more developed than TsKBS-1's submission. The latter's design for Battleship B, on the other hand, was considered superior, and from this point onward the two battleship designs would be handled by separate design bureaux and take very different development paths. The smaller battleship was declared 'wrecked' (that is, sabotaged by supposed enemies of the state) in August 1937 and was replaced by Project 64, armed with 356mm guns; by early 1938 it had grown to 48,000 tons, at which point it was cancelled in favour of building more Project 23 ships.

In theory KB-4's design, having won the competition, should have led directly to a technical design that would form the basis for the construction of the ship but, despite being judged the better design, the Navy was far from satisfied. It preferred siting the aviation facilities amidships, it disliked the long bow, which was unarmoured at its forward extremity, it wanted all the reduced armour thicknesses restored to the original specifications, and it demanded heavier deck protection: instead of 135mm with 180mm only over the magazines, it wanted a uniform deck of 180mm over the entire citadel. To accomplish all this the Navy was willing to boost the displacement to 46–47,000 tons. The result of these changes was a revised TTZ, issued on 26 November 1936 (Table 2, col A).

The Navy's new demands placed KB-4 in an impossible position: B G Chilikin, the bureau's chief constructor, argued that to fulfil all the requirements would require a ship of not less than 53,900 tons standard displacement. In an attempt to square the circle, KB-4's designers took the radical step of chopping eight meters off the forward hull. This eliminated the long unprotected bow, but it also led to a blunter lines, so the machinery power had to be boosted to 225,000shp to maintain the required 30-knot speed. KB-4 also concentrated the secondary battery in four triple turrets rather than the specified six twins.

Design KB4 November 1936: GA Plans

Inboard Profile

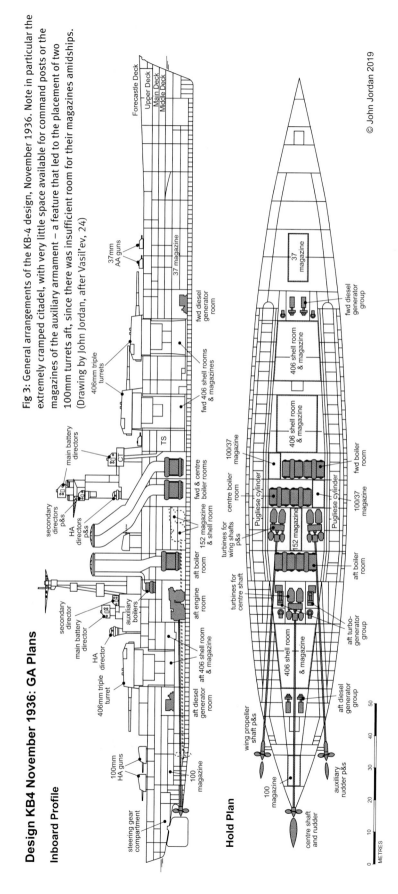

Fig 3: General arrangements of the KB-4 design, November 1936. Note in particular the extremely cramped citadel, with very little space available for command posts or the magazines of the auxiliary armament – a feature that led to the placement of two 100mm turrets aft, since there was insufficient room for their magazines amidships. (Drawing by John Jordan, after Vasil'ev, 24)

© John Jordan 2019

Hold Plan

Design KB4 November 1936: Protection

Profile

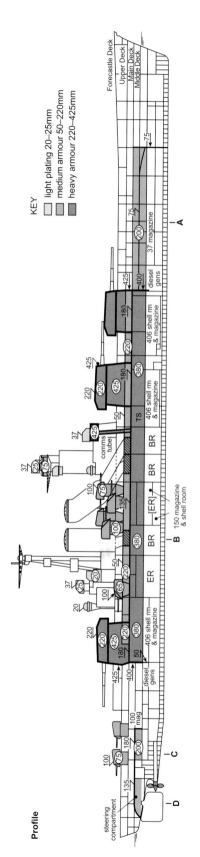

KEY
- light plating 20–25mm
- medium armour 50–220mm
- heavy armour 220–425mm

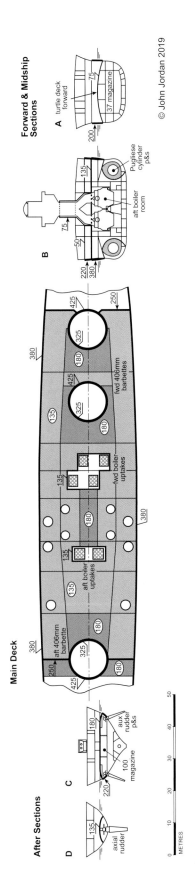

Forward & Midship Sections

A turtle deck forward

200

75

37 magazine

B

135

50

75

220

380

Pugliese cylinder p&s

aft boiler room

© John Jordan 2019

Main Deck

380

425

250

325

180

380

425

325

135

180

fwd 406mm barbettes

135

fwd boiler uptakes

180

135

380

135

aft boiler uptakes

380

aft 406mm barbette

250

425

325

180

After Sections

D

C

135

180

220

100 magazine

aux rudder p&s

axial rudder

METRES

0   10   20   30   40   50

Fig 4: The protection scheme of KB-4's design, November 1936. The sloping sides of the main battery barbettes are curious, as this would tend to improve the penetrating power of incoming shells by making their impact closer to the normal. Note also the relatively long unprotected bow, a feature the Navy wanted eliminated. (Drawing by John Jordan, after Vasil'ev, 26)

And once again it was forced to reduce armour thicknesses: the desired 180mm armour deck went by the board. Despite these measures the displacement came to more than 48,000 tons. Many of the other features of the competition design were retained, including the aviation facilities and the pair of 100mm turrets on the quarterdeck. Long, curved uptakes were still required for the forward boiler rooms (see Fig 5 and Table 2, col B).

In reviewing the progress of the design in April 1937, the Navy asked KB-4 to work out a parallel design limited to 47,000 tons and in strict accordance with the TTZ. This was designated Variant II, the larger design being Variant I. Both variants were presented to the Defence Committee on 4 July 1937, although it is unlikely that Variant II was a complete design. The shipbuilding industry's representatives came expecting a fight with the Navy over the need for more tonnage to meet the requirements, but in the interim the Navy's leadership had come to the same conclusion. Recent reports indicated that both Japan and Germany would soon begin construction of battleships displacing 50–52,000 tons, so the Navy was now more than willing to increase the size of Project 23. As a result, it was soon agreed by all parties that displacement should be 55–57,000 tons. A new TTZ was therefore issued (Table 2, col C), and work began on Variant III.

Two photographs of a model of Variant I. This model was probably used at the 4 July 1937 session of the Defence Committee and shows the arrangements of the superstructure and auxiliary armament. The crude forms of the 152mm and 100mm turrets probably reflect the fact that these mountings had yet to be designed. (Boris Lemachko collection)

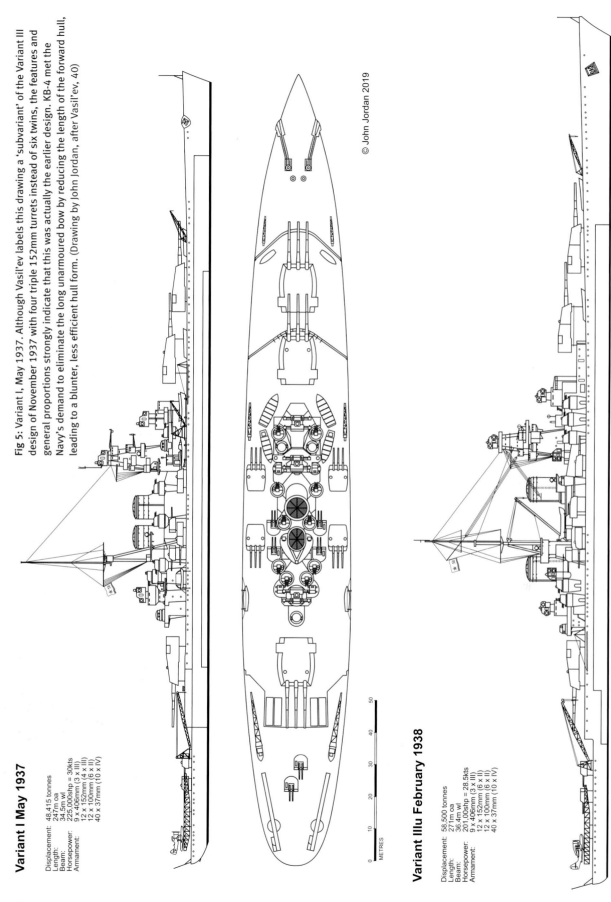

Fig 5: Variant I, May 1937. Although Vasil'ev labels this drawing a 'subvariant' of the Variant III design of November 1937 with four triple 152mm turrets instead of six twins, the features and general proportions strongly indicate that this was actually the earlier design. KB-4 met the Navy's demand to eliminate the long unarmoured bow by reducing the length of the forward hull, leading to a blunter, less efficient hull form. (Drawing by John Jordan, after Vasil'ev, 40)

© John Jordan 2019

Variant I May 1937

Displacement: 48,415 tonnes
Length: 247m oa
Beam: 34.5m wl
Horsepower: 225,000shp = 30kts
Armament: 9 x 406mm (3 x III)
12 x 152mm (4 x III)
12 x 100mm (6 x II)
40 x 37mm (10 x IV)

0 10 20 30 40 50
METRES

Variant IIIu February 1938

Displacement: 58,500 tonnes
Length: 271m oa
Beam: 36.4m wl
Horsepower: 201,00shp = 28.5kts
Armament: 9 x 406mm (3 x III)
12 x 152mm (6 x II)
12 x 100mm (6 x II)
40 x 37mm (10 x IV)

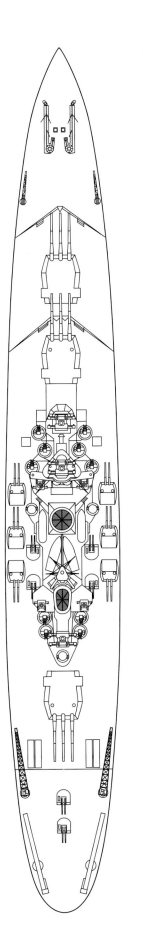

© John Jordan 2019

Fig 6: Variant IIIu of February 1938. Many features from KB-4's original November 1936 design are still in evidence – in particular the aviation arrangements and the two 100mm turrets on the quarterdeck. This version still has ten quad 37mm mountings, but two would soon be eliminated to reduce the 'congestion' of the superstructures. (Drawing by John Jordan, after Vasil'ev, 44)

Variants III & IIIu

It was at about this time that an orgy of destruction fuelled by Stalin's paranoia swept through the Navy and industry, further complicating the work of the design bureaux. Both Muklevich of the shipbuilding industry and Orlov of the Navy were arrested and executed; many naval officers and constructors – including the head of KB-4, L S Grauerman, and the Navy's chief technical adviser to the design bureau, E P Libel – were purged. This ongoing slaughter must be kept in mind when evaluating the progress of the battleship design; everyone in industry and the armed forces was working under the constant threat of death or imprisonment in the Gulags.

Nevertheless, the work did continue, and by November 1937 Variant III was completed (see Table 2, col D). Once again, however, the Navy, now headed by M V Viktorov, was unhappy with the design, especially the speed, which had fallen off by about half a knot despite a modest increase in horsepower. There were also concerns about the inordinate length of the wing shafts. This was the result of a rearrangement of the machinery;

A model of Variant IIIu. This 1:100-scale model was shown at the 27–28 February 1938 meeting of the Defence Committee. Several photographs of this model were published in the late 1980s and early '90s, leading to the erroneous conclusion that it represented the final configuration of the design. (Boris Lemachko collection)

the positions of the boiler and engines rooms had been reversed, so that the foremost machinery compartments were the engine rooms for the wing turbines. This unusual arrangement reduced the length of the boiler uptakes, but the shaft runs were thereby increased.

Moreover, the Navy now demanded a major change to the horizontal protection. Full-scale bombing trials carried out in the Black Sea demonstrated that 500kg high-explosive (HE) bombs would penetrate both the 40mm forecastle deck and the 50mm upper deck before exploding against the 180mm main deck; armour fragments would then spall off into the vital spaces below. The Navy therefore wanted to move the heavy armour deck to the upper deck and turn the main deck into a thin splinter deck. This meant the 220mm upper belt would have to be replaced with an upward extension of the 380mm main belt, increasing its height from 3.7m to 6.4m (a similar change had been made during the preliminary design of the *King George V* class[9]).

This and other changes were incorporated into the next design iteration, Variant IIIu ('u' for *uluchshennyi*, 'improved'; see Fig 6 and Table 2, col E), which was presented to the Defence Committee at a meeting on 27–28 February 1938. The shipbuilding industry's main spokesman was B G Chilikin, the lead constructor, while the Navy's representative was its deputy commissar, I S Isakov. There was broad agreement that the design could be taken as a basis for starting construction, but there were several sharp disagreements between the industry and the Navy over details. The most bitter argument revolved around industry's desire to eliminate the two 100mm turrets on the quarterdeck to reduce a trim by the stern; deleting these turrets along with their heavily-protected magazines was seen as the best way to resolve – or at least mitigate – the problem without reworking the entire design. Isakov argued for the retention of these turrets, but the Defence Committee eventually approved their deletion.

Another disappointment involved the Navy's hopes to move the aviation facilities amidships, which were stymied by the 'congestion' there – a problem that had already forced the Navy to agree to a reduction in the number of quad 37mm mounts from ten to eight. To some degree this congestion was caused by an excessive concern about the blast effects of the 406mm guns, which led to the 152mm, 100mm, and 37mm batteries being squeezed into a small area.

As usual, Stalin had the last word:

> I V Stalin summed things up; he stood up from his armchair and paced around the hall, calmly and persuasively expressing his thoughts on the questions raised during the course of the discussion. He proposed approving the technical design as presented ... Regarding the Navy's demand for the additional gun turrets, he proposed to the People's Commissar of the Navy that this question be reviewed in two months ...[10]

And that settled matters for the moment.

The Defence Committee had largely sided with the shipbuilding industry at this meeting, but only a week later, on 7 March, it approved a series of changes recommended by the Navy. The most consequential of these was a change to the deck armour; instead of a 140mm upper deck with a 60mm splinter deck below, the upper deck was to be 155mm and the splinter deck reduced to 50mm; other more or less minor adjustments to the armour protection made the change weight-neutral. Another question settled was the underwater protection: although tests to determine the most effective method were still in progress, the decision was made to use the Pugliese system in order to prevent further delays.

Two months later the Navy faced another disappointment when trials with a self-propelled 1:10-scale model of the Project 23 hull revealed that the maximum speed of the ship would be only 27.5 knots at normal draught, and 28.5 knots when the machinery was forced. The poor performance was blamed on the propeller design, which had been selected to avoid the cavitation problems that had plagued the destroyer leader *Leningrad*.

By now the displacement of the design stood at 57,576 long tons (58,500 tonnes), whereas the Anglo-Soviet naval agreement, which had finally been signed on 17 July 1937, specified a 35,000-ton limit. However, on 6 July 1938 a protocol to the agreement permitted the signatories to build battleships of up to 45,000 tons, reflecting the recent invocation of the 'escalator' clause of the Second London Naval Treaty. Two days later, the Soviet government informed Great Britain of the forthcoming keel-laying of the lead ship of the class, *Sovetskii Soiuz*; while the ship's armament was accurately reported, her displacement was given as only 44,190 tons to maintain the appearance of complying with the treaty's terms.[11] Although the final design was still far from ready, the ship was laid down at the Ordzhonikidze Works on 15 July.

The Final Design

The Defence Committee had decreed that the design was to be completed by 1 June 1938, but this had to be pushed back to 1 September. That deadline also passed; the design was finally submitted only on 13 October (see Fig 7 and Table 3), and even then it was still not complete – the fire control systems were still under development, as were the stabilised AA directors. KB-4, which had earlier been so concerned about the ship's trim by the stern that it had insisted on deleting the two 100mm turrets on the quarterdeck, now made things worse by shifting the after main-battery turret 6.39m towards the stern. This improved the arcs of fire of the main battery and allowed the secondary and heavy AA batteries to be better spaced out, but the change was not sufficient to make room for the aviation facilities amidships.

KB-4 also recommended a change in the belt armour. The bureau noted that, while the protection scheme was based on engaging a target 40 to 50 degrees off the bow, on these bearings the enemy's shells would strike the belt

Table 3: **Final Technical Design, 1938–1941**[1]

Displacement:	
standard	59,150 tonnes design; 60,190 tonnes actual (est)[2]
normal	62,155 tonnes design
full load	65,150 tonnes design; 67,370 tonnes actual (est)[2]
Length	269.4m oa, 260.0m wl
Beam	38.9m max, 36.4m wl
Draft	9.36m standard, 9.78m normal, 10.10m full load
Armament:	
main guns	9 x 406mm (3 x III); 100rpg
secondary guns	12 x 152mm (6 x II); 190rpg
HA guns	8 x 100mm (4 x II); 300rpg
light AA	40 x 37mm (10 x IV); 1800rpg
Catapults	1
Aircraft	4 x KOR-2
Protection:	
main belt	420–406–390–375–380mm at 5°
upper belt	180mm
belt forward	220mm
transverse b/hds	285–230mm fwd, 365–180mm aft
forecastle deck	25mm
upper deck	155mm
middle deck	50mm
lower deck fwd	100mm
406mm turrets	495mm faces, 230mm sides, 410mm backs, 230mm roof
406mm barbettes	425mm
152mm turrets	100mm faces, 65mm sides, 65mm backs, 100 roofs
152mm barbettes	100mm
100mm turrets	65mm faces, sides and backs, 100mm roofs
37mm turrets	25mm faces, sides, backs, roofs
U/w protection system	Pugliese & 'American', 7.0–8.2m deep
Machinery:	
boilers	six three-drum type
engines	three sets geared turbines
horsepower	201,000shp (231,000shp forced)
speed	28 knots (29 knots forced)
Endurance:	
normal load	oil 5,280 tonnes = 5,960/6,300nm at 14.5kts (winter/summer)
deep load	oil 6,440 tonnes = 7,260/7680nm at 14.5kts (winter/summer)
Complement	1,784

Notes:

[1] The 'final' technical design was approved in October 1938, but changes continued through 1941; the data here reflect the status of the design as of 1941.

[2] The estimated displacements were calculated by A M Vasil'ev based on changes to the design as of 1941.

Source: Vasil'ev, 51, 81, 87, 89, 90.

at angles closer and closer to the normal as the hull narrowed forward. In fact, calculations indicated that the 380mm belt could be penetrated out to 102 cables (20,400 yards) abreast the forward turret, whereas over the machinery it was proof down to 75 cables (15,000 yards). The bureau therefore proposed varying the thickness of the belt, decreasing it over the machinery to 375mm and increasing it over the forward magazines; over the after magazines the original thickness of 380mm was retained. The weight of armour would remain exactly the same, but it would provide roughly equal belt protection along its entire length.

KB-4 also raised an issue with the rudder arrangements. From its earliest stages the design had featured three rudders – a large centre rudder and two smaller side rudders, all positioned in the wake of one or other of the propellers. The centre rudder was farthest aft and its steering gear had only light splinter protection, whereas the gear for the two side rudders was under heavy armour. Tests with the 1:10 scale model of the battleship's hull showed that if the centre rudder became jammed, the side rudders could not overcome its effects, and the ship would turn in circles. KB-4 therefore recommended deleting the centre rudder; it was estimated this

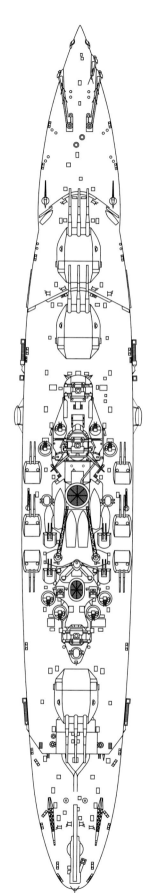

Fig 7: The technical design as of July 1939. The ship has almost reached its final form, the only changes still to be implemented being the restoration of two twin 100mm mountings – now in splinter shields rather than turrets – on the quarterdeck, and the addition of wooden planking on the forecastle deck. The rudder and propeller arrangements are based on written descriptions rather than drawings, and are therefore somewhat conjectural. (Drawing by John Jordan, after Vasil'ev, 47)

© John Jordan 2019

Sovetskii Soiuz: Final Design July 1939

Displacement: 59,150 tonnes
Length: 269.4m oa
Beam: 36.4m wl
Horsepower: 201,000shp = 27.5kts
Armament: 9 x 406mm (3 x III)
12 x 152mm (6 x II)
8 x 100mm (4 x II)
36 x 37mm (8 x IV)

would increase the ship's turning radius from 3.18 ship-lengths to 4, but subsequent trials showed that the actual turning radius at full speed would be 4.5 ship-lengths.

The Navy expressed its views on KB-4's design and proposals in a report dated 22 November 1938. It accepted the graduated belt armour scheme, but instead of arranging the plates horizontally in two strakes, as advocated by the shipbuilding industry, it wanted them arranged vertically in a single strake. Regarding the trim by the stern, it noted that the elimination of the centre rudder and its armour protection would help, and it was willing to trade one of the two catapults for two 100mm mountings in splinter shields (rather than enclosed mountings), with only fifty rounds per gun, stowed in ready-use lockers (rather than in a heavily-protected magazine). But the Navy was very unhappy that the ship's speed was now 27.5 knots at natural draught rather than 28.5 knots, and demanded that it be increased to at least 28 knots.

By the summer of 1939 KB-4 had incorporated most of these changes into the design, even managing to increase the speed by half a knot thanks to a new propeller design. But it adamantly refused to add the two 100mm mounts on the quarterdeck. This 'final' technical design was approved by the Defence Committee on 13 July 1939.

However, this was not the end of the design changes. The new Navy Commissar, Admiral N G Kuznetsov, who had been appointed in April 1939 – his three immediate predecessors had been purged in succession – was determined to get back those 100mm guns on the quarterdeck, and finally on 14 January 1941 the Defence Committee agreed. How they were to be sited, given the single cata-pult on the centreline, is not clear. Another change Kuznetsov insisted upon was that the weather decks have wood planking, which had not originally been included in order to save weight. However, Kuznetsov argued that it was essential for habitability, and in February 1941 his proposal was adopted at a cost of an additional 243 tons of weight; the planking was to be removed in wartime. Given this and other additions, the leading historian of

the design estimates that by 1941 the ship's displacement would have amounted to 60,190 tons standard, 67,370 tons full load.[12]

General Features

The hull was of riveted construction, and the framing was of the 'mixed' type: within the citadel it ran longitudinally, but at the bow and stern it was laid transversely. The frame spacing within the citadel was 1.42m (although it was reduced to 0.71m in places bearing heavy loads), while at the bow and stern it was 0.9m. The hull was divided into 33 main watertight compartments by trans-verse bulkheads. The metacentric height of the final design was 3.11m at standard displacement, 3.31m at normal load, and 3.49m at full load. For weights, see Table 4.

The hull form was very full, with a block coefficient of 0.657 – for comparison, *Yamato's* was 0.596. This resulted from the requirement to maintain a deep under-water protection system abreast the forward magazines, but it meant that very high power was required to achieve even modest speeds. This was further exacer-bated by the choice of a three-shaft propulsion plant, which led to very high shaft loading and a loss of propul-sive efficiency.

The cost per ship was estimated at 1.18 billion rubles, but one authority has suggested that the actual cost would have been 1.5–1.8 billion each, based on cost overruns on other Soviet ships of the period.[13]

Armament

The 406mm guns were one of the few unqualified successes in the design of the ships, with the first gun passing its trials at the proving ground near Leningrad in 1940 (for gun data, see Table 5; for the layout of the various control positions, see Fig 8). The guns were to have elevation limits of -2° to +45° with a fixed loading angle of +6 degrees. This led to a varying rate of fire, depending on the angle of elevation: 2.5 rounds per

Table 4: **Weights**

Design	KB-4	Variant IIIu	Tech Design	Final Tech Design
Date	Oct 1936	Feb 1938	Nov 1938	June 1939
Hull	14,969 tonnes	18,144 tonnes	19,385 tonnes	20,188 tonnes
Armour	17,165 tonnes	23,499 tonnes	23,306 tonnes	23,306 tonnes
Armament	8,121 tonnes	11,468 tonnes[1]	8,653 tonnes	8,547 tonnes
Ammunition	1,758 tonnes	—	1,953 tonnes	1,920 tonnes
Machinery	2,876 tonnes	3,517 tonnes	3,742 tonnes	3,727 tonnes
Crew & supplies	590 tonnes	659 tonnes	642 tonnes	642 tonnes
Margin	451 tonnes	1,200 tonnes	820 tonnes	820 tonnes
Standard displacement	45,930 tonnes	58,420 tonnes	58,500 tonnes	59,150 tonnes
Fuel, water, lubricants	5,100 tonnes	6,042 tonnes	6,000 tonnes	6,000 tonnes
Full load displacement	51,030 tonnes	64,460 tonnes	64,500 tonnes	65,150 tonnes

Note:
1 Includes ammunition.

Source: Vasil'ev, 52.

The 406mm gun on a test mounting at the proving grounds near Leningrad. Trials of the first gun were carried out 6 July to 2 October 1940 and were deemed successful. The trials gun was subsequently replaced by another, and from 29 August 1941 this gun was used to bombard German positions during the siege of Leningrad. A total of twelve barrels were manufactured by the Barrikada Works in Stalingrad. (Boris Lemachko collection)

minute (rpm) up to 14 degrees, slowing to 1.73rpm at higher angles. The turrets weighed 2,087 tonnes and rotated on 150 ball bearings; there were also 204 vertical rollers to receive the horizontal thrust when the guns fired. The ammunition outfit was 100 rounds per gun (rpg). The main battery was controlled by three command-rangefinder posts (*komandno-dalnomernye posty*, or KDP, equivalent to the Royal Navy's director

control tower), each of which was equipped with two 8-metre stereoscopic rangefinders – one for measuring the range to the target, the other for ranging on the ship's own shell splashes (scartometry). In addition, the main-battery turrets were each fitted with 12-metre rangefinders.

The 152mm/57 guns had elevation limits of -5° to +45°, with a fixed loading angle of +8 degrees. Rate of fire varied due to the fixed loading angle, with the

Table 5: **Guns of Project 23**

Mounting designation	MK-1	MK-4	MZ-14/B-54*	46-K
Gun designation	B-37	B-50	B-54	—
Calibre	406.4mm/50	152.4mm/57	100mm/56	37mm/67.5
Barrel length	20,720mm	8,950mm	5,795mm	2,510mm
Barrel weight	136,690kg	11,999kg	2,503kg	65kg
Barrel life	300 rounds	450 rounds	750 rounds	2,000–3,500 rounds
Elevation limits	−2° to +45°	−5° to +45°	−8° to +85°	−10° to +85°
Rate of fire	1.7–2.5rpm	4.8–7.5rpm	16rpm	160–180rpm
Crew (per mounting)	100	32	17/18[1]	13
Projectiles and performance				
Weight of projectile	1,108kg	55kg	15.8kg	0.76kg
Propellant charge	309.4kg	35kg	30kg[2]	1.5kg[2]
Muzzle velocity	830m/sec	950m/sec	900m/s	915m/sec
Maximum range	45,670m	30,210m	22,000m	5,000m

Notes:
[1] Turret/shielded mountings.
[2] Weight of cartridge.

Sources:

Vasil'ev, 62.

A V Platonov, S V Aprelev and D N Siniaev, *Sovetskie boevye korabli 1941–1945 gg*, vol IV, Tsitadel' (St Petersburg, 1997), 43, 45, 48, 50.

A B Shirokorad, *Entsiklopediia otechestvennoi artillerii*, Kharvest (Minsk, 2000), 937, 960, 976, 988.

Sovetskii Soiuz Final Technical Design: Superstructures

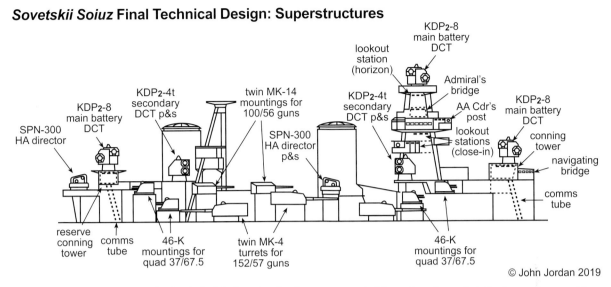

© John Jordan 2019

Fig 8: The superstructures of the final technical design, showing the arrangement of the various fire control directors and command posts. (Drawing by John Jordan, after Platonov, Aprelev and Siniaev, *Sovetskie boebye korabli*, vol. IV, 28)

maximum rate of 7.5rpm being possible at angles up to +16 degrees; it dropped to 4.8rpm between 30–40 degrees. The ammunition outfit was 190rpg. The guns were controlled by four KDPs, one on either side of the superstructure abreast the tower mast and a further pair abreast the after funnel; they were each equipped with two 4-metre rangefinders.

Initially the final technical design had a heavy AA battery of eight 100mm/56 guns in four twin turrets, two mounted on either beam above the secondary battery; at the Navy's insistence it was augmented in January 1941 by the addition of two twin mounts in open shields on the quarterdeck. All had elevation limits of -8° to +85° and a rate of fire of sixteen rounds per gun per minute. The ammunition outfit was 300rpg for the turrets, and 50rpg for the guns on the quarterdeck. They were controlled by three stabilised directors, one on either beam abreast the forward funnel and one on the centreline at the after end of the superstructure; each of the directors was equipped with a single 4-metre rangefinder.

The 1939 design called for thirty-two 37mm/67.5 guns mounted in quadruple mountings, all located on the superstructures. Elevation limits were -10° to +85°; rate of fire was 160–180 rounds per barrel per minute. The ammunition outfit was 1800rpg.

The 1939 design had a single catapult on the quarterdeck on the centreline, but the decision to restore the two aft 100mm mounts would probably have required some rearrangement of the aviation facilities. The hangars were located under the barrels of the after 406mm guns; in the earlier designs, the aircraft would have been lifted out of the hangars by cranes through hatches in the forecastle deck, but in the final design the hangar opened directly onto the quarterdeck, the aircraft being moved on rails. Two KOR-2 flying boats could be stowed in the hangars, with a third in the open just forward of the catapult and another carried on it.

There were four 900mm searchlights and four 450mm signalling lights.

Protection

The Soviet designers rejected the 'all-or-nothing' armour scheme, instead developing a protection system based on the following principles:

– citadel armour, protecting the main battery, its magazines, and the propulsion plant
– armour fore and aft of the citadel, ensuring the ship's ability to remain afloat
– local protection for various command posts, auxiliary armament, etc.

The result was an extremely complex system, employing no fewer than twenty-five different plate thicknesses; it can best be understood by studying Fig 9. The weight devoted to armour was necessarily enormous – at 23,370 tonnes, it slightly exceeded that of the *Yamato* class (23,262 tonnes). Yet because of the extensive protection devoted to the hull fore and aft, the armour of the citadel was somewhat lighter than that of its Japanese contemporary.

The protection was based on on the battleship being struck by a nominal 406mm armour-piercing shell weighing 1,000kg fired with a muzzle velocity of 880m/sec, and the armour scheme assumed an enemy located 40–50 degrees off the bow, perhaps reflecting the tactical situation during the approach to battle. The final design was expected to have an immune zone extending from 84–88 cables (16,800–17,600 yards) to 155 cables (31,000 yards). However, the manufacture of armour plate was a lost art in the Soviet Union, and as a result there were issues with the quality of the plates; had the ships been completed their level of protection

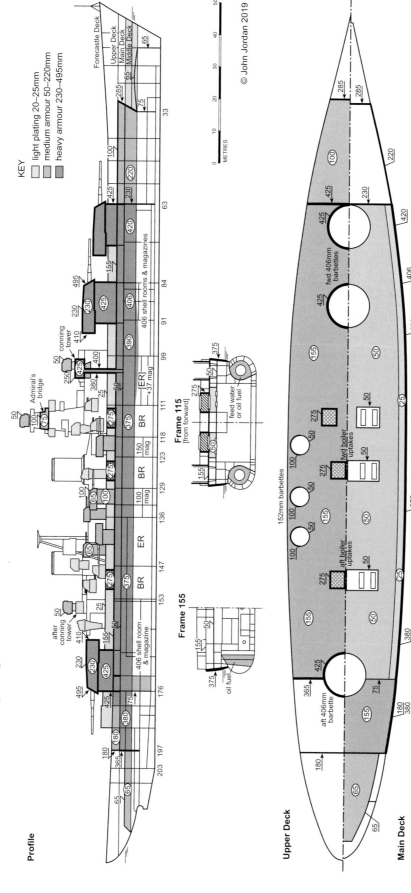

Fig 9: The protection scheme of the final technical design. Note the extensive use of light armour, the varying thicknesses of the main belt abreast the forward magazines, and the peculiar 'step down' of the main belt abaft turret No 3, intended to reduce the design's trim by the stern. The cross-sections illustrate the Pugliese system, used to protect most of the citadel, and the so-called 'American' multi-bulkhead system, used aft where the hull form was unsuited to the Pugliese system. (Drawing by John Jordan, after Vasil'ev, 67)

The hull of *Sovetskii Soiuz* under construction at the Orzhonikidze (Baltic) Works in Leningrad, December 1939. By this time many of the transverse bulkheads had been installed, and the framing for the Pugliese underwater protection can be seen clearly. (Boris Lemachko collection)

A close-up view of the framing for the Pugliese cylinder in *Sovetskii Soiuz*, taken on 9 December 1939. (Boris Lemachko collection)

might well have been less than the thicknesses employed would suggest.

The main belt extended from frame 62 to frame 170; it had a length of 148.4m, equivalent to 57 per cent of the waterline length, and was inclined at 5 degrees. For the majority of its length it was 6,275mm high, of which 1,770mm was to be immersed at full load displacement. However, for 32 metres (frames 88–110) in the vicinity of the magazines for turret no 2 the belt bulged downward, increasing its immersion at its lowest point to 2,670mm, or 900mm deeper than the rest of the belt. This was because model tests had revealed that at high speeds the trough of the bow wave would expose the hull below the armour belt in that region.[14]

As already noted, the specifications called for fighting an enemy on forward bearings. This led to a graduated scheme of belt armour to maintain equivalent protection along the length of the citadel. The foremost section of the belt was the thickest, 420mm, and extended for a length of 11 metres (frames 62–72); next came a 17.5-metre section of 406mm plating (frames 72–84.5), then a 390m section (frames 84.5–94.5). Over the central portion of the citadel the belt was 375mm (frames 94.5–157). It was then slightly increased, to 380mm, abreast the magazines for the after turret (frames 157–170). Above the belt the hull sides were protected by 25mm plating.

Over the citadel the forecastle deck was 25mm, the upper deck was 155mm, with the 50mm main deck below it acting as a splinter deck. The horizontal protection was also expected to resist 500kg HE bombs.

A peculiarity of the protection was a continuation of the belt aft by a 380mm section, the top of which was at main deck, rather than upper deck level. This was a weight-saving measure intended to help ameliorate the design's heavy trim by the stern. The side above this extension was 180mm. As a result of this odd arrangement the main armour deck had to drop to the middle deck, so there was a 365mm transverse bulkhead between the barbette for turret No 3 and the ship's sides. The after end of this lower extension was protected by a 180mm transverse bulkhead between the upper and main decks, and a 360mm bulkhead below the main deck. The forward end of the citadel was closed off by a 230mm transverse bulkhead.

Forward of the citadel the waterline belt was continued by a 220mm extension, which terminated at a steeply sloped (30°) transverse bulkhead, 285mm reducing to 250mm at the lower deck, and continued to the inner bottom by a vertical 75mm bulkhead. This belt extension was topped by a 100mm upper deck. Forward of the bulkhead there was a 20mm 'ice belt' extending to the bow, to enable the ships to transit the Northern Sea Route from the White Sea to the Pacific; it had a 65mm turtleback deck meeting its lower edge at the sides. Aft, there was a 65mm turtleback deck extending almost to the stern, but no ice belt.

The main-battery turrets were protected against 406mm AP shells on their faces and roofs, and against 406mm HE shells on their sides; their faces were 495mm, backs 410mm, the sides and roofs 230mm. There were 180mm shields that elevated with the guns to cover the embrasures, and 60mm internal bulkheads between the guns. The barbettes had 425mm armour, but they were nevertheless considered a weak point – their circular shape meant that there was always a small possibility that the vertical projection of a shell's line of flight could strike them at a 90-dgree angle.

The armour of the secondary and heavy AA guns was intended to protect them against 152mm HE shellfire. The 152mm turrets had 100mm faces and roofs, 65mm sides and backs, and 100mm barbettes (reduced to 65mm on their inboard sides). The 100mm enclosed mountings had 65mm faces, sides, and backs, with 100mm roofs and 100mm barbettes. The two 100mm open mountings on the quarterdeck had 25mm splinter shields, as did the 37mm guns.

The conning tower was rectangular in plan and had 425mm sides, which were also slightly inclined, probably at 5 degrees. There was a 'reserve command post' below the after main-battery director, but it was given only 20mm protection. The flag bridge high in the tower mast had 75mm sides and a 100mm roof, while the three main-battery directors had 50mm protection. The secondary and AA battery directors had 20mm splinter protection. Several small command posts in the tower mast also had 20mm armour.

The funnel uptakes were protected by 275mm plating between the upper and forecastle decks, with armour gratings over the openings in the deck. Below the upper deck the uptakes were protected by 50mm, while above the forecastle deck the uptakes of the forward boiler rooms were also given 50mm protection, to prevent the command spaces in the forward superstructure from being

'smoked out' in the event they were damaged in battle.

The underwater protection was designed to withstand torpedoes with a 750kg TNT-equivalent warhead; the ship was designed to be able to remain afloat with any five adjacent compartments flooded, or with the unarmoured above-water sides destroyed and three torpedo hits on one side, or two under-bottom torpedo hits. The depth of the side protection was 8.2 metres amidships, reducing to 7.5 metres aft and 7.1 metres forward. From frames 64 to 154 – an extent of 123 metres – the Pugliese side protection system was used. The cylinder had a diameter of 3.15m and its walls were 7mm thick; it was to be immersed in fuel oil or water. The main semi-circular holding bulkhead was 35mm, with a flat 10mm watertight bulkhead inboard of it. At the after end of the citadel, for a length of 33 metres (frames 154–170) the hull lines and propeller shaft runs made it difficult to fit the Pugliese system, therefore a so-called 'American' multi-bulkhead system was used, with four bulkheads each 20mm thick. A degree of bottom protection was provided by raising the magazines above the inner bottom.

Machinery

The machinery was arranged with alternating engine and boiler rooms, but their order was inverted compared to the usual layout (see Fig 10).[15] The forward engine rooms, which had a narrow centreline compartment between them, housed the turbine sets for the two wing shafts; directly abaft these was the forward boiler room with two boilers, then a space for 152mm and 100mm magazines, then the second boiler room with two more boilers, then another space for magazines and the fire control station for the 152mm battery, then the engine room for turbines powering the centre shaft, and finally the after boiler room with another two boilers. This arrangement resulted in very long shaft runs: 106m for the outboard shafts and 79m for the centre shaft.

There were three impulse/reaction turbine sets with single reduction gearing. They were each rated at 67,000shp (201,000shp total) with natural draught and 77,000shp (231,000shp total) with forced draught, the latter to be maintained on a two-hour trial. A contract was concluded with Brown Boveri & Cie, based in Baden, Switzerland, for the manufacture of four sets of turbines and technical assistance in establishing the production of further units at the Stalin Works in Kharkov. The turbines made by the Swiss firm were intended for the *Sovetskaia Rossiia*, while the first Kharkov units were earmarked for *Sovetskii Soiuz* and *Sovetskaia Ukraina*. Brown Boveri completed all four turbine sets, one of which was a working model to be used as a pattern for the manufacture of further units in the USSR. Three sets – including the working model – were delivered to Arkhangelsk in 1940–41, but the fourth set was still in Switzerland when the war broke out. As for the Kharkov turbines, none were ever completed.

There were six three-drum boilers in three boiler rooms, with a working pressure of 37kg/cm² (525psi) at

One of *Sovetskii Soiuz's* barbettes under construction. The roller path can be seen, with the large ball bearings already in place. Each of the battleship's turrets was to be supported by 150 of these bearings, which had a diameter of 206.2mm. (Boris Lemachko collection)

Sovetskii Soiuz Final Technical Design: Machinery

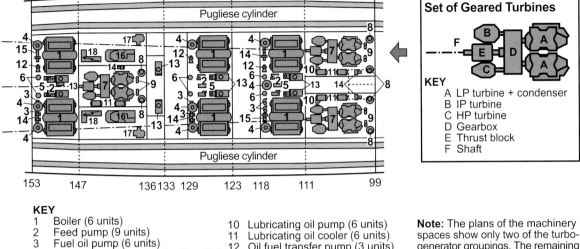

KEY
1 Boiler (6 units)
2 Feed pump (9 units)
3 Fuel oil pump (6 units)
4 Boiler room ventilation fan (9 units)
5 Feed water heater (6 units)
6 Fuel oil heater (6 units)
7 Set of geared turbines (3 units)
8 Circulating pump (6 units)
9 Condenser pump (6 units)
10 Lubricating oil pump (6 units)
11 Lubricating oil cooler (6 units)
12 Oil fuel transfer pump (3 units)
13 Feed water evaporator (8 units)
14 Fire pump (6 units)
15 Air compressor (2 units)
16 1300kW turbo-generator (4 units)
17 Auxiliary condenser (4 units)
18 Switchboard (4 units)

Note: The plans of the machinery spaces show only two of the turbo-generator groupings. The remaining two groupings were directly above the forward engine rooms, with the magazine and hoists for the 37mm guns between.

© John Jordan 2019

Fig 10: The machinery layout of the final technical design. The engine and boiler rooms were reversed from the usual arrangement, with the engine rooms forward of the boiler rooms. This eliminated the massive uptakes under the forward superstructure, but it made for very long shaft runs: 106m for the wing shafts. Note that only two of the four turbo-generators (16) and their associated auxiliary condensers (17) are shown, the forward units being located in a compartment above the forward engine rooms. (Drawing by John Jordan, after Vasil'ev, 74)

370°C (700°F). A prototype boiler was designed by KB-4 in 1936 and built by the Baltic Works, but due to various delays it was lit for the first time only on 21 July 1938, and even then the delays continued due to defects in workmanship and design. Moreover, it failed to achieve its intended steam output. The first boiler was supposed to undergo trials ashore in the fourth quarter of 1941, but the German invasion on 22 June 1941 put an end to these plans and the trials never took place.

The 1939 technical design specified a speed of 27.5 knots, 28.5 knots with forced draught; the improved propellers developed by NIVK were expected to add half a knot to the speed (28 knots, 29 knots forced). The propellers were three-bladed, the wing screws having a diameter of 5m and the centre propeller 4.8m. At full speed the wing shafts were to revolve at 247rpm, the centre shaft at 250rpm.

In addition to the main boiler plant, there were three auxiliary boilers for harbour service and other uses, located on the upper deck directly beneath the mainmast, outside the citadel.

Electrical power was provided by four main turbo-generators, each with a capacity of 1,300kW working at 230V; there were also four diesel generators, each rated at 650kW. Two of the turbo-generators were located in compartments outboard of the engine room for the centre shaft, the other two on the main deck just forward of no 1 boiler room. The diesel generators were located in pairs forward of turret no 1 and abaft turret no 3. The electrical system provided both direct and alternating current.

Construction and Fates

Construction of *Sovetskii Soiuz* began at Leningrad in July 1938; *Sovetskaia Ukraina* was laid down at Nikolaev three months later, and work on *Sovetskaia Belorussiia* and *Sovetskaia Rossiia* began in a new enclosed facility at Molotovsk, near Arkhangelsk, in December 1939 and July 1940 respectively (see Table 6). All four were scheduled to be delivered in 1945, while a fifth ship, *Sovetskaia Gruziia*, was to be laid down in 1941 at the Baltic Works.[16] However, work on the first four ships was plagued by delays in the issuing of working drawings, and in the delivery of steel and of armour. Moreover, much of the material proved to be defective. Further progress was hindered by the unusually severe winter of 1939–40, and air-raid blackouts during the Winter War with Finland (November 1939 – March 1940), which prevented night work. The planned delivery dates of the ships were repeatedly pushed off into the future.

The situation with *Sovetskaia Belorossiia* was even worse; in 1940 it was discovered that 70,000 of the rivets used were made of inferior steel. A decree of the Council of People's Commissars dated 19 October 1940 cancelled her construction outright; it also put an end to plans for laying down of the fifth ship.

Table 6: Construction Dates (Actual and Projected)

Name	Sovetskii Soiuz ('Soviet Union')	Sovetskaia Ukraina ('Soviet Ukraine')	Sovetskaia Belorussiia ('Soviet Belorussia')	Sovetskaia Rossiia ('Soviet Russia')
Hull number	S-299	S-352	S-101	S-102
Shipyard	No 189 Ordzhonikidze (Baltic) Works, Leningrad	No 198 Marti Yard, Nikolaev	No 402 Molotovsk	N. 402 Molotovsk
Laid down	15 July 1938	31 October 1938	21 December 1939	22 July 1940
Planned launch date	June 1943	June 1943	—	3rd quarter 1943
Planned delivery	1945	1945	—	1945
Constructors	Nikifor Fedorovich Muchkin	Pëtr Ivanovich Ermalaev	Sergei Pavlovich Kirilov	Sergei Pavlovich Kirilov
Date scrapping ordered	29 May 1948	24 March 1947	19 October 1940	24 March 1947

Note: It was planned to lay down a fifth ship, *Sovetskaia Gruziia* ('Soviet Georgia') in 1941 at the Baltic Works, but this was cancelled due to the outbreak of war with Germany.

Source: Vasil'ev, 115.

After the German invasion on 22 June 1941 a series of government decrees halted all work on the ships and their equipment. At that time *Sovetskii Soiuz* was 21.19 per cent complete, with 15,818 tonnes erected on the slipway; *Sovetskaia Ukraina* was 17.98 per cent complete, with 13,001 tonnes on the slipway, and *Sovetskaia Rossiia* was 5.04 per cent complete, with 2,125 tons erected. *Sovetskaia Ukraina* fell into German hands when Nikolaev was occupied in August 1941; she had been slightly damaged by the retreating Soviet forces, and during the war the Germans stripped about 30 per cent of her materials for other purposes. When the Germans withdrew they further damaged the hull with explosives. Both *Ukraina* and *Rossiia* were cancelled by a government decree dated 24 March 1947.

Sovetskii Soiuz lingered a little longer. During the siege of Leningrad some of her material had been used to in defensive works, but she was still judged 19.5 per cent complete at the end of the war. Some consideration was given to finishing her to an updated design, but the shipbuilding industry opposed the idea and the Navy was lukewarm at best; she was ordered broken up on 29 May 1948. A small section of her hull was launched in 1949 for use in underwater protection experiments, but Stalin's death in March 1953 put an end to all thoughts of capital ship construction, and it too was discarded.

Acknowledgements:

I owe a special debt to John Jordan, whose observations on the drawings proved invaluable in understanding the development of the design. Sergei Vinogradov provided much of the material used in this article, and as always my wife Jan Torbet applied her superb editorial skills to improve the text.

Sources:

This article is based principally on A M Vasil'ev, *Lineinye korabli tipa 'Sovetskii Soiuz'*, Galeia Print (St Petersburg, 2006). A number of other publications provided background or details, including:

Gribovskii, V Iu, 'The "Sovetskii Soiuz" Class Battleship', *Warship International*, vol XXX, no 2 (1993), 150–69.

Krasnov, V, 'Linkory tipa "Sovetskii Soiuz"', *Morskoi sbornik*, 1990, no 5, 59–63.

Markelov, V P, 'Sozdanie zashchity sovetskikh linkorov v predvoennye gody', *Tsitadel'*, no 11 (2004), 103–29.

Molodtsov, S V, 'Stalinskie linkory', *Briz*, 1996, no 9, 9–22.

Zubov, Boris Nikolaevich, *Zapiski korabel'nogo inzhenera: Razvitie nadvodnogo korablestroeniia v Sovetskom Soiuze*, Kliuch (Moscow, 1998).

Endnotes:

[1] I V Stalin, *Sochineniia, Gosudarstvennoe izdatel'stvo politicheskoi literatury* (Moscow, 1952), vol 13, 284–99.

[2] Vasil'ev, 4; Molodtsov, 9.

[3] Molodtsov, 17.

[4] Zubov, 126.

[5] Note that at this time the Soviets probably believed that *Littorio* was a 35,000-ton ship.

[6] Vasil'ev, 15.

[7] Enrico Cernuschi and Vincent P O'Hara, 'The Breakout Fleet: The Oceanic Programmes of the Regia Marina, 1934–1940', *Warship 2006*, 90–2. For details of UP.41, see William H Garzke Jr and Robert O Dulin Jr, *Battleships: Allied Battleships in World War II*, Naval Institute Press (Annapolis, 1990), 308–9.

[8] Vasil'ev, 17.

[9] David K Brown, *Nelson to Vanguard: Warship Design and Development 1923–1945*, Chatham Publishing (London, 2000), 29.

[10] Zubov, 125.

[11] Vasil'ev, 113; John Roberts, editorial, *Warship* vol III, no 12, Conway Maritime Press (London, 1979), 217.

[12] Vasil'ev, 50 table.

[13] Vasil'ev, 114.

[14] Markelov, 109–10.

[15] Note that the description in Stephen McLaughlin, *Russian and Soviet Battleships*, Naval Institute Press (Annapolis, 2003), 394, is inaccurate.

[16] Vasil'ev, 113; Evan Mawdsley, 'The Fate of Stalin's Naval Program', *Warship International*, vol. XXVII, no 4 (1990), 400–405.

THE IJN SUBMARINES OF THE *I 15* CLASS

In technical terms the fleet submarines of the *I 15* class were one of the Imperial Japanese Navy's most impressive achievements. However, they failed to make the anticipated impact on the Pacific War due to conceptual flaws and strategic errors in their employment. **Kathrin Milanovich** writes about the origins of the design, provides a detailed technical description of the boats and their equipment, and outlines their subsequent war service.

Following the termination of the naval armaments limitation treaties, the Imperial Japanese Navy (IJN) began the design and construction of three new types of submarine in response to the requirements of the fleet attrition strategy (*Zengen suishō sakusen*), which aimed to reduce the US Pacific Fleet to a force strength comparable to that of the Combined Fleet. The common feature of these submarines was that they were to combine the endurance of the cruiser type (*Junsen*) with the high surface speed of the large 'fleet' type (*Kaidai*). Because of differences in their tactical missions they were divided into 'A' (*Kō*), 'B' (*Otsu*), and 'C' (*Hei*) models (*gata*). Six, twenty-nine, and eleven boats were built respectively, demonstrating the priority accorded by the Naval General Staff (NGS) to the Model 'B'. After the completion of the first boats in the series a number of modifications were incorporated, so the class can be divided into four sub-groups (see Table 1). In this article only the first two groups will be described.

Generally speaking, the Model 'B' was slightly smaller and more simply fitted than the Model 'A' flagship type, and was judged to be well suited to its tactical purpose. Despite the large dimensions of these boats, steering and manoeuvrability were good in both the surface and the submerged conditions, diving time was comparatively fast (50 seconds), and the aircraft facilities were a marked improvement on those of earlier IJN submarines, thereby eliminating a perceived weakness. However, the wartime scenario for which they were designed failed to materialise, and the attrition strategy proved to be an illusion. Volume 98 of the official Japanese War History (*Senshi sōsho*) *History of the Submarine* states that the IJN submarines had fewer torpedoes than those of other nations because 'their targets were *warships*', and it was thought that this would mean fewer opportunities to attack.

The most spectacular success was undoubtedly that of *I 19*, which sank the US carrier *Wasp* and the destroyer *O'Brien* and damaged the battleship *North Carolina* on 15 September 1942 with a single torpedo salvo. Another successful boat was *I 26*, which torpedoed the carrier *Saratoga* on 31 August 1942 and sank the damaged AA cruiser *Juneau* on 13 November of the same year.

Requirements and Design

Following the First World War and the signing of the Washington Treaty of February 1922, the IJN pinned great hopes on the submarine, whose role was to compensate for its inferiority in capital ships, with the 5:5:3 ratio later being extended in 1930 to the cruiser, destroyer and submarine categories. Rather than

Table 1: Building Programme

Group	Basic design	Building programme	Hull numbers	Completed boats	Total	Principal differences
1	S 37	Circle 3 (1937)	Nos 37–42	I 15, I 17, I 19, I 21, I 23, I 25	6	–
2	S 37	Circle 4 (1939)	Nos 139–152	*I 26–I 39*	14	Safety equipment reduced to two emergency escape chambers
3	S 37B	Emergency (1941)	Nos 370–375	*I 40–I 45*	6	–
4	S 37C	Supplementary	Nos 627, 629, 631	*I 54, I 56, I 58*	3	Two diesel generators for battery charging

Of these 29 boats only one, *I 36*, survived. The remaining 28 boats were lost as follows:

1941	1942			1943			1944			1945		
Avail Dec	Add	Lost	Avail Dec	Add	Lost	Avail Dec	Add	Lost	Avail Dec	Add	Lost	Avail Dec
7	10	3	14	8	10	12	1	11	2	0	1	1

focusing on the destruction of mercantile traffic, submarines were to take on operational tasks in support of the fleet. The primary missions of the new generation of submarines were (i) to patrol and reconnoitre the US Navy bases at Pearl Harbor and on the West Coast of the United States and (ii) to wage a war of attrition against the enemy main force as it crossed the Pacific and advanced into the area intended by the IJN for the 'decisive fleet battle'. In order to fulfil these tasks large submarines with superior characteristics, particularly great range and high surface speed, were required.

Two types of large fleet submarine were developed. The first, intended for the extended observation of the enemy bases, was a cruiser type (*Junsen*) derived from the German *U-Kreuzer* of 1916–18, and had a displacement (standard) of 1,900 tons, a range of 20–24,000nm at 10 knots and a maximum speed 18–20 knots. The second, whose primary mission was the reduction of the enemy main force during its transit to the Western Pacific, was developed independently as a large fleet type (*Kaidai-gata sensuikan*, shortened to *Kaidai*) and needed a high maximum speed in order to enable it to shadow the enemy ships and to make repeated torpedo attacks. Standard displacement was around 1,500 tons and range 10–14,000nm at 10 knots, while maximum speed on the surface was in excess of 20 knots (23–24 knots for the latest boats of the KD6 type).

When the basic requirements for both the *Junsen* and *Kaidai* classes were fulfilled, the further goal was the combination of the long range of the former with the high surface speed of the latter. In order to overcome the most serious defect of the submarine for patrol, observation and tracking operations, the IJN developed submarine-based reconnaissance floatplanes, which were trialled and developed in the *Junsen* type. A key feature of the new fleet submarines of the 'B' type, in addition to high speed and long range, was the ability to operate aircraft.

Hull Form and Construction

The boats had a double hull: an inner pressure-resistant hull with external framing (considered advantageous by the Japanese designers), and an outer non-resistant shell. The designers made great efforts to reduce surface as well as underwater resistance, as high surface speed was a primary requirement; the outer shell was streamlined, and projections and sharp knuckles were eliminated as far as possible.[1] The outer hull casing was given marked sheer forward in order to improve seaworthiness.

The length of the cylindrical pressure hull was 86.8m; diameter was 5.7m. It was constructed of 20mm Ducol steel (DS) and was designed to withstand the water pressure at a depth of 100m. The pressure hull contained all the control systems for handling the submarine and also the crew accommodation. The external framing method was used because it favoured deep diving; it also saved space and weight within the pressure hull. This method had been adopted for the construction of the later German submarines, and testing had demonstrated its

superiority compared to the internal framing that had been used in the early U-boats. It also had the advantage of permitting the use of equal frame spacing. The early boats of the 'B' type had 70mm Z frames, but from immediately before the beginning of the Pacific War 100mm Z frames were used.

The Ducol steel used for the construction of the hull and the frames had a superior resistance to mild steel (MS); however it could not be welded using the then-standard Japanese technique of electric welding. Also, the use of electric welding had been seriously curtailed as a consequence of the 4th Fleet Incident,[2] so the designers were compelled to use a riveted structure. Riveting improved hull strength, but in the event of the detonation of depth charges close to the submarines the rivets were loosened, thereby impairing the integrity of the fuel tanks; leakage of heavy oil then revealed the position of the submerged submarine to the enemy. This was avoided by the adoption of electric welding for the fuel tanks in the boats built during the war; at the same time the thickness of the plating was increased from 6mm to 7mm.

The last boats of the first (S–37) series had some of the outer hull plates welded on land – ie a limited prefabrication method was adopted – and after April 1941 electric-welded tank frames and plates as well as brackets for fitting the superstructures were employed.[3] For the second series (S–37B: *I 40–45*) block building of tanks and hull parts was applied, and after a conference convened to discuss submarine hull structure in April 1943 tanks were all-welded, and the outer hull blocks prefabricated using this method; welding was also adopted for parts of the pressure hull; however, these later boats are not covered in the present article.

Above the centre section of the main pressure hull there was a second pressure-resistant cylinder of reduced diameter that formed the conning tower. It was encased within the outer plating to form the bridge structure. At the forward end of the bridge fairing there was a cylindrical pressure-resistant aircraft hangar that housed a single small reconnaissance floatplane. After the outbreak of the Pacific War plate thickness was increased from 12mm DS to 20mm over 60% of the upper part to protect the hangar against strafing, and the thickness of the hangar door was increased from 14mm MS to 22mm. The door was operated by a hydraulic piston with a crosshead located outside the pressure hull; unfortunately there was no arrangement to prevent leakage of oil from the cylinder.[4]

Table 2 gives the construction data for the 20 boats built to the basic S–37 design.

General Arrangement

The pressure hull was divided horizontally into an upper and lower half by a single deck with the exception of the engine room for the main diesels, which occupied the full height of the pressure hull. The general arrangement from bow to stern was roughly as follows:

Table 2: **Building Data and Fates**

Name/No list	Builder	Laid down	Launched	Completed	Lost	Removed from
I 15 [No 37]	Kure NY	25 Jan 1938	07 Mar 1939	30 Sep 1940	Missing since 03 Nov 1942; sunk 02 Nov N of San Cristobal US DD *McCalla*.	24 Dec 1942
I 17 [No 38]	Yokosuka NY	18 Apr 1938	19 Jul 1939	24 Jan 1941	Sunk 19 Aug 1943 40nm SE of Noumea US aircraft and New Zealand corvette *Tui*.	01 Dec 1943
I 19 [No 39]	Mitsubishi Kobe	15 Mar 1938	16 Sep 1939	28 Apr 1941	Missing since 19 Nov 1943 Central Pacific; sunk 25 Nov 50nm W of Makin US DD *Radford*.	30 Apr 1944
I 21 (II) [No 40]	Kawasaki Kobe	07 Jan 1939	24 Feb 1940	15 Jul 1941	Missing since 27 Nov 1943 S of Tarawa; sunk US carrier-based aircraft.	30 Apr 1944
I 23 (II) [No 41]	Yokosuka NY	08 Dec 1938	24 Nov 1939	27 Sep 1941	Missing since 24 Feb 1942 S of Oahu.	30 Apr 1942
I 25 [No 42]	Mitsubishi Kobe	03 Feb 1939	08 Jun 1940	15 Oct 1941	Missing since 24 Aug 1943; probably sunk by US DD *Patterson*.	01 Dec 1943
I 26 (*I 27*) [No 139]	Kure NY	07 Jun 1939	10 Apr 1940	06 Nov 1941	Missing since 25 Oct 1944; probably sunk by US DE *Richard M Powell*.	10 Mar 1945
I 27 (*I 29*) [No 140]	Sasebo NY	05 Jul 1939	06 Jun 1940	24 Feb 1942	Sunk 12 Feb 1944 NW of Addu, Maldives (1°25'N/77°22'E) RN DD *Paladin* + *Petard*.	10 Jul 1944
I 28 (*I 31*) [No 141]	Mitsubishi Kobe	25 Sep 1939	17 Dec 1940	06 Feb 1941	Sunk 17 May 1942 SSE of Truk (6°30'N/152°00' E) US SS *Tautog*.	15 Jun 1942
I 29 (*I 33*) [No 142]	Yokosuka NY	20 Sep 1940	29 Sep 1941	27 Feb 1942	Sunk 26 Jul 1944 Balintang Channel (20°10'N/121°50'E) US SS *Sawfish*.	10 Oct 1944
I 30 (*I 35*) [No 143]	Kure NY	07 Jun 1939	17 Sep 1940	28 Feb 1942	Sunk 13 Oct 1942 mine 3nm E of Singapore (1°15'N/103°55'E).	15 Apr 1944
I 31 (*I 37*) [No 144]	Yokosuka NY	06 Dec 1939	13 Mar 1941	30 May 1942	Missing since 14 May 1943; sunk 13 May US DDs *Edwards* & *Farragut* off Attu.	01 Aug 1943
I 32 (*I 39*) [No 145]	Sasebo NY	20 Jan 1940	17 Dec 1940	26 Apr 1942	Sunk 24 Mar 1944 south of Wotje (08°30'N/170°10'E) US DD *Manlove* & SC/PC 1135.	10 Jun 1944
I 33 (*I 41*) [No 146]	Mitsubishi Kobe	21 Feb 1940	01 May 1941	10 Jun 1942	Sunk by accident 13 Jun 1944 Inland Sea (Iyo-nada).	10 Aug 1944
I 34 (*I 43*) [No 147]	Sasebo NY	09 Jan 1941	24 Sep 1941	31 Aug 1942	Sunk 13 Nov 1943 SSE Penang (218° 18km Muka Lighthouse) torpedo RN SS *Taurus*.	05 Jan 1944
I 35 (*I 45*) [No 148]	Mitsubishi Kobe	02 Sep 1940	24 Sep 1941	31 Aug 1942	Sunk 22 Nov 1943 off Tarawa (1°22'N/172°77'E) US DDs *Meade* & *Frazier*.	30 Apr 1944
I 36 (*I 47*) [No 149]	Yokosuka NY	04 Dec 1940	01 Nov 1941	30 Sep 1942	Survived; handed over to Allied forces and sunk off Gotÿ Islands 1 Apr 1946.	30 Nov 1945
I 37 (*I 49*) [No 150]	Kure NY	07 Dec 1940	22 Oct 1941	10 Mar 1943	Missing since 06 Dec 1944 *Kaiten* operation against Palau; sunk 19 Nov US DD Conklin & DE McCoy Reynolds in Kossol-Passage.	10 Mar 1945
I 38 [No 151]	Sasebo NY	19 Jun 1941	15 Apr 1942	31 Jan 1943	Missing since 07 Nov 1944; sunk 12 Nov 85nm S of Yap US DD *Nicholas*.	10 Mar 1945
I 39 [No 152]	Sasebo NY	19 Jun 1941	15 Apr 1942	22 Apr 1943	Sunk 26 Nov 1943 W of Makin (3°10'N/171°55'E) US DD *Boyd*.	30 Apr 1944

Note: (I) or (II) means first or second submarine with same number; number in parentheses (I 27) is the original number, which was changed during construction; hull number is in square brackets beneath.

IJN Type B1 Submarine: Profile & Plan

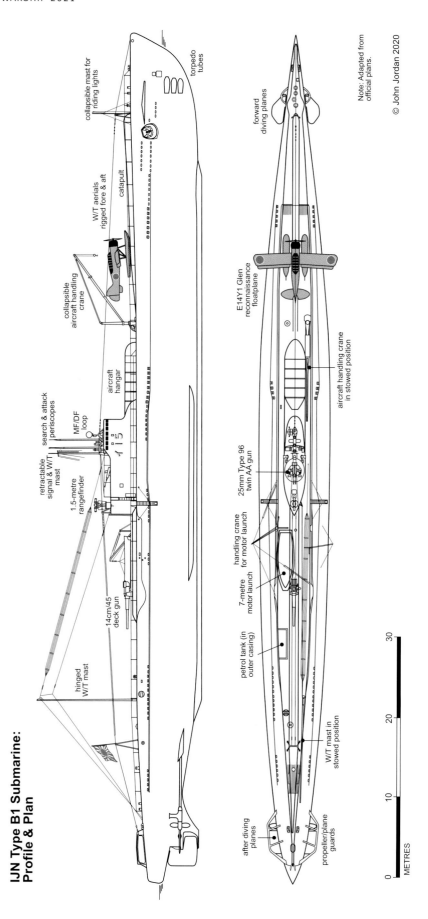

torpedo tubes

collapsible mast for riding lights

W/T aerials rigged fore & aft

catapult

collapsible aircraft handling crane

aircraft hangar

search & attack periscopes

MF/DF loop

retractable signal & W/T mast

1.5-metre rangefinder

14cm/45 deck gun

hinged W/T mast

forward diving planes

E14Y1 Glen reconnaissance floatplane

aircraft handling crane in stowed position

25mm Type 96 twin AA gun

handling crane for motor launch

7-metre motor launch

petrol tank (in outer casing)

W/T mast in stowed position

after diving planes

propeller/plane guards

Note: Adapted from official plans.

© John Jordan 2020

0   10   20   30
METRES

Editor's note: The drawings have been adapted from the official plans of *I 15* reproduced in *Nihon Kaigun Kantei Zumen Shū*, published by the Society of Naval Architects of Japan (Tokyo, 1975). The profile view shows the tall hinged mast with its prominent W/T aerial prism in the raised position, and the aircraft and boat handling cranes deployed. Before the boat dived the mast and cranes were lowered and stowed in recesses in the upper casing (see plan view and the sections on page 33) to create a smooth waterflow around the hull, thereby maximising underwater speed and reducing self-noise. The profile view shows the 7-metre motor launch suspended from the crane, above the level of the casing where it was normally stowed. The plan view also shows a derrick deployed to starboard abaft the bridge fairing; this would have been used to embark ammunition and stores.

Although the plans of *I 15* have been grouped together in the book, they appear to have been drawn at slightly different stages in the construction process. Note that there are small differences in the arrangement and number of free-flood holes in the upper casing in the profile and plan views. The number of free-flood holes clearly proved to be insufficient for rapid diving, as the boats as completed had a double row running for most of the length of the outer casing (see the photos of the launch of *I 17* and of *I 29* as completed). Also, while the original outboard plans clearly show the Yokosuka E14Y1 'Glen' (embarked following its entry into service in the late summer of 1941) atop the catapult, the aircraft atop the catapult in the inboard profile is its predecessor, the Watanabe E9W1 'Slim', a biplane which entered service in 1938. (Somewhat bizarrely, the broken outline of the aircraft depicted in the hangar is the 'Glen'). Note that the W/T aerial which ran from the forward end of the bridge fairing to the bow would have had to be detached to enable the floatplane to be assembled and launched.

IJN Type B1 Submarine:
Inboard Profile & Sections

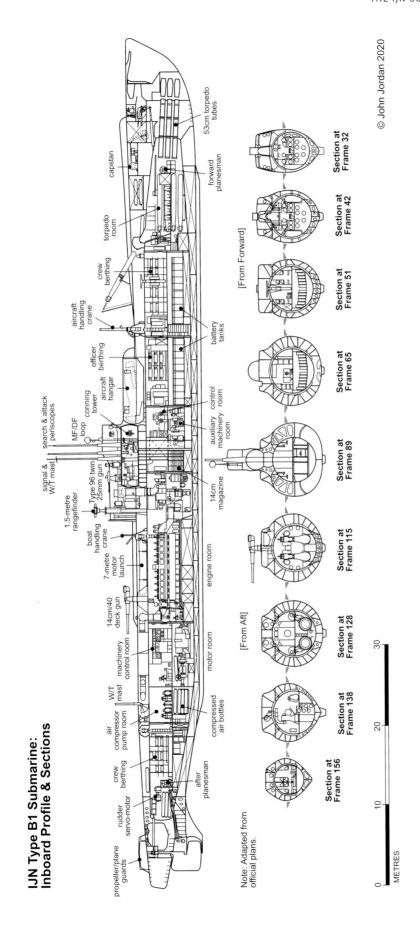

© John Jordan 2020

53cm torpedo tubes

capstan

forward planesman

torpedo room

crew berthing

aircraft handling crane

battery tanks

officer berthing

aircraft hangar

search & attack periscopes

MF/DF loop

conning tower

auxiliary control machinery room

control room

signal & W/T mast

1.5-metre rangefinder

Type 96 twin 25mm gun

14cm magazine

boat handling crane

7-metre motor launch

14cm/40 deck gun

engine room

motor room

machinery control room

W/T mast

air compressor pump room

compressed air bottles

crew berthing

after planesman

rudder servo-motor

propeller/plane guards

Section at Frame 32

Section at Frame 42

Section at Frame 51

Section at Frame 65

Section at Frame 89

Section at Frame 115

Section at Frame 128

Section at Frame 138

Section at Frame 156

[From Forward]

[From Aft]

Note: Adapted from official plans.

METRES

0   10   20   30

Editor's note: Particularly prominent in this inboard profile of *I 15* are the two massive Kampon No 2 Model 10 10-cylinder, 2-stroke diesel engines, which occupy almost the full height of the pressure cylinder. Each had a theoretical rating of 7000bhp, and when operating at their designed 350rpm they were capable of driving these boats at up to 23 knots on the surface. The compartment for reserve torpedoes forward was on two levels; the plans show stowage for up to seven Type 95 Model 1 53cm torpedoes on the upper level, and six torpedoes on the lower level (see Sections frames 42 and 32). Berthing for the crew was in separate compartments fore and aft.

The single floatplane was stowed broken down in the pressure-tight cylindrical hangar forward of the bridge fairing; the wings and floats were detached and stowed separately to the sides. Assembly normally took in excess of 30 minutes; the aircraft was then launched by the catapult, which was angled upwards towards the bow both to improve lift and to take the aircraft clear of the swell. The 7-metre motor launch is shown in its stowed position on the port side of the upper hull casing. When lining up an attack the search and attack periscopes were operated from the pressurised conning tower, the periscopes being lowered into vertical cylindrical housings in the pressure hull when dived.

- upper torpedo room with four Type 95 Model 2 torpedo tubes and up to seven reserve torpedoes (above); lower torpedo room with two torpedo tubes of the same type and six reserve torpedoes (below)
- forward crew berthing (above) and forward battery tank (below)
- officers' accommodation (beneath the hangar – above) and after battery tank (below)
- pressure-resistant cylinder integrated into the bridge structure housing the conning station (forward) and W/T office (aft); main pressure hull with the control centre (above) and the auxiliary machinery room (below)
- main engine room housing the two No 22 Model 10 diesel engines
- machinery control room (above) and the motor room (below), the latter housing the two main electric motors and some compressed air bottles (the boat was controlled from here when submerged)
- air compressor pump room (above) with banks of compressed air bottles (below)
- after crew berthing
- steering gear compartment.

The main ballast tanks, trim and buoyancy tanks (including the unusual negative buoyancy tank), compensation tanks and heavy oil fuel tanks were arranged in the usual way. Most were between the inner and outer hulls, but some important tanks were located inside the pressure hull below the batteries, and beneath the after crew's berthing spaces. The tanks were filled via large and small Kingston valves; high- and low-pressure air was used for the blowing out.

Also located between the inner and outer hull were the mechanisms for the aircraft crane (forward) and the W/T mast (aft), with the exhaust pipes and silencers for the diesel engines directly in front of the latter (see inboard profile). The capstan and the anchor cable compartment were above the torpedo rooms, and further forward was the mechanism for the forward diving planes and the line-handling equipment. The after control surfaces comprised two horizontal diving planes and a vertical rudder. A 7-metre motor launch powered by an 8hp engine was stowed to port abaft the conning tower between the pressure hull and the after deck.

In the first boats up to five sets of escape equipment for the crew were fitted, but the boats built in wartime had simplified fittings to save time and materiel. These boats had only two emergency escape chambers with double hatches and an air lock. They were located close to the forward and after crew spaces.

Machinery

The development of a high-performance diesel engine was key to high surface speed and the decisive criterion for success or failure of the design. The diesel was designed in the Fifth Division of the NTD by a specialised

The submarines of the *I 15* class as designed combined the endurance of the cruiser type (*Junsen*) with the high surfaced speed of the large fleet type (*Kaidai*). This photo shows the lead boat *I 15* on 15 September 1940, running at 23.6 knots on the surface.
(Hans Lengerer collection)

team. A prototype was successfully tested in early 1935 and series production followed.

The two-cycle, double-acting, ten-cylinder engine with air injection developed 7,000bhp and weighed 93 tonnes. The engine combined the best design features of the Swiss Sulzer and the German MAN and Lachman DA diesels (see Table 3 for technical data). The control station was at the forward end of the engine and was fitted with the HP air compressor control, the engine start-up and stopping levers, a fuel throttle with Vernier adjustment and the engine reversing gear.[5] The fuel injection pump was located below the control station, from which the plunger lift control mechanism was operated. All gauges and indicators, including the piston cooling monitor box, were mounted above.[6] The scavenging air was supplied by a detached electric motor driven by a rotary-type two-stage blower. The motor power required at maximum speed was 750hp, revolutions ranging from 1750rpm to 2700rpm. Blower capacity was 15m³/sec with a discharge pressure of 0.7–1.0kg/cm².

All the necessary pumps except the fuel supply pump were independently mounted. Three lubricating oil pumps[7] (one per engine plus one stand-by) were located in the lower flats. The stand-by pump could be cross-connected and used as a fuel transfer pump. Two large lubricating oil coolers were arranged between the two large sea water pumps and lube oil pumps. One DeLaval-type oil purifier was provided for both engine lube oil pumps.

The propellers had three blades and rotated at the same speed as the engines. The tail shaft diameter was 300mm

Table 3: Kampon No 2 Model 10 Diesel

Cycle	2-stroke
No of cylinders (A)	10
Horsepower (B)	7,000bhp
Shaft revolutions	350rpm max
Cylinder diameter	470mm
Stroke	530mm
B/A	700
Ave effective pressure	6.18kg/cm²
Maximum pressure	5.13kg/cm²
Distance between cylinders	780mm
Length	10,920mm
Height above middle of shaft	3,678mm
Depth below middle of shaft	1,720mm
Width of the engine seating	1,590mm
Weight (W)	93 tonnes
W/B	13.3kg/hp

Note: Data for horsepower vary between 5,650bhp and 7,000bhp depending on the source. The data given here are from Ohara Nobuyoshi's 'Marine Engines of the Imperial Japanese Navy' in *Sekai no Kansen* 10/1973, 114. On the other hand, eight Japanese source state uniformly 12,400bhp for the Kampon No 22 Model 10 engines of the *I 15* class (the *I 40* class had Kampon No 1 Model 10, the *I 54* class Kampon No 22 Model 10 engines), so effective performance may have been 6,200bhp per unit.

and a single thrust bearing and shaft friction brakes were fitted on either side. There were positive jaw couplings between the main engines and main motors.

The engine was not without its defects. In rough weather conditions, water came in through the muffler due to the low exhaust pressure and speed had to be reduced, thereby impairing the combat power of the submarine. Also, the diameter of the two air supply pipes (800mm) was too small, and this resulted in an excessive air flow speed at times and also a reduction in pressure when the air flow changed, which created problems for the crew.[8] Surface speed ranged from slow speed (6 knots with 105rpm) to full speed (*c*23 knots with 350rpm) and full astern was 14 knots. Battle speeds Nos 1–4 were 16–22 knots with 235–340rpm. With the maximum fuel oil stowage of 814 tonnes range was 14,000nm at 16 knots.

The DC electric motor belonged to the Type Special Model 5 and had an output of 1,000hp at 220V x 3300A. Maximum overload was 1,250hp at 220V x 4170A for 10 minutes. One motor was mounted on each of the two shafts, and the motors were also used to generate electricity. Four main motor air coolers were provided and separate motor-driven fans used to circulate the cooling air in order to limit maximum temperature to 90 degrees (field coils and commutators). Underwater speed was 8 knots with the maximum power of 2,000hp, and endurance 60nm at 3 knots, which corresponded to 20 hours in the submerged condition.

The energy for the motors used for underwater operation was supplied by 240 battery cells Type No 2 Model 15 in the boats belonging to the first group, and Type 1 No 1 Model 13 in the boats of the second group. The former delivered 25% more energy than the latter with 240V x 10,000A for eight hours. For operations in the South Pacific battery cooling was considered essential in order to maintain capacity, and two coolers each rated at 25,000Kcal were mounted. In contrast to the submarines of the first group, authorised as part of the Third Fleet Replenishment Program of 1937, the coolers used Freon gas instead of CO_2; this constituted an improvement, although capacity remained the same. The use of cooling motors also improved habitability. A single DC diesel auxiliary generator rated at 450kW was mounted in order to charge the battery. This was considered insufficient, and quicker charging was deemed essential; the boats of the last group were therefore fitted with two generators of the same type.

Torpedo Tubes and Torpedoes

The 'B'-type submarine had six torpedo tubes arranged as two vertical banks of three at the bow. The type used in the boats of the first sub-group was the Type 95 Model 1; the boats of the second series had Model 2. This torpedo tube was developed in parallel with the Type 95 oxygen-propelled torpedo; the later model was an improved Model 1 but technical details are lacking. The torpedoes were fired using compressed air.

The Type 95 constituted the last stage of the development of the IJN's submarine torpedo tubes. Domestic development began with the Type 15 mounted on *Ro 57* (completed in 1922), but this tube had various defects that ranged from the manual opening and closing of the muzzle door to other more serious issues, namely: alerting the enemy to the firing of a torpedo (and hence the position of the submarine) by the huge splash on the surface that resulted from the large volume of air escaping from the tube behind the torpedo; and the difficulty experienced in rapidly adjusting trim, which meant that the bow broached the surface when all four torpedoes were fired.

In order to remedy the defects of the Type 15, the Type 88 was developed; a prototype was completed in 1928 and the tube was formally adopted in 1931 after extensive testing in *I 65*. It was equipped with a piston connected to the after door of the tube. When the launch valve was opened and air admitted into the tube the piston was pushed forward and stopped automatically when the torpedo had reached the required speed. The air behind the piston was automatically discharged via a valve into a container inside the pressure hull and the piston was returned to its original position by the water pressure in the tube following the launch of the torpedo. In this way concealed firing became possible. However, this torpedo tube also had its weak points: the rubber seals were sometimes damaged by the shock of the piston and, depending on the intensity of the shock, the tube itself, which was a high-pressure silicon bronze ('Silzin') casting, tended to develop fine cracks, with the risk of catastrophic failure.

The IJN made serious efforts to resolve this problem once and for all, and developed the Type 95 torpedo tube, which did not require a piston to expel the torpedo. In developing this tube the IJN and Kobe Seikoshō[9] cooperated closely, but many studies and experiments were necessary before, in 1936, Sasebo NY was ordered to undertake extensive tests using *I 64*. Key features of the new torpedo tube were an air cut-off valve (*dankiben*), a 'harmonising' valve (*chowaben*),[10] and an automatic air/seawater separation valve that stopped the flow of seawater following the expulsion of the air into the hull container. This was achieved by locating an intermediate tank for seawater separation inside the chamber of the tube. One end of the air pipe was connected to the tank, the other end opened into the hull to form the air channel. When the air used for launch passed into the hull after accelerating the torpedo, sea water flooded into the tank, which was fitted with a float that when raised closed the air channel. *Kaigun Suirai Shi* (page 245) notes that the principle was very simple in theory, but that many difficulties had to be overcome before reliable function could be guaranteed. As usual, the IJN considered it to be a revolutionary development, and this view is often repeated in secondary Japanese sources.[11]

Seventeen Type 95 torpedoes, propelled by 100% pure oxygen, were normally embarked: six in the tubes and eleven as reserve. The oxygen torpedo had been developed for destroyers and cruisers as the 61cm torpedo Type 93.[12] The Type 95 developed for use by submarines was essentially the same torpedo but with a reduced diameter of 53.3cm (21in). The general view of most of the COs of the first class submarines was that its performance largely outweighed the disadvantages in handling and preparation attributed to the use of pure oxygen.

The second boat of the B type, *I 17*, on the day of her launch at Yokosuka Navy Yard on 19 July 1939. The floatplane hangar and the conning tower are lacking their streamlined outer casing, and the catapult and deck gun have yet to be fitted. (Hans Lengerer collection)

Circuit Diagram, Type 95 Torpedo Model 1

A vessel containing air at a pressure slightly above (10kg/cm^2 greater) that in the oxygen vessel is added to the oxygen circuit shown above. Working from the nose to the tail there is first an oxygen charging stop valve and charging connection at the forward end of the vessel for charging the oxygen vessel with 100% oxygen. From the after end of the oxygen vessel there is a lead through the delivery stop valve and non-return valve to the 'first air vessel'. From this vessel the connection goes to the air stop valve and charging connection and from there to the group. The group is connected to the main reducer which in turn is connected to the generator. No

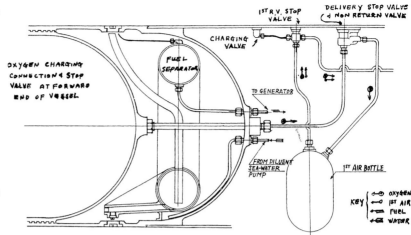

oxygen was discharged from its vessel until the pressure in the 'first air vessel' had fallen below that in the oxygen vessel. The torpedo started on natural air and the oxygen concentration was gradually increased to 100%. Thus no explosion at the moment of ignition could take place. (USNTMtJ 0–01–1, Fig 15 page 88)

Once the superiority of early models of the 61cm Type 93 torpedo became apparent, the NTD ordered the Nagasaki Weapon Production Factory (Nagasaki Heiki Seisakushō) to design and test produce two 53cm torpedoes using the same main medium (oxygen) for the propulsion system. NTD requirements were for a range of 9,000 metres at 49 knots and a 400kg explosive charge. Initially designated 'trial torpedo B', tests were carried out on land in the factory and then at the Dozaki torpedo range. Despite the IJN's experience with the Type

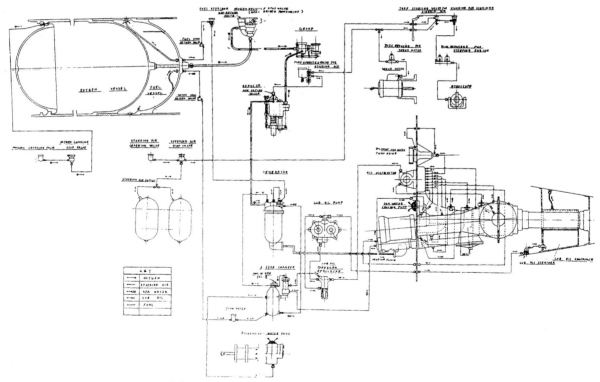

Circuit Diagram, Type 95 Torpedo Model 2

The oxygen vessel had an oxygen charging stop valve and charging stop valve and charging connection at the forward end of the vessel for charging with 100% oxygen. Oxygen from the vessel passed through the mechanically operated oxygen delivery stop valve through the group to the reducer and thence to the generator. Air from the steering air vessel (centre left) flowed through the steering air stop valve to the tube-operated starting valve. From this valve, one lead went to the water flap-operated valve attached to the group, another to the disc reducer and then to the servomotor; a third lead went to an air strainer and then to the disc reducer and to the gyroscope. (USNTMtJ 0–01–1, Fig 16 page 89)

93, a number of problems had to be resolved before series production could be started in 1937. These can attributed principally to the smaller diameter of the torpedo, which meant that the dimensions, and sometimes the configuration of the components had to be modified, and account taken of the smaller capacity of containers. Like the Type 93 Model 1 torpedo this type also employed a 'first air vessel' circuit to prevent an explosion at the moment of ignition.[13] This means that the torpedo was started on natural air and the oxygen concentration was gradually raised to 100%. The circuit is shown diagrammatically in the accompanying figure. The sequence of operations is listed in the table in comparison with that for the Model 2 torpedo.

The 'first air vessel' had a capacity of 7 litres, reduced from 13.5 litres in the Type 93 torpedo. Since the volume of this vessel was small, the pressure soon fell below that of the oxygen vessel, and as soon as the oxygen delivery stop valve was opened, oxygen passed into the first air vessel and enriched its charge. Leakages often occurred at the joint of the pipe leading to the oxygen vessel.[14] The rapid decline in pressure was not a problem in the larger Type 93 Model 1, as the pressure could be checked while the torpedo was in the tube and the air topped up if necessary, but in the smaller Type 95 this was more difficult. A leak would result in an explosion at the start of the run that might result in the loss of the submarine. This prompted studies of another start-up method, leading to the adoption of the 'steering air vessel' circuit shown diagrammatically in the accompanying figure.

Table 4: Type 95 Torpedo

Sequence of Operations

	'First Air Vessel' Circuit (Type 95 Model 1)	'Steering Air Vessel' Circuit (Type 95 Model 2)
1	Open 'first air vessel' stop valve by hand when torpedo is in the tube.	Tube starting valve is opened and air is admitted to servomotor, gyro and pilot valve in group.
2	Open oxygen delivery stop valve by hand when torpedo is in the tube. (Special geared spanners to give slow opening are used, and the valves are opened just before firing.)	Torpedo is fired and water flap goes aft admitting air to reducer and generator.
3	Tube operated starting lever is pulled aft to operate safety mechanism on group. Air is supplied to gyroscope, servomotor and steering engine by opening a second valve in the circuit.	Engine revolves and opens main oxygen delivery stop valve fully in 60 revolutions and closes steering air valve suddenly after 40 revolutions.
4	Torpedo is then fired and flap goes aft and opens small valve on group.	Range is set on group in the normal way; the valve closes when the range is run off.

Note: The method of starting by 'steering air' was considered superior to the 'first air vessel' method not only with regard to safety, but it did not suffer the mechanical disadvantages of its predecessor.

Characteristics

	Type 95 Model 1	Type 95 Model 2
Range	9,000m at 49 knots	5,500m at 49 knots
	12,000m at 45 knots	7,500m at 45 knots
Deflection at extreme range	250m left / 170m right	130m left / 90 right
Depth setting	5m ± 0.5m	5m ± 0.5m
Total weight / displacement	1,665kg / 1,345kg	1,730kg / 1,345kg
Total length	7,150mm	7,150mm
Weight of warhead	405kg (Type 97)	550kg (Type 97)
Length / shape of the warhead	1,530mm / hemispherical	1,750mm / streamlined
Oxygen vessel volume	386 litres (215kg/cm^2)	220 litres (200kg/cm^2)
Fuel: type / volume	Kerosene / 50 litres (displaced by water)	[as Model 1]
Engine	Horizontal reciprocating two-cylinder, double-acting with bore 130mm, stroke 160mm and expansion rate 1.5	[as Model 1]
Horsepower	330hp with 1,150rpm	430hp with 1,250rpm
Afterbody / propellers	4/6 fins, 4 blades / 482mm and 439mm diameter fwd and aft respectively	[as Model 1]
Wake	Scarcely visible	[as Model 1]

Note: These performance characteristics and the wakeless running made the Type 95 the best submarine torpedo in the world.

This photo of *I 26* is dated October 1941. Built at Kure NY, *I 26* was commissioned on 6 November 1941, and the date of the photo and the high bow and stern waves suggest that it must have been taken during one of the speed trials. *I 26* torpedoed the US carrier *Saratoga* (CV-3) on 31 August 1942, 200nm southeast of Guadalcanal, and sank the light AA cruiser *Juneau* (CL-52) when retiring to Espiritu Santo on 13 November 1944. Her CO chose *San Francisco* (CA-38) as his target, but after the torpedoes had been fired *Juneau* masked the heavy cruiser and was sunk instead. Personnel losses were very heavy, and the loss of the five Sullivan brothers caused the US Navy to issue a ruling under which close relatives were prevented from serving in the same ship. (Hans Lengerer collection)

'First air' and 'steering air' was combined into a single container with a capacity of *c*30 litres, and manual actuation of the oxygen delivery stop valve was changed to automatic opening. These modifications were developed by the Torpedo Division of Kure NY for the Type 93 Model 2 torpedo in 1943 and communicated to the Nagasaki Weapon Factory, which subsequently applied the technology to the Type 95 submarine torpedo Model 2.

However, this was not the only difference between the two models. In late 1943 or early 1944 it was decided that warhead weight needed to be increased from 400kg to 550kg. However, this could not be realised without impairing other characteristics, and the Model 2 was fitted with an oxygen vessel of reduced capacity to compensate for the increased length and weight of the warhead. Also, the warhead was given a streamlined configuration modelled on that of Italian torpedoes in place of the hemispherical warhead of the Model 1, which was thought to have caused premature detonation in the Type 93 torpedo.[15] Following the Japanese advance into the Southwest Pacific in April 1942, submarines had reported premature detonations in steadily increasing numbers. However, *Kaigun Suirai Shi*

(page 31) points out that this information was received very late by the Kure NY and the Nagasaki Weapon Factory was not informed, so remedial measures were taken very late.[16]

Torpedo Fire Control

For fire control there was a Type 92 submarine FC computer Mod 1 and two Type 14 Mod 1 calculators. The latter was a very simply-constructed instrument – in essence little more than an alidade. Developed for surface vessels, one has to question the value of fitting such a primitive system in a comparatively sophisticated modern submarine.[17]

In 1931 the IJN purchased a torpedo fire control computer from Vickers (UK), and presented it to the Japan Optical Co (Nippon Koyaku KK), Mizunokuchi, Tōkyō, for study with an order to produce a prototype. Testing on land took place at the Submarine School (*Kaigun Sensuigakkō*) and the computer was then trialled on board a submarine. Testing resulted in some improvements, after which the director was formally adopted as the Type 92 submarine fire control computer Mod 1 (*92 shiki sensuikan hōiban kai 1*). Development was completed in 1933 and series production began in the following year; The Type 92 was to become the standard torpedo FC system for submarines, with 162 units supplied including Mod 2.[18]

The introduction of this system revolutionised torpedo fire control in IJN submarines. The individuals responsible for its design and construction were Lt-Cdr (later Captain) Norita Kiyoshi, Engineer Kasunori Sunnosuke of the Second Division of the NTD, and Engineers Ninomiya, Mimaki, and Yamada of the Japan Optical Co. These men were also responsible for the development of the automatic target bearing indication system fitted to 8-metre, 9-metre and 10-metre periscopes, which provided continuous transmission by selsyn of the bearing angle of the periscope to the FC computer when the periscope was raised. This was a key component of the Type 92 FC system; development was likewise completed in late 1933.

After the outbreak of the Pacific War production had to be increased to cope with the new programmes of submarine construction, and in late 1943 the Japan Optical Co was instructed to accelerate mass production of the Type 92 FC computer Mod 2 without lowering capabilities. In Report 0–32 Japanese Torpedo Fire Control, page 1, the US Navy investigation team summarised its findings thus:

> The Japanese spent considerable effort on the design and manufacture of torpedo fire control equipment. The various units are well constructed and function with good accuracy. Their submarine torpedo data computers and auxiliary equipment are more simplified and less accurate than US equipment ... Target designation systems as well as torpedo indicating and firing panels were not well developed in submarine installations ...

The same report describes the Type 92 torpedo FC computer Mod 2 (pages 10–17), and draws the following conclusion:

> [It] is a central computing instrument ... designed so that the necessary firing data with no dead time may be obtained by one operator. To accomplish this, the gyro angle is calculated for a future time (chosen arbitrarily from 0 to 40 seconds) and five seconds before the time to fire a bell begins to ring. The torpedo is launched when the bell stops ringing.[19]

Gun Armament and Aviation

The single deck gun, which was mounted abaft the conning tower, was a 14cm/40 11th Year type. The 14cm gun was the largest calibre employed for submarine guns in the IJN, and fired a projectile weighing 38kg; the magazine held 158 rounds. Range was 15,000 metres with an elevation of 30°. A rangefinder with a base length of 1.5 metres, probably Type 96, was mounted on a fully retractable pedestal in the after part of the conning tower. The general plans of *I 37* show that the rangefinder was later replaced by a single Type 97 12cm binocular mounting.

For anti-aircraft fire a Type 96 25mm twin MG was

This starboard broadside photo of *I 29* shows the conning tower, the aircraft hangar – which might better be termed 'aircraft cylinder' because the reconnaissance plane could be stored in broken-down condition only – and the Type Kure No 1 Model 4 catapult, which was powered by compressed air and had a total length of 19m, mounted directly forward of the hangar. (Hans Lengerer collection)

mounted on the after part of the bridge. The early boats had magazine stowage for 1,100 rounds per gun (total: 2,200), but this proved insufficient and was later doubled – the boats of the 4th series had 2,200rpg. Small arms comprised four Type 38 rifles and sixteen 14th Year type pistols with a total 4,850 and 2,464 rounds respectively.

The single floatplane embarked when the first boats were completed was the Watanabe E9W1 Type 96, superseded from late 1941 by the Yokosuka E14Y1 'Glen'. The principal aircraft facilities comprised a cylindrical hangar, a crane, and a catapult for launch. In earlier submarines these had been located abaft the conning tower, but this was found to have a number of serious disadvantages: the floatplane could not be launched when the boat was underway, and the turning circle was greater when submerged. They were now located forward of the conning tower for the first time. The plane was stowed broken down inside the hangar cylinder, and the Type Kure No 1 Model 4 catapult was mounted atop the bow casing directly in front of the hangar. The catapult was specifically designed for submarines; whereas the catapults mounted on surface vessels were powder-operated, the Kure No 1 Model 4 submarine catapult used compressed air. Total length was 19m, effective length 15m, and acceleration 26m/sec. Aircraft with a maximum weight of 1,600kg could be launched.

For assembling and dismantling the aircraft there was a sort of turntable between the hangar and the catapult. The collapsible crane used to hoist the floatplane on board was normally stowed in a recess to starboard of the hangar. Locating the hangar and catapult forward of the conning tower proved a success: the aircraft could be catapulted off when the submarine was underway, and the turning circle in the submerged condition was not impaired. Trials in this class of submarine demonstrated that the aircraft could be assembled 10–15 minutes after surfacing and the engine started for warm-up. However, these results were obtained under training conditions; in service this procedure generally required between 30 and 60 minutes. Also, when the plane returned it took at least an hour before the boat was ready to submerge, and this proved to be excessive when operating in an area where the enemy had air superiority.

Periscopes and Electronic Equipment

The two periscopes were fitted on the centreline, one behind the other. The forward (attack) periscope was the Type 88 Model 3 Mod 1. It had a total length of 10 metres and could be raised or lowered by 5.5m.[20] The diameter of the tube that protruded above the surface was 31mm. Magnification ranged from x1.5 (low) to x6 (maximum), the respective fields of vision being 40 degrees and 9°30'. This periscope was used by day and was fitted with the automatic bearing angle transmitting system described above. The after periscope, designated Type 88 periscope Model 4 Mod 1, was used at night and also for navigation. It had a total length of

9 metres, a diameter of 90mm, and maximum magnification was x10; the field of vision was 40 degrees at low magnification (x1.5) and 6 degrees at maximum magnification (x10). Other optical sensors included Type 93 12cm binoculars fitted in watertight casings on the roof of the conning tower. With a magnification of x20 and a field of vision of 3 degrees, performance exceeded the world standard.

The underwater search equipment comprised a Type 93 hydrophone and a Type 93 Model 4 sonar. The latter was developed by the IJN and entered production in 1933. It was the first usable active underwater search sensor, but was difficult to operate and errors were large (±6° in direction, ±12% in range), so the sonar proved to be of little tactical value when installed in a submarine. Underwater signal equipment (*suichu shingo sōchi*) was fitted.

When these boats were completed IJN had no submarine radar. ESM sensors to detect metric and centimetric radar emissions were fitted, probably in late 1943, and in 1944 No 22 and No 13 radars were installed. The horn antenna of the former was mounted atop the hangar in a watertight casing weighing 840kg, making the total weight of the installation 2.14 tonnes. The antenna of the No 13 radar was fixed to the telescopic W/T mast behind No 2 periscope. The performances of these radars were inferior to those of their Allied counterparts, and the models fitted in the submarines proved less effective than the surface installations.

The wireless equipment comprised four Type 99 special No 3 transmitters, a Type 99 Tan No 3 transmitter, four special receivers, a wave length meter, a short wave mast, a long wave mast (housed in a recess when not in operation), an ultra short wireless telephone and a Type 0 No 4 long wave direction finder.

Complement

The designed complement was ten officers and 87 men, although this saw an increase in wartime – sources record complements of up to 100 officers and men.

Operational History

Owing to the limited space available, only the successes of *I 19* and *I 26* against warships will be recounted here. These accounts are extracts of those by Shibuya Ikuo, formerly chief engineer of *I 19*, and Hasegawa Minoru, CO of *I 26* with his former name Yokota Minoru, in Fukuda Ichirō's 923-page *History of IJN's Submarines* (*Nippon Kaigun Sensuikan-shi*, pages 685–89).

Report of the sinking of the carrier *Wasp*:

We were tasked with forming a patrol line in the Solomon Islands area to intercept and attack an enemy force advancing on Guadalcanal. Discretion was paramount; large warships such as carriers or battleships were the targets; attacks on smaller ships that might reveal our presence were to be avoided absolutely.[21]

Taken during trials in January 1943, this fine photo of *I 38*, built at Sasebo NY and commissioned on 31 January of the same year, shows how the aircraft stowage cylinder was faired into the conning tower. The gun abaft the conning tower was a Year 11 type 14cm/40-cal low-angle gun, the largest calibre fitted in IJN submarines; there was rarely an opportunity to use it after the early stages of the Pacific War. Note the Type 96 25mm twin MG atop the open bridge. (Hans Lengerer collection)

On 23 August an enemy plane attacked *I 19*, and from the 24th onwards the boat remained submerged from 0330 until around 1630 to avoid air attacks. When the boat surfaced at 1245 on the 25th an enemy plane, a cruiser and a destroyer were sighted but the targets were already outside effective torpedo range and made off at high speed. On the following day, at 1425, the sonar detected the presence of a group of ships that was subsequently identified by the periscope as comprising a carrier, a battleship, a cruiser and several destroyers. The enemy formation passed over *I 19*, which waited silently at depth. However, there was no chance to attack!

Then came the day of 15 September. At 0950 the sonar operator reported 'group noise', possibly indicating an enemy force! When the periscope was raised for the first time no enemy was seen. Later, at 1050, a task force comprising a carrier, a cruiser and several destroyers was sighted using the periscope; however range was too great. CO Kinashi Takakazu[22] changed course towards the enemy at his best speed. Then fate turned in favour of *I 19*. The enemy had been steering course WNW, but at 1120 changed to SSE and approached our boat. When the range was down to 900m, at 1145 six torpedoes were launched with a gyro angle of 50° right.

When diving to 80m depth four explosions were heard, and six minutes later the first depth charge detonation was heard, quickly followed by a second and a third ... I noted 85 detonations in all directions and at various depths. *I 19* changed course and depth constantly, and no detonation was close enough to cause damage.

Our boat surfaced at 2010 and reported to the HQ of the 6th Fleet:

> *I 19* detected the noise of a group of ships with sonar sensitivity stage 3 at 12°18'S/164°15'E at 0950, and sighted the ships using periscope at 1050, 45° to starboard at 15,000 metres: a carrier of the *Wasp* class, a cruiser and several destroyers. Six torpedoes were fired at 1145 and four explosions were heard. Some 80 DCs were dropped, so no effect could be confirmed.

At almost the same time *I 15*, which was operating in the vicinity of *I 19*, reported:

> Following an attack by our aircraft fire broke out on the enemy carrier. Heavy list to port. Two cruisers and several destroyers gave up rescue work at 1515 and retired to the South; the sinking of the carrier was observed at 1800.

It was later established that the drifting *Wasp* was sunk by American bombers. The sinking of the destroyer *O'Brien* and the damage to the battleship *North Carolina* were discovered only after the end of the Pacific War.

Report on damage to the carrier *Saratoga* and sinking of the cruiser *Juneau*:

On 31 August 1942 the carrier *Saratoga*, protected by a ring formation, was attacked east of Guadalcanal. A salvo of six torpedoes was launched, of which two hit.[23] The carrier was taken under tow by a heavy cruiser. After manoeuvring to a firing position and ordering torpedoes to be fired, the bow of a destroyer appeared in the periscope from port to starboard heading straight for *I 26*. 'Periscope down; dive, dive, dive; depth 100m!' was ordered. The destroyer sighted the periscope and gave the alert, followed by the dropping of depth charges. While diving, loud explosions were heard ... the crew was jubilant ... the numerous depth charge detonations had no effect on them.

On the morning of 13 November 1942, during the observation of the strait between the islands of San Cristobal and Guadalcanal, a group of enemy ships steering south was sighted. The CO launched three torpedoes[24] from the position marked in the sketch the sketch that accompanied the report against a ship later identified as the cruiser *San Francisco*, and heard one explosion. Ten years later it became clear that *I 26* sank the AA cruiser *Juneau*. It is thought that the torpedoes were launched prematurely at a ship that appeared in the crosshairs of the periscope.

Endnotes:

1   The power-operated doors for the torpedo tubes had a wave-like form to blend in with the hull lines, and the valves of the main ballast tanks were cast in such a way as to follow the curve of the outer plates, and were fitted with covers that also blended perfectly into the hull lines. These features reduced underwater resistance, albeit at the expense of increased construction times.

2   See Hans Lengerer, 'The Fourth Fleet Incident', *Warship 2013*, 30–45.

3   Navigation in northern waters had demonstrated that heavy seas damaged the upper plates of the tanks, loosening the riveting and causing leakages.

4   According to USNTMtJ Report S.01-1, 12, considerable leakage was observed in operation.

5   An unconventional type was used in which the camshaft was not moved longitudinally; instead a layshaft carried, on eccentrically operated arms, two sets of rollers for the injection nozzles and air starting valves. The turning of the layshaft, either hydraulically or manually, changed the roller contact from the ahead to the astern cams or vice versa within five seconds.

6   The piston was in three sections (upper and lower crowns and centre) and was equipped with five upper and five lower fire rings of the plain cast iron type. The piston was oil cooled via the conventional grasshopper gear and hollow piston rod. The oil entered the piston cooling chamber via the inner tubing and returned through the outer passage formed by the tube outer wall and piston rod inner bore.

7   Lubrication oil consumption was about 30 litres per tonne of fuel. Fuel consumption was approximately 50 tonnes per day for two engines at a speed of 20 knots.

8   The air was sucked in through a series of louvres spaced around the port and starboard top level of the open bridge. The supply pipes were provided with the necessary outboard valves and directed the air flow into the main induction pipe. The air entered the engine at its after end through a large opening in the inboard piping some 0.75m above the engine room deck. The inboard pipe was provided with a flapper valve near the large inlet elbow and tapered off into the bilge to form a drain line port and starboard.

9   The Type 88 tube had been developed by a team directed by the head of the Weapon Design Section of Kobe Seikoshō, Engineer Kumamoto.

10   In fact a pressure reduction valve.

11   The system was retro-refitted to the Type 15 tube and the Type 88 tubes were partially modified, but the IJN's request to replace all existing tubes with the Type 95 could not be realised due to cost and bottlenecks in production and ship/dockyard capacity.

12   See Jiro Itani, Hans Lengerer and Tomoko Rehm-Takahara, 'Japanese Oxygen Torpedoes and Fire Control Systems', *Warship 1991*, 121–131.

13   The solution to this problem was the foremost among the issues to be resolved. Others were: control of the temperature in the generator; the use of sea water as diluent; an increase of the structural strength of the engine; and improvements in the gyroscopic control.

14   In *Kaigun Suirai Shi*, 34, another problem resulting from the smaller diameter is mentioned: the difficulty of maintenance and repair of the oxygen admission valve.

15   When the US battleships *South Dakota* and *Washington* sank the IJN *Kirishima*, the cruiser *Takao* reported six premature detonations out of a salvo of eight torpedoes, and the remaining two torpedoes may have detonated in the wakes of the US battleships.

16   No date is given for this modification, which presumably was undertaken during the first half of 1943. The IJN was not the only navy to suffer from problems with its torpedoes; the US Navy and the German *Kriegsmarine* experienced problems with torpedoes that failed to detonate.

17   It was equipped with a 5cm telescope and the line of sight, which was determined by two upright wires, was adjusted for the correction and firing angles to obtain a firing solution. The submarine had to manoeuvre until the target appeared in the line of sight; the torpedo was then launched.

18   The Japan Optical Co began production of the Mod 2 in April 1943; production ceased in March 1945 after the delivery of 60 units.

19   Readers are referred to the illustrated and detailed description of the USNTMtJ report.

20   It was retracted into a large well 1.5m in diameter in which there was a platform that was picked up by the lower extension of the periscope as it was raised, and on which the operator rode up and down in the trunk.

21   Hasegawa (Yokota) also complains about the order at the beginning of the Pacific War to fire only a single torpedo at small warships and merchant vessels, and claims that the sinking of several merchant ships was not achieved due to this order.

22   When Cdr Kinashi, later CO of *I 29*, returned from Germany his boat was torpedoed by the US submarine *Sawfish* (SS 276) on 26 July 1944 in the Balintang Channel (20°10'N/121°55'E); Kinashi lost his life.

23   *Saratoga* was hit by a single torpedo, which detonated on the after part of the bulge to starboard, causing the after engine room to be flooded.

24   At the beginning of November a flying boat (taking off from Shortland) was supplied with fuel by *I 26* in the Coral Sea. When *I 26* advanced South to search for the enemy, she collided with an uncharted reef and damaged three of the lower torpedo tubes, so only three torpedoes could be fired.

THE DEVELOPMENT OF THE SMALL CRUISER IN THE IMPERIAL GERMAN NAVY (Part II)

In the second part of his design history of the German *Kleiner Kreuzer* ('small cruiser') **Dirk Nottelmann** continues the story by highlighting significant aspects of the second generation of these vessels, beginning with the *Magdeburg* class and ending with the projects drawn up at the end of the Great War.

The appearance of Admiral John Fisher's 'Dreadnought cruisers' made a further increase in the speed of German small cruisers more than just desirable. After learning, semi-officially, in 1907 of their design speed of 25 knots, the results of *Indomitable*'s initial trials leaked through to the German authorities in May 1908. Although initially no more than rumours, a variety of values between 26 and 28 knots were being quoted (she actually made 26.1 knots). This set off alarm bells, as the *Kolberg* class were designed for only 25.5 knots, although they made up to 26.8 knots on trial.

The Germans assumed that these new 'battle cruisers' (as they would be rechristened from November 1911) would, like their own large cruisers (*Grosser Kreuzer*), act as a fast wing of the battle fleet, or in support of small cruisers. This was not initially the case for the British vessels, and thus German small cruisers were unlikely to face them in home waters. Not knowing this, however, the designed speed of the next generation of German cruisers (1909/10 estimates) had to be adjusted according to the information gathered thus far; this was accordingly set at 27 knots – 1.5 knots more than the *Kolberg*s.

It came as no surprise that this determination required the designers to provide significantly more powerful machinery – a designed output of 25,000shp, *vice* 19,000shp in *Kolberg* – with the addition of an extra boiler in the forward boiler room bringing their total number to sixteen. Besides the provision of one more funnel to cope with the increase in the exhaust gases of the enlarged boiler plant, this meant that the foremost boiler room would now require additional volume to install two boilers side by side, which resulted in broader hull lines at this point. Consequently, to keep the forward lines as hydrodynamically efficient as possible in order to achieve the higher required design speed, waterline length had to be extended more than would have been necessary simply to accommodate one extra boiler. What would become the *Magdeburg* class would thus be 6 metres longer than the *Kolberg*s. In addition, the modified stem considered for the second batch of the *Kolberg* class (see Part I, 116) was finally adopted, while the forefoot was cut away, mainly to balance the anticipated deterioration in turning abilities stemming from the new ships' additional length.

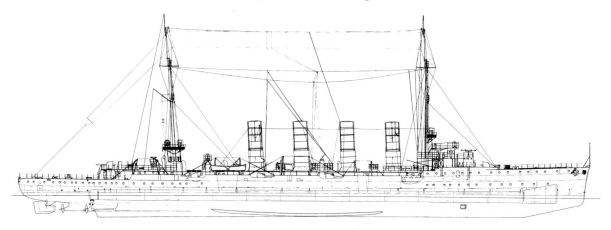

An as-fitted drawing of the small cruiser *Breslau* immediately after delivery. The new 'modern' profile is immediately apparent; it set the pattern for every small cruiser to come. (Drawn by the author)

To ameliorate increased stress moments that originated from the extra length of the hull in relation to the continuous height of its sides – the latter resulting from unavoidable military requirements – the designers decided to deviate from traditional practices. Thus far, the hulls of German naval vessels had been constructed with a combination of classical transverse and longitudinal framing. After lengthy calculations, which would eventually delay the keel laying of the first vessels for more than six months, it was decided to construct the hulls entirely on the longitudinal framing system. The time necessary for these calculations led to another confrontation between Tirpitz and the Kaiser, the latter always eager to accuse the RMA of building too slowly. Tirpitz pointed out that the Kaiser had been kept fully informed by the RMA about the ongoing work, from a first informal speech on 5 October 1908 to the requested decision in April 1909, in the wake of lengthy discussions with the prospective shipyards involved, in order to obtain the approval of the Emperor for the new system. Only after this approval, Tirpitz noted, had he been able to order that detailed plans and specifications be prepared, which – according to his preliminary estimates – would take at least six months. The Kaiser's options in April 1909 had thus been explicitly to choose between a repeat *Mainz*, to be completed more rapidly, or to wait another 10 months for the much-improved type he had insisted upon. As in most of these controversies, Tirpitz's

analysis and references to imperial speeches made long ago prevailed over the more subjective position of the Kaiser, to the benefit of all concerned. The new system proved to be completely successful and would be employed in all later German cruisers.

Longitudinal construction had the added benefit of considerable savings in hull weight, and these savings would be invested in another novel feature, stemming from the military necessity to counter the re-introduction of the 6in (15.2cm) gun in the recent British small cruisers of the *Bristol* (two guns) and *Weymouth* (eight) classes: the provision of a vertical armour belt to protect the waterline. Experience from the battles of the Yellow Sea and Tsushima, at which the largely unprotected sides of Russian vessels had been ripped open by Japanese high-explosive shell, also came into consideration. It was thus foreseen that German vessels, whose vertical protection thus far had comprised only a cofferdam filled with cork, might suffer similarly if faced with 6in HE shell. A waterline belt therefore appeared to be mandatory for future ships.

As was common practice in the IGN, a mock-up target was first erected at Krupp's proving ground. Part of the target was built using the classical construction method: 50mm armour plates bolted onto the skin plating of the hull with a 100mm wooden backing. The other part was more innovative in its structure, introduced not least to minimise weight – as always a key metric. In this, the skin

Ersatz-*Prinzess Wilhelm* (*Graudenz*) on the stocks of the Imperial Dockyard Kiel in 1912. The photo shows clearly the system of longitudinal framing adopted for these ships. (Author's collection)

An illustrious group of officers and civilians pose in front of the cruiser target at the Meppen proving ground: among them are Grand Admiral Tirpitz (in civilian clothes, 8th from right) and Gustav Krupp von Bohlen und Halbach, chairman of the Krupp works (4th from right). (Author's collection)

plating was dispensed with altogether so that the 60mm armour itself formed the outer plating of the hull; this method of construction naturally required the plates to be riveted seamlessly to each other. Both sections were fitted with a sloped armoured deck of 20mm thickness behind the 'belt'. During two consecutive trials the target was fired on using a variety of calibres and projectiles, ranging from 8.8cm HE, 10.5cm HE and shot, 15cm AP, HE and common, to 30.5cm HE. For the first trial the coal bunkers were left empty; in the second the upper bunkers were filled with coal. The results of both trials were considered satisfactory overall, with the 'classical' form of protection judged to be slightly inferior.

During the first series of trials it had been demonstrated that complete protection against nose-fused 10.5cm shell could be achieved, whereas the immunity zone against nose-fused 15cm shell extended down to about 3,400m. Unsurprisingly, the second series validated these results with regard to the nose-fused HE shell but demonstrated, on the other hand, the limitations of the system: 15cm AP shell could be kept out only at ranges beyond 3,900m. Regarding the effect of even larger calibres it was stated in the report that:

> The effect of 30.5cm common shells equipped with nose-fuses against the waterline strake is such that – at least at a distance of 10,000m or more – it will not necessarily always penetrate beneath the armour deck.

Nevertheless, the results were encouraging enough to introduce the 'integral' 60mm waterline belt of low-percentage nickel steel in the forthcoming classes of small cruiser.

A further significant deviation from established hull forms must be mentioned, which should, by a glance at the ships' profile, have indicated even to the less informed eye that a shift in policy had taken place: the omission of the poop deck. Retained mainly to accommodate the officer corps, its continued employment had to be questioned as soon as the offensive laying of mines came more into focus in the wake of experience gained from the Russo-Japanese War. While there is insufficient space to explore in detail the history of the German minelaying service, established at this very time, lack of available funding required that the two purpose-built minelayers *Nautilus* and *Albatross* were henceforth to be supported by fitting each new small cruiser to lay up to 120 mines. This can be seen as a strong indication of the anticipated range of operations for such vessels. As an interim measure the four vessels of the *Kolberg* class and some of their predecessors received provision for stowing up to 100 mines. However, in these vessels, the process of laying them over the sides via the after gangway ports on the main deck was always considered unsatisfactory. Accordingly all forthcoming small cruisers would have a flush upper deck aft of the forecastle to allow the fitting of removable mine rails extending to the stern.

In contrast to what had been done with the preceding *Kolberg* class, where the builders were directed as to which turbine installations to employ, the competing shipyards were (at least in theory) now given much more freedom in their choice of propulsion and auxiliary machinery – although in reality they were bound by subcontracts, patent rights and, later on, by renewed interference from the RMA. This extended to the number of shafts: they could offer twin-, triple-, or four-shaft arrangements at their own discretion. Only one parameter was laid down: maximum speed over the measured mile had to exceed 27.25 knots, while the average speed during the 6-hour run had to exceed 26.75 knots.

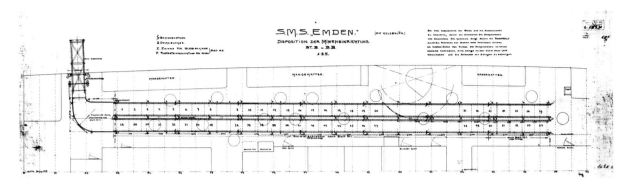

The optional fitting of a double row of mine rails on the main deck of the older small cruisers, in this case *Emden* (i). The mines were to be laid through the gangway port. (Author's collection)

As an example of the complex negotiations that ensued, one may cite parts of the correspondence between Vulcan and the RMA – initially with reference to the yard's offer to build Ersatz-*Bussard* (later named *Magdeburg*), and subsequently for the already-ordered Ersatz-*Falke* (*Breslau*). According to Vulcan's calculations, the most cost-effective option employed four shafts – which would eventually be chosen. The choice of the turbine system is interesting insofar as the Vulcan-preferred reaction-type AEG-Curtis system was chosen, which normally had the advantage of combining the HP and LP stages in one casing, driving two shafts. In this case, however, the stages were separated to produce a four-shaft arrangement: HP turbines drove the inner shafts, while the wing shafts were powered by the LP turbines. Despite this division into four propulsion units, the total weight for this more complex system was less than for the alternatives, coming in at 33 tonnes below the limit set by the RMA in the preliminary specifications. Vulcan went even further: they dared to guarantee an extra quarter-knot above the requirements set by the specifications for the full speed and the 6-hour trials. And they would eventually keep their promise.

The *Magdeburg* Class[1]

Ersatz-*Falke* was ordered from Vulcan in December 1909, when AG Weser of Bremen also received the order for Ersatz-*Bussard*. As the latter yard was linked with the Bergmann company of Berlin, whose newly-introduced turbine branch would be sold to Siemens a few years later, they offered a vessel propelled by three Weser-Bergmann turbines, another derivative of the Curtis type. Their lowest bid was for a three-shaft installation, which was accepted by the RMA, always open to the most economical solution. However, AG Weser was to pay dearly for its choice, as the construction of the turbines lagged badly behind schedule and it took five more months to complete the vessel compared to her Vulcan-built sister.

At the end of 1909, negotiations began for the second pair of vessels of the class, to be built under the 1910 estimates. Because of the long delay in ordering the first pair resulting from the adoption of longitudinal framing, the four vessels were actually ordered quite close together in time. Politics also contributed to this, as a worldwide shipbuilding crisis was underway, and this had hit many German yards hard; thus, all the major yards were lobbying the RMA to be involved in naval contracts. According to the files, Tirpitz was initially inclined to offer Schichau one of the vessels without competition, but changed his mind shortly afterwards – probably after an intervention by Hunold von Ahlefeld, a former colleague of the admiral at the RMA and now Director of AG Weser. Like other yards, Weser had offered to build a second vessel at a reduced price should the yard receive the contract for one of the initial pair. As the Weser yard, unlike Schichau, was not involved in the building of one of the 1910 battleships or large cruisers, it was in desperate need of work and stood on the verge of bankruptcy. This was a situation Tirpitz could not afford, as he needed the support of all the big shipyards to continue his 'plan', if only to keep competition alive. Thus Tirpitz decided to make a direct award of the first of the 1910 cruisers to Weser, albeit subject to a significant reduction in its original bid, down to a price that Germania had now offered – which was actually the same as the price for Vulcan's Ersatz-*Falke*.

There seems to have been an earlier understanding between the RMA and Weser that the new cruiser (Ersatz-*Cormoran*, later named *Stralsund*) would not follow Ersatz-*Bussard* in being propelled by the three-shaft Bergmann arrangement, but instead employ a three/four-shaft installation using Turbinia-Parsons turbines. The reasoning behind this is not entirely clear, but it seems that the RMA was unconvinced by the cruising mode of the three-shaft installation. However, the financially-pressed shipyard could not afford, after being forced by the RMA to reduce prices further, to pay licence fees to Turbinia-Parsons, and thus had to revert to the cheaper three-shaft Bergmann installation; no objections were forthcoming from the RMA.

Something of an enigma surrounds the way in which the contract for the fourth cruiser of the series was allocated, as it was given to the Imperial Shipyard Wilhelmshaven, thus taking away a precious order from the private shipyards in hard times.[2] The most likely reason for this move is that Ersatz-*Condor* (later

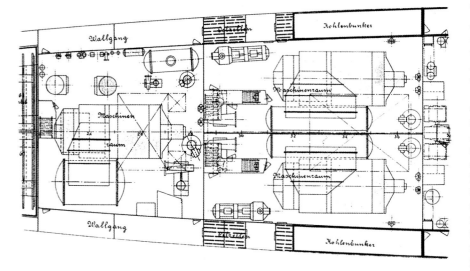

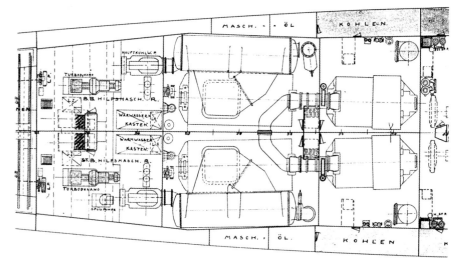

The three-shaft machinery arrangement of *Stralsund* and *Magdeburg* (above) compared to the two-shaft arrangement of *Strassburg* (below), which would become standard in the later German fleet cruisers. The casings for the Bergmann turbines adopted for *Stralsund/Magdeburg* housed the HP stage as well as the LP stage, whereas in the Navy-type turbines of *Strassburg* they were separated. The condensers for the wing turbines had to be placed close to the centreline bulkhead due to space constraints in the former, while in the latter the larger condensers required so much space that the wing compartments had to be reduced in width.
(Author's collection)

Strassburg) was to be used as a test-bed for a new Navy-type turbine, initially manufactured exclusively by the Imperial Shipyard. The yard had just expanded its machinery building plant for that purpose, and was thus eager to show its capabilities. The Navy-type turbine derived from a design by Brown, Boveri & Cie (parent of the Turbinia company), which had been evolved from a pure Parsons-type into a Curtis/Parsons hybrid in which the casings for the HP and the LP stages were separated in two engine rooms but drove a common shaft. During trials the vessels achieved the results tabulated in Table 1.

In addition to the delay caused by the late delivery of her machinery, *Magdeburg* was hampered during her short career by technical problems related to the Bergmann turbines. The closing report of her trials commented favourably on all aspects of their functioning, but just three months later both wing turbines were suffering severe leaks, so that *Magdeburg*'s maximum speed was reduced to 22 knots and even that speed could be attained for short sprints only. Twelve months later, in January 1914, her centre turbine failed and had to be removed for rebuilding. It had not been reinstalled by mobilisation in August, thus her operational speed was again reduced to about 22 knots on the wing turbines only, which limited her employment to the less important Baltic theatre. Interestingly, her 'machinery sister' *Stralsund* was initially unaffected by these problems but suffered her own breakdown of the

Table 1: **Performance on Trials**

	Magdeburg	*Breslau*	*Stralsund*	*Strassburg*
On the measured mile	29,900shp	33,470shp	35,515shp	33,740shp
	27.6 knots	27.6 knots	28.3 knots	28.3 knots
6-hour sustained trial run	25,100shp	28,870shp	27,032shp	25,650shp
	26.9 knots	27.2 knots	27.0 knots	26.9 knots

Breslau during her trials at high (but not maximum) speed in perfectly calm water. (Author's collection)

centre turbine in the summer of 1916. During August of that year, first her centre shaft and subsequently the whole turbine were removed. In this condition *Stralsund* operated on two shafts until August 1917, when the Imperial Shipyard Kiel installed the reconditioned centre turbine of *Magdeburg*, thereby restoring her full capability, and *Stralsund* became probably the IGN's most active small cruiser during the war.

Despite these technical issues, in the *Magdeburg* class the IGN received a very valuable reinforcement of its cruiser squadrons. They would prove their effectiveness and sturdiness during a whole range of deployments in peace and war, and offered a firm basis for the design and construction of future vessels.

The *Karlsruhe* Class

On 6 May 1910, members of the relevant departments assembled at the RMA to discuss the requirements for the 1911 cruiser. Initially, Tirpitz was willing to grant an increase in displacement of only 100 tonnes, as he had just been informed that the money available for small cruisers was only sufficient to cover the average cost of the 1909/10 ships. However, at the same time there were some pressing demands from the military side that it had not been possible to work into the previous class. The main issue was the narrowness of the belt. Tirpitz stated that it was imperative that flooding through large holes close to the waterline must be avoided in all circumstances. The second issue was whether any approved higher displacement should be used to enhance the armament. As usual opinion was far from unanimous. One faction, led by Rear Admiral Gerdes, head of the

weapons department 'W', advocated a mixed armament, comprising fully-enclosed 15cm quick-loading (QL) twin mounts fore and aft, supported by six 8.8cm anti-torpedo boat guns on the beam. Another faction, led by Captain Schrader, head of general department 'A', proposed a uniform installation of twelve 12cm QL. A third option was to group both the forward and the after 10.5cm QL in twin mountings. All of these options would have had the disadvantage of delaying construction by several months, as construction department 'K' would have to be involved to a greater degree than current resource levels permitted. In particular, over the following months it was fully occupied with the development of a diesel battleship. This was, according to Tirpitz, intended to avoid Germany losing the 'lead we have among the nations' during this anticipated revolution in propulsion engineering. Thus, in the end, only marginal improvements were worked into the 1911 cruiser, which is apparent from the wording of the official order for construction:

I decree that the small cruisers Ersatz-*Seeadler* and Ersatz-*Geier* shall be constructed according to the plans of Ersatz-*Bussard*, incorporating the following changes: displacement shall be about 4,880 tonnes, WL length shall be 139m, the height of the WL belt shall be extended up to 2.05m above the CWL, whereas the horizontal part of the armoured deck is to be 1.45m above the CWL. Freeboard is to be increased to 3.625m amidships and 7.8m at the bow. Bunker capacity is to be increased to 1,350 tonnes of coal. The choice of the turbines and the number of shafts is left to the discretion of the state secretary of the RMA.

Wilhelm IR[3]

Another new feature stemming from the freedom of choice granted by this comparatively open decree was the introduction of oil-fired boilers in cruisers. The advantages of oil as a combustion material had not gone unnoticed in the RMA; indeed it had been first employed as early as 1870, before it descended into obscurity for more than two decades. Shortly before the end of the 19th century trials were renewed, first as part of mixed firing, later as oil firing alone. However, it took more than another decade to develop a reliable boiler-burner combination that, in 1910, coincided with an offer made by the General Petroleum Industry joint stock company, granting the IGN oil at a price around twice that of coal – which was seen as favourable. It was then decided to equip every large vessel with a mixture of coal- and oil-fired boilers – employing an approximate 6:4 ratio of output for cruisers. Cautious as ever, the RMA did not apply this ratio to Ersatz-*Seeadler* (later *Karlsruhe*) and Ersatz-*Geier* (*Rostock*). In the latter ships the total number of boilers was reduced to fourteen, twelve of them coal burning, with the (oil-fired) forward pair doubled in size and fitted as double-enders; this was possible only because there was no need to leave space for stoking. Bunker capacity was enlarged to 1,300 tonnes of coal plus 200 tonnes of oil.[4] The designed power of the two Navy-turbine sets driving the two shafts, which would be employed in all future German small cruisers of RMA design, was raised to 26,000shp, resulting in a slightly higher design speed of 27.8 knots. However, as trials would reveal, there had been a miscalculation regarding the output of the twelve coal-fired boilers, and extra coal had to be fed into the furnaces to compensate for a grate area that was too small. This in turn affected the overall efficiency of the propulsion plant, resulting in the endurance of the new class being inferior to that of the *Magdeburg* class (5,000nm *vice* 5,800nm), despite the latter's smaller bunker capacity. Apart from this drawback, the machinery proved very satisfactory in operation: *Rostock* exceeded her designed horsepower by 17,000shp, attaining 29.3 knots on the measured mile, while *Karlsruhe* reached 28.4 knots with 37,450shp. Both would be the last small cruisers to be completed and run their trials in peacetime.

The *Regensburg* Class

The 1912 small cruiser again differed little from the preceding class even if, at first glance, its profile suggests otherwise. It was ordered during the difficult days of the introduction of another '*Novelle*' to the Fleet Law, which was ultimately approved despite reductions in overall funding. It was agreed that the cost of all vessels, in particular battleships, large cruisers and small cruisers, was to be kept down as far as possible. As a result, the changes from the *Karlsruhe* design were largely cosmetic. There was a further reduction in the number of boilers, from fourteen to twelve, allowing a reduction in the number of funnels from four to three while maintaining the requirement of exhaust ducts that were close to the vertical. The ratio between coal- and oil-fired boilers was

Karlsruhe leaving the inner port of Kiel, having passed *Blücher* to starboard and the fleet flagship *Friedrich der Grosse* to port. Externally, she differed little from the preceding class. (Author's collection)

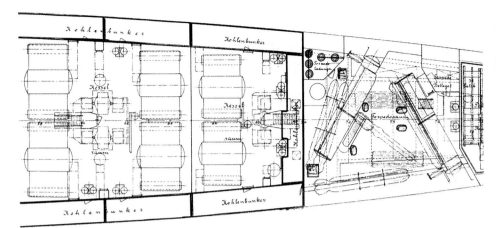

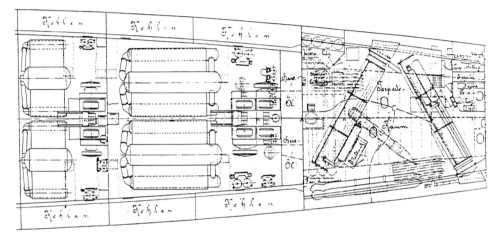

Two extracts from the plans of *Stralsund* (top) and *Karlsruhe* (bottom) showing the introduction of the oil-fired double-ended boiler in the latter. It is probable that the explosion that ultimately destroyed *Karlsruhe* originated from a fire in her athwartship oil bunker. (Author's collection)

also reduced, as the 'lost' boilers were coal burners, while both double-ended oil-fired units were retained. The installed power nevertheless, remained at 26,000shp for 27.25 knots. Both vessels exceeded their designed power: Ersatz-*Irene* (*Regensburg*) attained 38,500shp for 28.0 knots, while Ersatz-*Prinzess Wilhelm* (*Graudenz*) improved on this with 42,470shp for 29.2 knots on the measured mile. The difference in performance between the two ships can be attributed to the trials being run in different depths of water.

The most striking difference between the new cruisers and their immediate predecessors was the number of funnels and the height of their masts. *Regensburg* and *Graudenz* were both intended as flagships for the commodores of the torpedo boat flotillas, as the IGN had given up on the construction of a separate class of vessels for that particular purpose, with flotillas led tactically by boats of the standard series. To allow them to communicate with the fleet flagship and with HQ ashore long-range wireless telegraphy was required, secured by an increase in height of the masts of both ships by 7 metres compared to the 'standard' masts of earlier ships.

A further modification would serve as a stepping stone towards to the major next innovation in development: the superimposition of the after 10.5cm guns, actually achieved by moving one of the guns down from its former position atop the after superstructure to the upper

deck. This was a left-over from informal discussions that had suggested that the vessels might be equipped either with ten 13cm[5] or with eight 15cm guns. In the case of the eight-gun variant, superimposition was seen as mandatory in order to provide a sufficient number of guns in the broadside. Even though this up-gunning came to naught for the time being, and remained a far cry from the mooted twin mountings for the *Karlsruhe* class, it nevertheless set the stage for a more effective distribution of the main artillery and greater things to come.

The Small Cruiser of 1913

The development of the 1913 small cruiser marked a significant waypoint in the history of the IGN that has tended to go unnoticed and unremarked upon by most historians. Since the introduction of the small cruiser, the fundamental nature of the type had been settled and improvements had been made in small increments only, and at the beginning it was far from clear that the 1913 ships would differ significantly from their predecessors. Initially the main premise for cruiser 'K' was stated as follows:

Changes in displacement and costs compared to the Ersatz-*Irene* type shall be kept to a minimum as far as possible.

A fine image of the recently-commissioned *Graudenz*, taken on 17 August 1914 at her builder, the Imperial Dockyard, Kiel. Her tall masts are prominent as are her 'stepped' aftermost 10.5cm guns. (Author's collection)

However, events moved rapidly, especially as the 1913 estimates gave a little more financial freedom, as only two battleships/large cruisers were to be laid down in accordance with the Fleet Law. Thus, in May 1912 the Kaiser had ordered Tirpitz to report, *inter alia*, about the intended future armament of small cruisers. The subsequent report was drafted by Rear Admiral Scheer, then head of 'A'. He now openly opted for the 15cm solution, already discussed in previous years, as a response to the first reliable information regarding the most recent British cruisers, the *Chatham* class, which were reported to be fitted with a waterline belt, the extent and thickness of which was so far unknown. Initially this solution was rejected by the Kaiser and the fleet command, whose principal concern was the unhandiness implicit in larger vessels; in their view a 13cm calibre would suffice. However, Tirpitz was of the view that this calibre did not possess sufficient 'punch' against any form of side armour. Thus, the 13cm variant would eventually, apart from a brief oral mention, not even be presented to the Kaiser during the annual autumn report at Rominten, and the omission was apparently not even noticed, the 15cm variant being accepted. Additionally, for the first time, four 5.2cm anti-aircraft guns or *Ballon-Abwehr-Kanonen* (BAK) were proposed, to deal – proactively, as the British did not possess operational dirigible airships thus far – with the anticipated aerial threat. These guns were to be mounted in pairs on a common platform between the bridge and the first funnel.

At this point it is worth remembering the long struggle for the introduction of the 15cm calibre into small cruisers. Apart from the statements in various later publications – led by Tirpitz himself – regarding the rationale behind the decisions either to retain the 10.5cm gun or to mount larger calibres, the final explanation for employing the latter – the introduction of side armour into British small cruisers – was in the end simple. Despite our knowledge that he had on occasion expressed a personal preference for a larger calibre than 10.5cm ever since 1894, Tirpitz would, nevertheless, choose to cloud the whole process in his memoirs:

Regarding the small cruisers, a criticism had been made that they had been too lightly armed. ... As long as the effective range of torpedoes had been comparatively short, that is until 1910/11, and torpedo boats had to close considerably to effectively deliver their projectiles, the small cruiser's 10cm [*sic*] gun was not only sufficiently efficient for engaging hostile torpedo boats but was even superior to a 15cm gun because it could be mounted in greater numbers and its rate of fire was higher than that of the larger calibre.[6]

Other official sources mention the higher costs of the larger calibre, which had precluded its introduction since 1909, and so indirectly – and perhaps unintentionally – blame Tirpitz again for not securing sufficient funds for shipbuilding purposes. One must also not forget that, in the wake of the gunnery trials on *Mainz* mentioned in Part I of this article (*Warship 2020*, 117), the fleet Commander-in-Chief, Admiral von Holtzendorff, as well as the leader of the I. Scouting Group, Rear Admiral Bachmann, had argued for the retention of the 10.5cm calibre.

There were other innovations included in this groundbreaking class of vessels. The first was a return to trainable deck-mounted torpedo tubes of the kind last fitted in *Gazelle*. Nevertheless, submerged broadside tubes were retained, thereby resulting in a doubling of the torpedo armament. The upper deck mounts were installed as far forward as possible so as not to interfere with the mine rails aft, in case the full load of up to 120 mines was embarked.

Another innovation was the introduction of the 'Föttinger transformer' in *Wiesbaden* (Ersatz-*Gefion*) which, for the first time, opened up the possibility of reducing the speed of rotation, and hence increasing the efficiency, of the propellers.[7] On the measured mile in the Little Belt *Wiesbaden* attained an output of 39,531shp.

This resulted in a speed of only 27.4 knots; however, if corrected for deep-water running this would correspond to a maximum speed of about 29 knots. *Frankfurt* (Ersatz-*Hela*) is reported to have reached 42,500shp, although the maximum speed has not been recorded. Her highest confirmed speed value was attained on a run at deep draught on 22 September 1915, giving 26.7 knots at an unrecorded power, which corresponds with her sister's performance but may not reflect the maximum horsepower of which she was capable. Unfortunately these two sisters, although present at Jutland (with fatal consequences for *Wiesbaden*), and subsequent vessels derived from them, had few chances to display their full potential in ship-to-ship engagements on comparable terms. Accordingly, their true value must remain uncertain. At least one of their characteristics was, however, noted early on – indeed, before the 1913 vessels were launched and the 1914 vessels finally approved – that would influence discussions regarding future construction, and which will be dealt with below.

The 1914 Small Cruiser and Beyond[8]

The contract for the 1913 vessels had not even been signed when 'K' approached 'A' in March 1913 regarding any changes to the 1913 design that might be desirable for the 1914 small cruiser. 'K' had already put forward five proposals of which the most notable were the enlargement of the accommodation for both officers and men, a larger calibre for the BAK and – rather surprisingly – a submerged twin torpedo tube mounting. 'A' was quick to respond that, in general, the 1913 type should be retained, although small variations, such as the exten-

Images of *Wiesbaden* are rare. Here she is seen running her trials at high speed on 12 September 1915. The large trough aft indicates the shallow depth of the water. Just visible below the second funnel is the upper-deck 50cm torpedo tube. (Author's collection)

sion of the deckhouses or the substitution of two 8.8cm BAK for four 5.2cm BAK might be feasible. After the usual hiatus over the summer, the project was presented to Tirpitz for his approval before being taken to Rominten in September to secure the approval of the Kaiser. Tirpitz endorsed the decision to adhere to the existing general design, although necessary changes (which were not yet fully agreed) would raise the displacement by a further 150 tonnes. Shortly before final approval was given in April 1914 another proposal brought the agreed design to its limits, when 'K' opted for the replacement of the linoleum cladding of the main deck with wood for maintenance reasons. Linoleum had been found susceptible to excessive wear during coaling. Its substitution by teak planking 55mm thick added an extra 40 tonnes in weight and an extra 40,000 Marks in cost. In April 1914 the Kaiser signed the order of construction for the last German cruisers designed in peacetime, once more in a rather peremptory fashion:

> I decree that for the small cruisers Ersatz-*Gazelle* and Ersatz-*Niobe* the plans of the Ersatz-*Gefion* class shall be retained, subject to the following changes: WL length shall be 145.8m, maximum beam shall be 14.32m, design draft shall be 5.28m aft, 4.83m forward, displacement shall be about 5,300 tonnes.
>
> Depending on the results of the tests that are currently proceeding either four 5.2cm or two 8.8cm BAK will be mounted.
>
> The choice of the turbines and the number of shafts is left to the discretion of the state secretary of the RMA.
>
> Bunker capacity is to be 1,340 tonnes of coal and 500 tonnes of oil.
>
> Given at Achilleion, Corfu, 24 April 1914.
>
> Wilhelm IR

Just two months later these specifications would be changed as, during the trials of *Strassburg*, it had been found that at high loading of the boilers the generated steam tended to condense at the turbine nozzles, which endangered the engines. Thus, the new vessels needed superheaters to dry the steam and thus avoid the problem in the future; this required a further 20 tonnes of displacement, which was beyond design limits. Consequently the bunker capacity for oil had to be limited.

In the meantime, following the first presentation of the draughts for the 1914 cruisers at Rominten and their final approval in April, 'K' pressed for work to begin on outlining the design for the 1915 small cruiser. The reason for this haste is hard to see, as the department was

not overly busy with new designs at the time. However, the main point of concern was the current small cruiser's speed deficit in relation to the latest large cruisers/battle-cruisers of other nations, principally the UK, although similar ships were now being built for Russia and Japan. The rule of thumb, as laid down by 'A', remained that 'the small cruiser must possess at least the same speed as the modern large cruiser, while carrying the highest possible bunkerage',[9] and the IGN was now confronted with known or anticipated figures of 30 knots for the UK's *Tiger*[10] and 29 knots for Russia's *Borodino* class.[11] The situation was exacerbated by an understanding that the latest British small cruisers (the *Arethusa* class) were significantly faster than previous vessels.[12]

Accordingly, 'A' proposed increasing the design speed of the future small cruiser to at least 28.5 knots. Since, for budgetary reasons, a major increase in the size of these vessels had to be avoided, the necessary extra power output would have to be achieved by an increase in the ratio of oil-fired to coal-fired boilers. However, this substitution of fuels was not to go as far as in the Royal Navy, which relied entirely on oil for its newest small cruisers: a certain number of coal-fired boilers should be retained, sufficient for cruising at around 15 knots. To support the resulting increase in oil stowage, it was seen as necessary to perform flammability tests on oil bunkers using 15cm HE shells – and even larger calibres. Still not settled was the question regarding the substitution of two 8.8cm BAK for the four projected 5.2cm BAK, as the new larger-calibre gun had only been ready for initial trials since November 1913. Although still seen as desirable, it was noted that this measure would probably add yet another 40 tonnes to the design.

During the subsequent interdepartmental discussions, in July 1914 'K' questioned the necessity for increased speed, as all the relevant information from abroad had come from unconfirmed oral sources. Even more seriously, owing to the small size of the vessels, a switch in emphasis towards speed would significantly degrade habitability and the suitability of the type for the dual fleet/overseas role that had underpinned the small cruiser type since its inception. The only solution to this dilemma would be either to introduce a special smaller 'fleet cruiser' similar to the British *Arethusa* class, or to increase displacement by about 1,500 tonnes for the 'standard' cruiser. Either way, the verdict was that even without major modifications the 1915 cruiser would need at least another 100 tonnes of displacement, just to compensate for the already acknowledged shortcomings of the 1914 vessels. A further 50 tonnes would be required for the proposed change in relation to the ratio

Table 2: Estimated Performance & Cost

	Displacement	6h trial speed	Maximum speed	Cost
1914-type	5300t	27.25kts	27.5kts	8.7m Marks
1915 – I	5400t	27.25kts	27.5kts	8.8m Marks
1915 – II	5600t	27.75kts	28.0kts	9.2m Marks
1915 – III	5700t	27.75kts	28.0kts	9.3m Marks

The first of the 1914 cruisers, *Königsberg* (ii), navigating the River Weser on 24 August 1916. Except for the enlarged bridge structure and the raised forefunnel, there are few changes from her immediate predecessors. (Author's collection)

of coal- and oil-fired boilers. 'K' thus provided three options to reflect the various boiler options and necessary changes in the configuration of the bunkers. Variant I would be the 'natural' follow-on, incorporating only the 'necessary' changes, Variants II and III would fit eight coal- and six single-ended oil-fired boilers: Variant II would have the oil bunkers arranged alongside the boiler rooms, while Variant III would retain the classical bunker arrangement of stowing oil in the double bottom. Estimated performance and cost of the different variants is tabulated in Table 2.

However, no 'official' 1915 cruiser would actually be built. The 1914 cruisers had been ordered in June 1914, just two months after their approval. Ersatz-*Gazelle* (to be named *Königsberg* [ii]) would be laid down by AG Weser in early August 1914, but Ersatz-*Niobe*

(*Karlsruhe* [ii]), built by Imperial Dockyard Kiel, had to wait until 5 May 1915 before eventually being laid down owing to the excessive workload at that yard, which was responsible for mobilisation-related support in the Baltic after war had commenced.

The ordering of the second pair of 1914 cruisers, Ersatz-*Nymphe* (*Emden* [ii]) and Ersatz-*Thetis* (*Nürnberg* [ii]), is shrouded in an obscurity that is exacerbated by gaps in the surviving documentation. The ships were ordered in August 1914, in spite of not legally being 'due' as replacements for the old *Nymphe* and *Thetis* until 1915 – the Fleet Laws were very explicit in stipulating that a vessel could be replaced only after the determined lifespan or in the event of loss. As the IGN had already lost four small cruisers in August alone, *Magdeburg* in the Baltic, together with *Mainz*, *Cöln* and

Table 3: **The Wartime Successors to the 1914 Ships**

Replacement	To be named	Shipyard	Ordered	Laid down
Ersatz-*Leipzig*	*Leipzig*	AG Weser	10 Apr 1915	August 1915
Ersatz-*Ariadne*	*Cöln*	Blohm & Voss	14 Apr 1915	August 1915
Ersatz-*Nürnberg*	*Wiesbaden*	Vulcan	10 Apr 1915	October 1915
Ersatz-*Emden*		AG Weser	10 Apr 1915	August 1915
Ersatz-*Mainz*	*Rostock*	Vulcan	10 Apr 1915	November 1915
Ersatz-*Cöln*		AG Weser	10 Apr 1915	August 1915
Ersatz-*Dresden*	*Dresden*	Howaldt	7 Aug 1915	May 1916
Ersatz-*Magdeburg*	*Magdeburg*	Howaldt	7 Aug 1915	June 1916
Ersatz-*Königsberg*	*Frauenlob*	Imp Dyd Kiel	20 Aug 1915	December 1915
Ersatz-*A*[1]		Imp Dyd Kiel	20 Aug 1915	September 1916

Note:

[1] Ersatz-*A* is sometimes referred to in print as Ersatz-*Karlsruhe*, but this is not reflected in the relevant files; nor is there an indication of her being ordered as an addition to strength, although this might be indicated by her letter designation – although the use of 'Ersatz-' is odd. Among the large cruisers there is a similar case of Ersatz-*Friedrich Carl* versus Ersatz *A*.

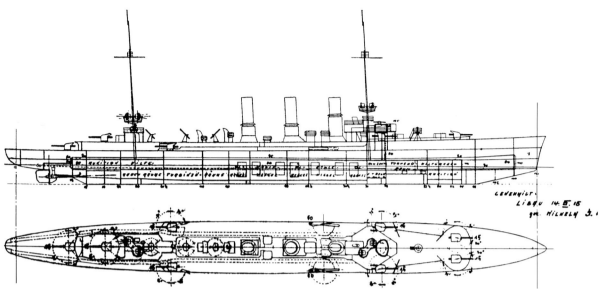

The sketch for the Ersatz-*Leipzig* class which gained the Kaiser's approval at Libau on 14 December 1915. Two significant features, later to be changed, are the position of the mainmast abaft the third HA gun (BAK), as well as the initial retention of submerged torpedo tubes forward of the foremost boiler room. (Author's collection)

Ariadne at Heligoland, their immediate replacement would legally have been possible. However, it is puzzling that only two vessels should then have been ordered, and under designations that referred to ships not yet legally replaceable rather than two of the lost cruisers. On the other hand, there is some indication that the August 1914 'order' was only a a 'promise' to the yards, since Ersatz-*Nymphe* would actually not be laid down by AG Weser before December – which could be easily passed off as 'almost' 1915 – while Ersatz-*Thetis* had to wait even longer, until 8 May 1915, before being laid down at Howaldts, the contract for the latter cruiser not being signed before 13 January of that year.

Before looking at the important question of the rearmament of the older small cruisers, which would be raised as early as January 1915, we will consider the design of the actual successors to the 1914 vessels, of which only two would eventually commission before the end of the war. On 22 March 1915 Tirpitz presided over another meeting of the RMA's departments convened to consider the ordering of replacements for small cruisers lost thus far during the war. The discussions mainly concerned numbers, rather than the design of the ships themselves, as any major deviation from the 1914 cruiser would lead to a loss of precious time. In the end it was decided that, of the nine replacement vessels, six would be ordered immediately to the same general specifications as the 1914 cruisers with a few modifications, none of which would exactly match one of the three '1915' proposals mentioned above. These included a change in the coal to oil ratio (to 8:6, with two extra boilers: maximum oil bunkerage was doubled, but that for coal reduced by 18 per cent), while the orders for the remaining three would be held back for about three months in order to allow for the possible inclusion of the latest lessons from the war.

One major lesson that would be worked into the 1914 design, including the vessels already laid down, was the raising of the 15cm mountings abreast the bridge from the upper deck to the forecastle deck. This resulted from the experience gained with *Strassburg* during her cruise with the 'Detached Division' to South America, which found that, especially in this forward position, the upper deck guns could be very wet, a problem exacerbated by their being sponsoned out. Additionally, it was observed that neither of these guns in the 1913 cruisers was able to fire straight ahead. Thus, this change was applied to the four original 1914 cruisers and the war-loss replacement vessels.

Experience gained through the war led to further changes, in particular the abandonment of the submerged torpedo flat that had long been a feature of small cruisers. There always had existed the fear of sympathetic detonations of the torpedo warheads in the event of a hit, either by mine or torpedo, and despite the mixed practical experience in this matter, its deletion was proposed in February 1916 by the chief of the Scouting Forces, Rear Admiral Hipper, following the loss of *Prinz Adalbert* and *Bremen*. Instead, two additional trainable G/7 torpedo tubes were added on the upper deck of the war-loss replacement vessels, and these were later uprated to enable them to launch the newly-introduced 60cm H/8 'battleship torpedo'. This would not have any negative impact on stability until the new mounts were covered by optional armoured hoods. A further modification was the reduction of the originally envisaged four 8.8cm BAK to three.

After the first trials of *Königsberg* (ii) in September

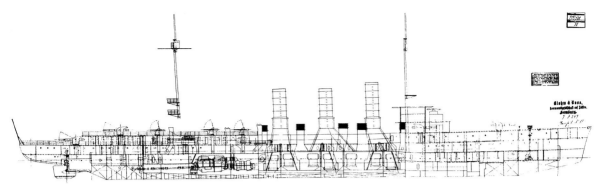

This excerpt from the booklet of plans for Ersatz-*Ariadne* show that there were still some uncertainties as to her final configuration. While the submerged torpedo tubes had already been suppressed, the position of the mainmast had yet to be finally decided upon. Not visible in this profile are the newly-introduced anti-rolling-tanks in this ship. (Author's collection)

1916, Hipper as well as the weapons department 'W' requested yet another change, in this case regarding the 15cm guns abreast the bridge. Both wanted a reversal of the decision to mount the guns on the forecastle, for reasons of flash and splinter protection of the guns and the adjacent superstructure. During a decisive interdepartmental meeting on 24 September, it was agreed almost unanimously that, despite the advantages of the present design as seen from the design department, the arguments from the fleet should prevail. However, given the different levels of progress in the work on the various vessels, and also the different priorities accorded to them in the interim, the order for the modification could not apply to all of them. Thus, the three still-unfinished 1914 cruisers, as well as Ersatz-*Ariadne*, were too far advanced to be rebuilt, while the remaining five vessels ordered in April 1915, then scheduled to be completed in 1917, owing to their having been assigned Priority I, were to be

modified under the said order. A decision on the four vessels of the August 1915 batch remained open for the time being. The RMA was not overly concerned by this development, because it would allow direct comparison between the two configurations of the war-loss replacement design. At that date nobody anticipated that only two of these vessels – one of each configuration – would eventually be commissioned.

The actual process of ordering the two new batches of cruisers was unusual, to say the least, for the otherwise strictly formal IGN. On 23 February 1915, Tirpitz had informed the Kaiser for the first time about the RMA's intention to continue with the 1914 design, subject to a few alterations. The Kaiser gave oral consent and, on the basis of this, in April the first batch of six vessels was ordered by the RMA, followed by the second batch in August, while detailed design work was still proceeding. Finally, in 14 December 1915, the Kaiser

A fine broadside view of the last small cruiser of the IGN, *Dresden* (ii). She marks the ultimate development of the type, with the positioning of the second gun at upper deck level, and the mounting of the four 50cm torpedo tubes on the same level. While the mounting of the third BAK abaft the mainmast can at least be verified for *Cöln* (ii), the same cannot be said for *Dresden*. (Author's collection)

Table 4: **Characteristics**

	Magdeburg	Karlsruhe	Regensburg	Wiesbaden	Königsberg	Cöln
Estimates	1909/10	1911	1912	1913	1914	war fund
No of ships	four	two	two	two	four	ten
Displacement (des)	4,564t	4,902t	4,912t	5,150t	5,300t	5,600t
Length oa	138.7m	142.2m	142.7m	145.4m	150.6m	155.5m
Beam (max)	13.3m	13.7 m	13.7m	13.9m	14.2m	14.2m
Draught (aft/fwd)	5.1/4.5m	5.2/4.5m	5.2/4.5m	5.2/4.7m	5.3/4.8m	5.3/4.8m
Engines	2/3/4-shaft turbines	2-shaft turbines	2-shaft turbines	2-shaft turbines	2-shaft turbines	2-shaft turbines
Horsepower (des)	21,600shp (22,800shp – *Breslau*)	26,000shp	26,000shp	29,000shp	29,000shp	29,000shp
Boilers	16 coal-fired	12 coal-fired 2 oil-fired	10 coal-fired 2 oil-fired	10 coal-fired 2 oil-fired	10 coal-fired 2 oil-fired	8 coal-fired 6 oil-fired
Speed (des)	26.75kts (27.0kts – *Breslau*)	27.25kts	27.25kts	27.25kts	27.25kts	27.25kts
Bunkerage	1,200t coal	1,300t coal 260t oil	1,280t coal 370t oil	1,280t coal 470t oil	1,340t coal 500t oil	1,100t coal 860t oil
Endurance[1]	5,820nm (12kts)	5,000nm (12kts)	5.500nm (12kts)	4,800nm (12kts)	4,850nm (12kts)	5,400nm (12kts)
Armament	twelve 10.5cm	twelve 10.5cm	twelve 10.5cm	eight 15cm two 8.8cm BAK	eight 15cm two 8.8cm BAK	eight 15cm three 8.8cm BAK
	two 50cm TT	two 50cm TT	two 50cm TT	four 50cm TT	four 50cm TT	four 60cm TT
Armour Belt	60mm	60mm	60mm	60mm	60mm	60mm
Protective deck	40/20/20mm	40/20/40mm	40/20/20mm	40/20/20mm	40/20/20mm	40/20/20mm

Note:

[1] Endurance was never specifically determined in advance. Consumption was measured during trials over several different speeds and an extrapolating graph drawn afterwards; this would represent trials condition only. The values given here are averages at a specified speed given in brackets.

signed a formal order of construction that covered all ten vessels. It would be the IGN's final order for small cruisers.

Rearmament

While the design work for the war-loss replacement vessels was going on, 'K' was faced with another task. All engagements by German small cruisers during the first six months of the war had given ample evidence that they were undergunned, whether their opponents had been British or Russian. Interestingly, the first push in the new direction came in December 1914 from the Admiralty Staff, which wanted to continue the war on trade by employing the new cruisers *Pillau* (the former Russian *Muravev-Amurskiy*, seized while under construction by Schichau at Danzig), which had just commissioned, and *Regensburg*, due to commission in January 1915. As the former generated the majority of her power through oil-fired boilers, her potential dependence on a proper oil supply was regarded as too high, and discussions therefore focused mainly on *Regensburg*. In this context, Admiral von Pohl strongly argued for an upgunning to compensate for the now-acknowledged inferiority of the artillery *vis-à-vis* British small cruisers, and proposed replacing the forward two 10.5cm guns and the after mounting on either side amidships by four 15cm guns in total. A few days later, Tirpitz initially agreed to the change, but with only the forward pair and the after 10.5cm guns replaced by single 15cm weapons.

This decision caused a wave of opposition from the various responsible departments and front-line commands. All were united in their concerns regarding the control of such a mixed armament, the disadvantages of which were considered to outweigh the putative benefits of possible single hits by the larger calibre. In the end, Pohl withdrew his proposal; however, following his appointment as C-in-C of the High Seas Fleet that same month, he immediately requested a complete rearmament from 10.5 to 15cm for all of the 'fast' small cruisers. Of the earlier vessels, only *Bremen* and *Lübeck* would receive 15cm guns – and then only two, replacing their forward and after pairs of 10.5cm weapons – owing to weight limitations and their exclusive use in the Baltic. The RMA responded positively to Pohl's proposal, proposing a six-gun 15cm fit, supplemented by two 8.8cm guns on each broadside. Pohl, in return, argued for the addition of a seventh gun aft instead, omitting the relatively worthless 8.8cm mounting. In this, he was strongly supported by the leader of the II. Scouting Group and all its commanding officers.

The main obstacle to a quick conversion of all relevant vessels was the limited number of available guns. In February 1915, the stock of 15cm guns amounted to a full outfit of just two cruisers armed with seven guns; the question therefore arose of which vessels should be prioritised. It was decided to base the initial studies on *Stralsund* and *Strassburg*, the oldest and slowest of the side-armoured ships. If additional weights of 50–60 tonnes were accepted, which would increase draught by

Following her reconstruction, *Lübeck* can be seen here carrying 15cm guns on the forecastle and the quarterdeck, while retaining six 10.5cm at the level of the main deck. (Author's collection)

4–5cm, their conversion was seen as both technically possible and most desirable from the military point of view. In a display of rare unanimity, all involved ultimately approved a uniform armament of seven 15cm L/45 SK with no intermediate calibre – in the case of the older vessels without even any BAK, if necessary. The order of these initial conversions would depend on when each vessel was due for a major refit or repair in the normal course of operations.

In July 1915, Rear Admiral Hipper endorsed briefly the proposal of *Rostock*'s CO to mount an eighth gun on the forecastle on the larger vessels – *ie* with a similar layout to *Wiesbaden* – to strengthen ahead fire. However, as noted by 'K', this would be impossible for *Rostock* herself, as she had emerged from her builders already 100 tonnes overweight. Eventually, all converted side-armoured cruisers received the seven-gun arrangement

except *Breslau*, which got eight but in an arrangement that limited her broadside to five guns. She was the only modern cruiser ever to carry a mixed armament, when between January and July 1916 she mounted two 15cm and ten 10.5cm guns because, despite the demands of the Black Sea theatre, priority for deliveries to the Ottoman Empire was seen as low. This was further manifested during her subsequent conversion to the intended full complement of eight 15cm guns, which lasted many months, only completing on 30 April 1917. Of all the planned conversions, only one would never be begun: *Rostock* was scuttled on 1 June 1916 following damage during the Battle of Jutland. All of the vessels that had already undergone conversion missed the battle: *Stralsund*, for instance, was still under conversion at the time, so we will never know whether the rearmament would have had any impact on the course of the battle.

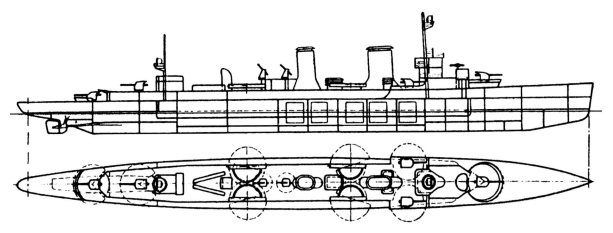

This had been for a long time the only surviving sketch of a pure 'fleet cruiser' project by the IGN, designated 'Projekt 1a'. It dates from 1916 and represents the smallest of a total of six proposals. (Author's collection)

Conclusion

These two articles have aimed to outline key elements in the design history of the German *Kleiner Kreuzer*, subject to the limited space available. As we have seen, their design was hampered from the beginning by the dual fleet/overseas role for which they were intended. Each resulting compromise led to weak points, the most significant of which has long been considered – with the benefit of hindsight – to be the comparatively small calibre of the main guns, which had to be rectified at some cost after the outbreak of the Great War. However, from the perspective of the time, the greater number of smaller guns had its positive points as well. Only the (re)introduction of side armour in their British adversaries made the small calibre truly obsolete, but the RMA reacted to this relatively quickly – despite the objections of some frontline commanders. With the 1914/15 cruisers, the vessels had shown the customary steady growth in size, with the result that a split in the lines of development was seriously discussed: large cruisers for overseas deployment, smaller units for fleet work, but a decision was postponed until after the end of the war. The last class of cruisers, though remaining for the most part uncompleted, were fine vessels able to compete with any foreign adversary of comparable size. They were but one more example of the IGN having at least reached parity with her opponents in the field of naval construction.

Sources:

In general the article draws mainly from files archived in the Bundesarchiv-Militärarchiv (BA-MA) at Freiburg im Breisgau, of which the most relevant used for direct citation are named below:

BA-MA RM2/1614 – *Fortgang der Neu- und Umbauten.*

BA-MA RM3/374 – *Inbaugabe* Ersatz-CORMORAN.

BA-MA RM3/894 – *Bau Kleiner Kreuzer BRESLAU.*

BA-MA RM3/3697 – *Bau Kleiner Kreuzer.*

BA-MA RM3/3698 – *Bau Kleiner Kreuzer.*

BA-MA RM3/23148 – *Schlussbericht der Schiffsprüfungskommission S.M.S. MAGDEBURG.*

BA-MA RM20/105 – *Akten betreffend Schiffsbauten September 1892 – August 1932.*

Bauer, Gustav, '*Über moderne Turbinenanlagen für Kriegsschiffe*', Jahrbuch der Schiffbautechnischen Gesellschaft – Zehnter Band, Julius Springer (Berlin, 1909).

Besteck, Eva, *Die trügerische 'First Line of Defence'*, Rombach Verlag (Freiburg, 2006).

Blohm & Voss (ed), Planbuch Kl. Kr. Ersatz-*Ariadne*, (Hamburg, 1916).

Föttinger, Hermann, '*Die hydrodynamische Arbeitsübertragung besonders durch Transformatoren, ein Rückblick und Ausblick*', Jahrbuch der Schiffbautechnischen Gesellschaft – 31. Band, Julius Springer (Berlin, 1930).

Gröner, Erich, *Die deutschen Kriegsschiffe 1815–1936*, JF Lehmanns Verlag (Munich, 1937).

Jacobsen, Hermann, *Die Entwicklung der Schiesskunst in der Kaiserlich Deutschen Marine*, Reichswehrministerium (Berlin, 1928).

NN, *75 Jahre Marinewerft Wilhelmshaven*, G Stalling (Oldenburg ,1931).

Reichs-Marine-Amt (ed), *Deutsche Kriegsflotte – Bd IV: Kleine Kreuzer*, Reichs-Marine-Amt (Berlin, 1907–1915).

Reichs-Marine-Amt (ed), *Bauvorschrift für die Hauptmaschinen (zwei Turbo-Transformatorsätze auf zwei Wellen) und Kessel mit zugehörigen Hilfsmaschinen, Zubehör und Inventar des Kleinen Kreuzers 'Ersatz-Gefion', M Teil I - Hauptmaschinen*, Reichs-Marine-Amt (Berlin, 1913).

Reichs-Marine-Amt (ed), *Bauvorschrift für den Schiffskörper des Kleinen Kreuzers 'Ersatz-~~Leipzig~~ Nürnberg, Mainz', S*, Reichs-Marine-Amt (Berlin 1915).

Schulz, Bruno, *Die Ölfeuerung*, W. Knapp (Halle/S., 1925).

Tirpitz, Alfred von, *Erinnerungen*, Koehler (Leipzig/Berlin, 1942).

Endnotes:

[1] Internally the class was designated *Breslau* class but, as *Magdeburg* had been launched three days in advance of *Breslau*, it became known publicly as the *Magdeburg* class.

[2] In the end the Navy profited from this choice as *Strassburg* emerged as the least expensive of the four ships.

[3] BA-MA RM2/1614 – fol 52.

[4] Another official source states as much as 260 tonnes of oil.

[5] The 13cm calibre had originally been developed by the Army for fortifications on shore.

[6] Tirpitz, *Erinnerungen*, 509f.

[7] Limitations of space preclude any detailed reference to this interesting device.

[8] The history of the requisitioned ex-Russian cruisers *Pillau* and *Elbing*, as well as the minelaying cruisers *Brummer* and *Bremse* are omitted here, as they are outside the line of development of the German small cruiser. For them and a comprehensive account of all the cruising vessels of the IGN, see A Dodson and D Nottelmann, *The Kaiser's Cruisers, 1871–1918*, Seaforth Publishing (Barnsley, forthcoming).

[9] BA-MA RM3/3698 – fol 61.

[10] Actual design: 28 knots (29 knots on trials).

[11] Actual design: 26.5 knots (28 knots with forcing).

[12] They were designed for 28.5 knots, rather than the 25.5 knots of the preceding *Birmingham* and *Chatham* classes.

THE ITALIAN AIRCRAFT CARRIER *AQUILA*

To follow his article on the interwar Italian aircraft carrier projects published in *Warship 2020*, **Michele Cosentino** provides a detailed account of the rebuilding of the former the liner *Roma* during the Second World War.

When Italy entered the Second World War on 10 June 1940, the existing law governing naval aviation stated that the Italian Navy, the *Regia Marina*, could operate only maritime reconnaissance seaplanes, both shore-based and embarked in battleships and cruisers. Even so, it was the Italian Air Force, the *Regia Aeronautica*, that controlled and administered naval aviation, which was considered an ancillary force for the Navy; the *Regia Aeronautica* owned all combat aircraft, including a newly-established torpedo-bomber unit.

At the outbreak of war, naval aviation comprised twenty-four land-based squadrons with 237 reconnaissance seaplanes, but only 163 were declared combat-ready. The squadrons were distributed along the coasts of the Upper and Lower Tyrrhenian Seas, the Lower Adriatic, the Ionian Sea, the Sicilian and Sardinian channels, the Aegean archipelagos, and the Central Mediterranean. Maritime reconnaissance missions included: surveillance off national coasts and enemy bases (Malta and Alexandria, and initially also Toulon and Bizerte – Gibraltar was too far distant), and at choke points along the Mediterranean shipping routes; antisub-

A Ro.43 floatplane being readied for catapult launch from the battleship *Littorio* while the ship was moored at Naples in November 1941. (Franco Bargoni collection)

marine duties for naval groups and convoys; the patrol of maritime areas likely to be used for enemy operations; and photographic reconnaissance of enemy naval and air bases. The shore-based seaplanes were Cant Z.501 and Z.506 types, while major warships were equipped with the IMAM Ro.43: three were embarked on each of the two battleships of the *Littorio* class, and two assigned to each of the nineteen Italian cruisers of various types. In total, there were forty-four Ro.43 embarked seaplanes and as many in reserve. Their tasks were short- and medium-range reconnaissance, signalling the movements of enemy naval groups during the tactical phase of an engagement, spotting of gunfire and antisubmarine search.

Unfortunately there were no procedures for joint maritime operations to ensure proper coordination between Italian shore-based combat aircraft and warships, and communications between the two were virtually non-existent. A request for intervention from an at-sea task group had to travel along a complex chain of command before being executed, and therefore rarely produced results. The poor performance of shore-based and embarked aircraft of the ancillary naval aviation arm was another significant shortfall that adversely affected Italian naval operations from the earliest phases of the conflict. As a consequence of the action off Calabria (9 July 1940), the Italian C-in-C, Admiral Inigo Campioni, clearly addressed the urgent requirement for an aircraft carrier that could provide quick and effective support to naval surface forces. However, the Italian Navy Staff, led by Admiral Domenico Cavagnari, was not ready to order the construction of an aircraft carrier based on any of the designs and proposals put forward during the 1930s by the design department of the *Regia Marina*.[1]

Most of these proposals concerned the conversion of the steam liner *Roma*, and one involved a radical reconstruction to secure a real fleet aircraft carrier. The latter, however, was little more than a sketch, the aim being to produce a carrier capable of operating as many aircraft as possible and featuring new machinery with sufficient power to enable the ship to accompany the battle fleet. Little information is available about aircraft stowage and flying operations beyond a 218-metre flight deck and a two-tier hangar. The proposal was therefore similar in conception to the configuration of the aircraft carrier *Ark Royal*, commissioned in 1938 and featuring a similar double hangar.

In December 1940, Admirals Arturo Riccardi and Angelo Iachino replaced Cavagnari and Campioni respectively. Before leaving office, Cavagnari sent a memorandum to the Italian Prime Minister Benito Mussolini emphasising the role played by British embarked aviation during maritime operations and the poor performance of the Italian air and naval forces against enemy air threats.[2] Riccardi initiated two measures that aimed to change this imbalance: a feasibility study to address the conversion of the battleship *Impero* to an aircraft carrier,[3] and the commandeering of *Roma* for conversion to an aircraft carrier by Ansaldo. *Roma* had been decommissioned in Genoa in late 1939, and Riccardi's orders allowed the execution of some preliminary work, notably the disembarkation of all furniture and other equipment not

The passenger liner *Roma* in Genoa during the late 1930s. (Italian Naval Historical Office)

The study model of the battleship *Impero* converted to an aircraft carrier. (Italian Naval Aviation Museum, Grottaglie, Taranto)

needed for combat operations. However, this work was abruptly suspended in February 1941 when the *Regia Aeronautica* notified the *Regia Marina* that there was no aircraft, neither in service nor projected, suitable for embarkation in the new carrier. This attempt at obstruction obliged the Italian Navy to realign its focus on refining and improving the previous conversion plans for *Roma* so that they would be ready for possible future execution – the Italian Navy staff had optimistically allowed only eight to ten months for the conversion.

The tragic losses suffered off Cape Matapan on 28–29 March 1941 prompted Mussolini to overturn the operational and technical objections to an Italian aircraft carrier. The *Regia Marina* design department therefore tasked Major General Gustavo Bozzoni, Naval Engineering Corps, and Ansaldo to update *Roma*'s conversion plans in line with the preliminary study carried out in 1936,[4] as there was no time to design and build a new aircraft carrier from scratch. Although confident that the German *Kriegsmarine* would be willing to provide technical and material assistance for *Roma*'s conversion, the Italian Navy would need to manage the risk raised by a number of technical issues surrounding the construction of the German aircraft carrier *Graf Zeppelin*.

The green light for the conversion of *Roma* was given at a meeting of the Italian Joint Staff held in July 1941, at which the head of the *Regia Aeronautica* guaranteed the urgent initiation of studies for an embarked plane with folding wings. For its own part, the *Regia Marina*

had already established an initial requirement for 34 embarked fixed-wing Re.2000s so that at least 25 would be always operationally available. Inspector General Carlo Sigismondi, Naval Engineering Corps, suggested that *Roma* could be converted in eight to nine months, as the new machinery was ready; however, this ambitious schedule relied on the availability of the necessary steel and other construction materials. Sigismondi was optimistic, but his assessment was probably based on what had already been achieved by the Italian naval designers, with several refinements to the design.

As a steam liner, *Roma* had an elegant silhouette. She featured nine decks, two tall pole masts and two raked funnels of elliptical cross-section. The steel hull was divided into 13 watertight compartments in order to ensure buoyancy with two adjacent compartments flooded. However, this did not satisfy the Italian Navy requirements for major warships, which required survivability with three adjacent compartments flooded. After disembarking all unnecessary items, *Roma* was commandeered by the *Regia Marina* with a light displacement of 21,000 tonnes.

Work on *Roma* began officially on 15 July 1941, with the ship moored at a pier owned by the Officine Allestimento e Riparazioni Navi (OARN), a subsidiary of Ansaldo. All the original superstructures and the steam propulsion machinery were dismantled and disembarked, while a dense exchange of correspondence took place between the Italian Navy Staff and several other bodies involved in the undertaking so that the steel and other material needed for the conversion could be quickly procured. The plans for the aircraft carrier were finalised and approved by the *Regia Marina* design and technical departments, while orders and contracts were issued for the sub-systems and equipment related to flight operations.

The Initial Design

The design of the aircraft carrier, named *Aquila* ('Eagle') in February 1942, was quite different from the previous studies and proposals put forward for the *Roma* conversion. The design objective was a fleet carrier able to provide air defence for surface forces and to carry out air strikes against enemy ships, this latter task being performed by a squadron of attack planes that included fighter-bombers, torpedo-bombers or multipurpose planes. The design project was officially designated UP 188 by Ansaldo. It was based on two distinct but integrated lines of development, platform systems and aviation equipment, and there were two key drivers: number of embarked aircraft and speed. According to official papers from Ansaldo,[5] the first design was completed in September 1941: the final design took into account several important modifications resulting from German assistance (see below), and was completed in February 1942.

Design UP 188 featured a vessel with a flight deck running for the entire length of the hull. There were no catapults, and the forward edge of the flight deck was faired into the stem, thus creating a small downward step to facilitate take-off. The lower part of the hull was fitted with bulges. The island was positioned on the starboard side and included a single funnel, a structure housing the bridge and the main operational spaces, and a tower supporting the fire control directors and rangefinders.

Main dimensions were as follows:

Displacement	25,500t
Length pp	202.4m
Length wl	208m
Flight deck	223.5m x 26.2m
Beam (max)	28.8m
Depth (to flight deck)	23m
Draught (normal)	6.96m

Initially the embarked air wing was to comprise 34 fighters or 16 fighters and 9 bombers. Armament would include 135/45 low-angle (LA) guns, 65/64 high-angle (HA) guns, and 20/65mm machine guns. All of the gun mountings were to be located along the hull sides. To achieve the maximum speed of 30 knots, the design required a propulsion plant able to deliver 140,000shp.

The naval designers had five main objectives when planning the conversion:

– an increase in speed from 21.5 to 30 knots
– a flight deck long enough to allow take-off without catapults but landing with arrester gear

– a single enclosed hangar large enough to stow and maintain all embarked aircraft
– compact watertight compartmentation within a robust structure able to withstand the effects of an underwater explosion
– the accommodation of an adequate number of anti-aircraft weapons and their related magazines.

Following tank tests with a provisional hull form and the decision to adopt new machinery, the naval designers decided to fit a pair of 130m-long bulges that would also contribute to increasing underwater protection. The hull configuration was also refined by flaring its upper forward section to withstand head seas and by shaping its after section to support the cantilever structure of the flight deck, as in contemporary British carriers.

The major part of the conversion work was the dismantling of *Roma*'s superstructures and the construction of a flight deck to form the ceiling of a 5.5m-high hangar deck. The hull below the flight deck was divided into seven decks, designated 'A' to 'F' and Hold. Other important work was carried out to create adequate magazines, fuel and crew spaces. A number of major modifications were incorporated into the initial UP 188 design following the availability of systems and technical assistance provided by the German Navy.

Machinery

In parallel with the design work on the hull form, a study of the propulsion system was entrusted to Major General Francesco Modugno, Naval Engineering Corps. Extensive tests carried out at the La Spezia Naval Tank showed that 132,660shp was required to attain a speed of 29 knots, which led to the conclusion that a top speed of 30 knots was achievable with a total horsepower of 151,000shp distributed between four shafts.

Roma's original machinery (thirteen cylindrical boilers and four geared turbines) was to be replaced, but design and construction of new machinery would have taken longer than allowed by the projected conversion schedule. The solution was the adoption of the machinery previously assigned to *Paolo Emilio* and *Cornelio Silla*, two of the twelve light cruisers of the 'Capitani Romani' class, whose construction had been suspended in June 1940 and was subsequently cancelled. Their machinery, comprising four boilers and two sets of geared turbines per ship, was almost complete and was therefore selected for *Aquila*.[6] The aircraft carrier was fitted with eight *Regia Marina*-type boilers supplying superheated steam at 29 kg/cm^2 and 320°C. Each of the four sets of Belluzzo-type turbines comprised a high-pressure turbine and two low-pressure turbines working in parallel and driving the propeller shaft via single-reduction gearing.

The disposition of the machinery was developed in accordance with a new concept of reliability and effectiveness that called for the location of the boilers and the geared turbines for each shaft in a single machinery space. This was intended to facilitate command and control of each machinery unit, to simplify the placement of pipes for fuel, steam, feedwater, lubricant, etc, and to enable damage control to be carried out promptly. Space availability on *Aquila*'s lower decks made possible the implementation of this unit concept and allowed four independent machinery units, each comprising two boilers and a set of geared turbines. The machinery spaces were separated by double watertight bulkheads with an inner gap. This solution contributed to a greater structural resistance for the machinery installation, and limited the flooding of two adjacent spaces because the collapse of one bulkhead would not necessarily compromise the integrity of the adjacent space. The machinery spaces occupied the central part of the lower hull on three decks. Machinery Rooms 1 & 2 and 3 & 4 were separated by double watertight bulkheads, and the two forward machinery rooms (MR 1/2) were separated from the after rooms (MR 3/4) by the bomb magazine (see plan of Hold). The turbines in MR 1 and 2 drove the wing shafts, while the turbines in MR 3 and 4 drove the inner shafts.

In the final configuration there were six 250kW turbo-generators located above the turbines at the outer ends of MR 1 and 4 ('D' Deck). The turbo-generators were driven by steam supplied by the main boilers when the ship was underway, and by auxiliary boilers when in port. These were backed up by four diesel generators amidships, in a separate space on the same level directly above the bomb room.

German Assistance

During the project definition phase of *Aquila*, the *Regia Marina* asked the German Navy for technical and material assistance for the reconstruction, particularly in the field of air operations. The *Regia Marina* was anxious to obtain as much information as possible on the construction of *Graf Zeppelin*, as the Naval Staff was conscious that Italian industries did not have any experience in the design and manufacture of embarked aviation equipment. The *Regia Marina* also wanted a peer technical assessment of the carrier construction programme and an exchange of information on maritime flight operations. In November 1941 an Italian Navy/Air Force delegation travelled to Germany, met a number of government officials and visited *Graf Zeppelin*. The key decision that emerged from the report written by the delegation when it returned to Italy was to modify *Aquila*'s design to incorporate catapults and arresting gear provided by German companies.[7] Other modifications related to the lifts, the hangar, the electrical ring main, the fire control system and some minor items of equipment.

For aircraft launch, the *Regia Marina* had initially proposed to install two compressed air catapults, possibly a new model developed by Colonel Luigi Gagnotto, Naval Engineering Corps, and an Italian pioneer in this field. However, the willingness of the *Kriegsmarine* to transfer Deutsche Werke catapults

The launch trolley used for launching all aircraft embarked in Italian warships, including *Aquila*. (Giancarlo Garello collection)

superseded this decision. Installing Deutsche Werke-type catapults meant a redesign of the forward upper section of the ship, as well as adopting the same procedures for preparing and launching aircraft as planned for *Graf Zeppelin*. *Aquila*'s flight deck therefore ended some 18 metres from the bow and had a total length of 211.6 metres. It was covered with wooden planking and maximum beam was 26 metres amidships.

The German-conceived aircraft preparation and launch process was time-consuming and cumbersome. The aircraft was fuelled in the hangar and placed on a catapult carriage that in turn was installed on a small wheeled trolley used for manoeuvring inside the hangar and onto the lift. When the whole assembly was on the flight deck, the wheeled trolley was removed and the catapult carriage was mounted on the double rail track linked to the catapults. Following take-off the carriage was returned to the hangar. In order to speed up the process, the following aircraft would be launched using one of the carriages already stowed in the forward section of the hangar. A well-trained crew could in theory launch a flight of eight aircraft in four minutes, but the procedure was never tested. The only alternative option for *Aquila* was a flight deck without catapults, using a length of about 170 metres to launch a fighter-bomber; this procedure would involve exploiting the full length of the flight deck.

The Final Configuration:

Hull & Protection

In early 1942, the design of *Aquila* reached its final stage (for displacement and dimensions see the accompanying table).

Fuel stowage at full load displacement was 2,800 tonnes. This provided an endurance of 4,150nm at 18 knots and 1,210nm at 29 knots. However, the maximum capacity of the fuel tanks at deep load was 3,660 tonnes, endurance increasing to 5,550nm at 18 knots and 1,580nm at 29 knots. The fuel tanks were placed amidships, outboard of the machinery spaces: some tanks were located in a double bottom that ran from Frame 80

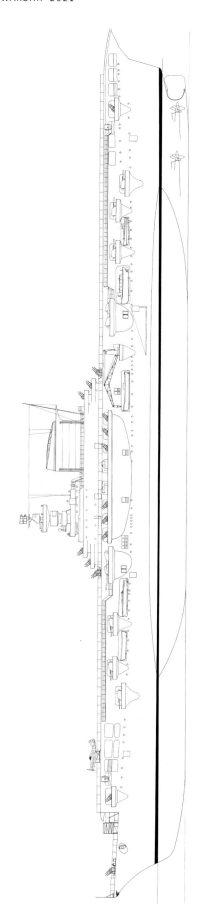

Port outboard profile of *Aquila*.

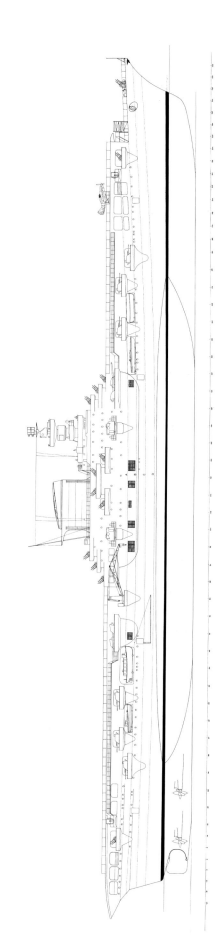

Starboard outboard profile of *Aquila*.

(INHO)

Aquila: Inboard Profile

Note: Adapted from the Ansaldo plans dated 30 September 1941.

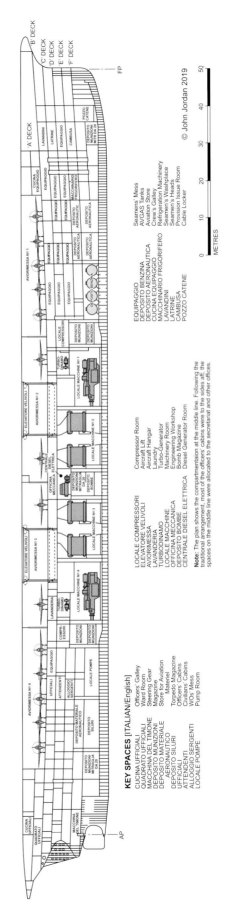

KEY SPACES [ITALIAN/English]

CUCINA UFFICIALI — Officers' Galley
QUADRATO UFFICIALI — Ward Room
MACCHINA DEL TIMONE — Steering Gear
DEPOSITO MUNIZIONI — Magazine
DEPOSITO MATERIALE AERONAUTICO — Store for Aviation Materiel
DEPOSITO SILURI — Torpedo Magazine
UFFICIALI — Officers' Cabins
ATTENDENTI — Civilians' Cabins
ALLOGGIO SERGENTI — WOs' Mess
LOCALE POMPE — Pump Room

LOCALE COMPRESSORI — Compressor Room
ELEVATORE VELIVOLI — Aircraft Lift
AVIORIMESSA — Aircraft Hangar
LAVANDERIA — Laundry
TURBODINAMO — Turbo-Generator
LOCALE MACCHINE — Machinery Room
OFFICINA MECCANICA — Engineering Workshop
DEPOSITO BOMBE — Bomb Magazine
CENTRALE DIESEL ELETTRICA — Diesel Generator Room

EQUIPAGGIO — Seamens' Mess
DEPOSITO BENZINA — AVGAS Tanks
DEPOSITO AERONAUTICA — Aviation Store
CUCINA EQUIPAGGIO — Crew's Galley
MACCHINARIO FRIGORIFERO — Refrigeration Machinery
LAVANDINI — Seamen's Washplace
LATRINE — Seamen's Heads
CAMBUSA — Provision Issue Room
POZZO CATENE — Cable Locker

© John Jordan 2019

METRES

Inboard profile of *Aquila*, showing the internal spaces at the middle line.

Note: The plan shows the compartmentation at the middle line. Following the traditional arrangement, most of the officers' cabins were to the sides aft, the spaces on the middle line were allocated to the secretariat and other offices.

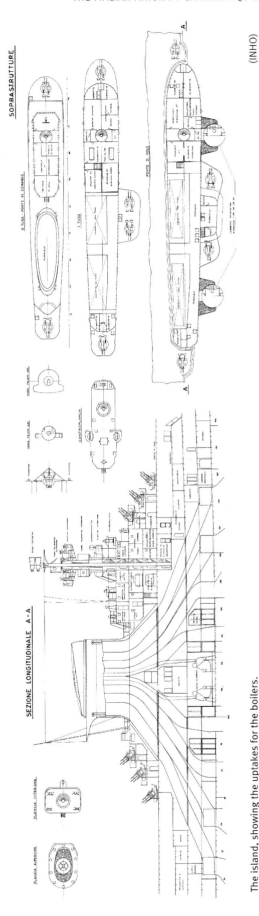

SOPRASTRUTTURE

(INHO)

SEZIONE LONGITUDINALE A - A

The island, showing the uptakes for the boilers.

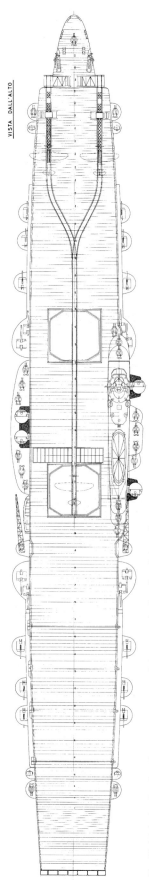

VISTA DALL'ALTO

The flight deck, showing the arrangement and the catapults for aircraft launch.

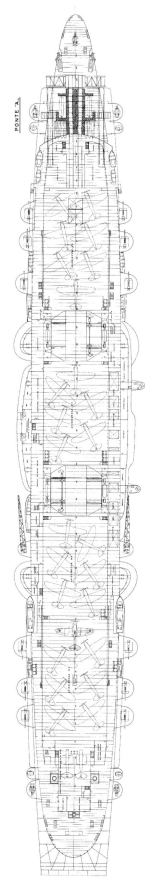

PONTE 'A'

'A' Deck (*Ponte A*), showing the accommodation of aircraft stowed in the hangar.

PONTE 'B'

'B' Deck (*Ponte B*), which was also the main (and hangar) deck.

(INHO)

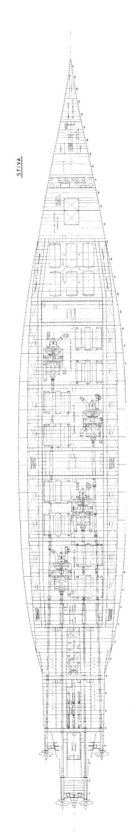

STIVA

The Hold (*Stiva*), which housed the four machinery units, each comprising two boilers and one set of single-reduction geared turbines. The gasoline tanks (labelled *deposito benzina*) were located in a compartment well forward.

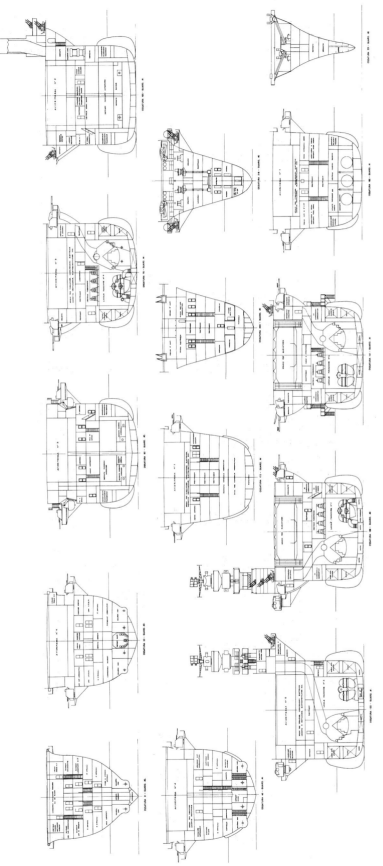

(INHO)

Section views.

A starboard stern view of *Aquila*'s lower hull in March 1942; one of the four propellers has already fitted. (Guido Alfano collection)

Characteristics

Displacement:	23,350t standard, 26,700t trial
	27,800t full load; 28,800t deep load
Dimensions:	
length	204.75m pp, 210.65m wl, 231.4m oa
beam	35.75m max, 29m wl
Moulded depth:	23.5m (to flight deck)
Draught (normal):	7.39m (fl)
GM (trials):	2.8m
Propulsion:	Four-shaft geared steam turbines,
	151,000shp = 30 knots
Endurance:	5,500nm at 18 knots; 1,580nm at 29 knots
Aircraft:	
normal	38 Re.2001/OR aircraft
maximum	51 Re.2001/OR aircraft
Armament:	8 – 152mm (8 x I)
	12 – 65mm (12 x I)
	132 – 20mm (22 x VI)
Protection:	
horizontal	60–80mm over magazines, fuel tanks
underwater	concrete-filled bulges
Complement:	1,532 officers and men

The starboard bulge of *Aquila* is filled with concrete. (Carlo Martinelli collection)

to Frame 145, while other tanks were located in the space created on either side by two long double longitudinal bulkheads. Some of the side tanks were used for boiler feed water, and fresh water for the crew was stowed in a portion of the double bottom.

The hull was divided into a Hold and seven decks designated A–F, some of them being non-continuous. 'A' Deck comprised two short sections fore and aft, and was interrupted by the upper part of the hangar, while 'B' Deck (main deck) was the hangar deck. The hangar floor was raised, and beneath it was a space 0.5m high that housed pipes, electrical cables, ventilation ducts and other similar systems; this solution provided a certain

Fitting out the flight deck of *Aquila*. (INHO)

degree of protection for the vital systems beneath against explosions or splinters. 'C' Deck was a continuous deck, while 'D', 'E' and 'F' Decks were interrupted by the spaces for the propulsion and auxiliary machinery. The exhaust ducts from the boilers and other machinery converged amidships and were combined into a single large funnel that occupied most of the island. This solution minimised the length of the island and maximised space on the flight deck. However, the cofferdam for the boiler ducts inside the central section of the hangar to starboard was 70 metres long, 3.5 metres wide and 10 metres high, thereby limiting the space available to stow aircraft. There were 19 watertight bulkheads, some of them double and most of them extending from the double bottom to 'D' Deck.

As for *Aquila*'s protection, the *Regia Marina* relied heavily on both the structural strength of the hull and the compartmentation. Underwater protection was provided by the double longitudinal bulkheads and the bulges. The latter were partially filled with reinforced concrete, a solution applied for the first time in *Aquila* and successfully tested ashore in scale models. The Pugliese underwater protection system, which comprised sealed cylinders surrounded by liquids at the lower corners of the hull,[8] could not be installed in *Aquila* because she was a conversion of an existing merchant ship and her hull was narrower than the battleships of the *Littorio* class. The compartment between the hull sides and the first of the two longitudinal bulkheads was filled with fuel. In the event of an underwater explosion, the bulge would collapse but the reinforced concrete would mitigate the effects of the explosion, while the residual energy would be absorbed by the double longitudinal bulkheads; the fuel stowed between them would distribute the residual energy over a larger surface instead of concentrating it at

a single point. Finally, the 5.5m stand-off between the outer skin of the bulge and the first internal longitudinal bulkhead would further decrease the residual pressure. As for horizontal protection, the upper decks could not be reinforced with armour plates because of stability issues. *Aquila* was therefore fitted with light armour (60–80mm) only over the magazines and avgas tanks, which were lower in the ship. The steering gear was protected by 30mm plating.

Flight Deck and Hangar

Following the adoption of German catapults the flight deck was 211.6 metres long and ended 10 metres aft of the fore perpendicular to make room for fitting the recovery equipment for the catapult carriages. The aftermost section of the flight deck had an 18-metre round-down to facilitate aircraft landing. The entire flight deck was integrated into the hull structure without supporting pillars at either end as in contemporary British carriers. The single hangar was 160 metres long, 18 metres wide and 5.5 metres high. It was divided into four sections, numbered 1 to 4 from fore to aft and fitted with fire-resistant bulkheads. Sections 2 and 3 were 25 metres long, while Sections 1 and 4 were 32 metres and 35 metres long respectively. There were two electrically-driven lifts on the centreline, between Frames 91–106 and Frames 132–147. Each of the lifts had an octagonal platform measuring 14m (L) x 15m (W) with a surface area of 182m² and a 5-tonne capacity. A rail track 72 metres long extended from the after lift to the forward end of the flight deck to facilitate the transfer of aircraft from the hangar to the catapults. In front of the forward lift the track divided into two arms, each of which connected with a catapult. To the sides of the hangar deck and 'A' Deck were the working spaces: the work-

A rare photo of *Aquila*'s hangar. (Giorgio Parodi collection)

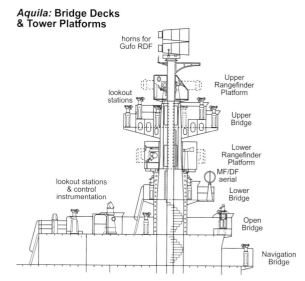

***Aquila:* Bridge Decks & Tower Platforms**

horns for
Gufo RDF

Upper
Rangefinder
Platform

lookout
stations

Upper
Bridge

Lower
Rangefinder
Platform

MF/DF
aerial

lookout stations
& control
instrumentation

Lower
Bridge

Open
Bridge

Navigation
Bridge

Note: Adapted from the Ansaldo
plans dated 28 January 1942.

© John Jordan 2020

Section of the tower, showing the platforms for the upper and lower fire control directors (*torreta telemetrica*) and, atop the housing for the upper rangefinder, the superimposed horn antennae of the Gufo radar.

shops, offices and storerooms were located on the port side, while most of the starboard side was occupied by the cofferdam for the boiler uptakes. The hangar deck was fitted with reinforced beams every two frames instead of every four frames, which was standard in the other sections of the hull.

Compressed air bottles for the catapults were housed directly beneath them. They were charged by eight electric- and four motor-driven compressors. The two Deutsche Werke catapults could launch a 2.5-tonne fighter at 140km/h or a 3-tonne bomber at 130km/h. The high pressure reservoir allowed each catapult to launch nine planes at the rate of one per minute, then needed to be recharged for 50 minutes.

Aquila was fitted with four arrester wires, officially termed 'braking cables' or 'brakes'. Manufactured by DEMAG, the system included a set of winch drums capable of stopping a 5-tonne plane landing at 120km/h. The arrester wires were placed at 30.1, 46.1, 55.6 and 66.6 metres from the after edge of the flight deck. A 14m x 4m hinged wind break was fitted in front of the after lift and served as an emergency crash barrier.

Island

In the initial design *Aquila*'s island was 42.5 metres long. However, new requirements led to a final configuration

A close-up of the funnel and the conning tower, seen from the starboard side. (INHO)

that featured a longer island (55m) with stepped deckhouses at either end on which the new six-barrelled 20/65mm MG (see below) were to be mounted. The lower deckhouse at the forward end of the island housed spaces devoted to flying operations. The second level housed the CO's sea cabin, the ready room for pilots and the main fire control station, and the third level the navigation bridge, the W/T office and other operational spaces. A central cylindrical housing provided ladder access to the upper level of the tower, where it tapered to an inverted cone. Around it were seated the main fire control platforms: above the main fire control director was a platform with lookout stations, and the tower was topped by a 5-metre rangefinder in a rotating housing.[9]

There was an antenna for the EC 3/ter *Gufo* ('Owl') radar atop the forward tower. It comprised two funnel-shaped quadrangular antennae that rotated through 360 degrees with the upper rangefinder turret. The need for two separate antennae, one for transmitting, the other for receiving, was due to the lack of efficient duplexer devices for concentrating these two functions into a single antenna. *Gufo* had an emission frequency of 400MHz and a maximum power of 10kW; its location in *Aquila* would theoretically ensure a detection range of 30,000m for surface targets, while an aerial target would be detected at up to 80,000m.

Armament

Like other modern foreign aircraft carriers, *Aquila* was equipped with both major-calibre guns for protection against surface attacks and anti-air defence weapons. The initial design included Ansaldo Mod 1934 152mm guns,

Workers on *Aquila* building one of the sponsons in 1942. (Carlo Martinelli collection)

The construction of a large sponson on the port side of *Aquila*'s hull. (Carlo Martinelli collection)

German-made 88mm HA guns, and Breda 37mm guns. The final armament was considerably modified. It comprised: eight OTO Mod 1938 135mm single shielded LA guns, mounted in pairs amidships on the hull sides, one level below the flight deck to ensure 360-degree coverage; twelve Ansaldo-Terni Mod 1939 65mm single HA mountings, located on sponsons fore and aft along the hull sides; and 132 Breda Mod 1935 20mm machine guns, installed as twenty-two six-barrelled mountings placed at the fore and after edges of the island and on other platforms, most of which were located amidships.

The OTO Mod 1938 135mm gun was the latest iteration of a weapon developed in 1937 as the secondary armament for the modernised battleships of the *Andrea Doria* class and as the main armament of the light cruisers of the 'Capitani Romani' class. Its rate of fire was 6–7 rounds per minute, while maximum range with a 32.7kg AP shell at 45 degrees elevation was 19,600m. The Mod 1939 65mm HA gun was developed by Ansaldo and Terni during the Second World War, and

prototypes were being completed. Its rate of fire with manual loading was about 20 rounds per minute. The mount was designed for an elevation of -10/+80 degrees. Range with a 4kg shell at 45 degrees elevation was 7,500m.

The Breda 20/65 machine gun was the most common anti-air weapon fitted in Italian warships during the Second World War. An air cooled and gas operated machine gun, it had a maximum range of 5,000m. Although considered an efficient weapon, its low magazine capacity of 12 rounds meant frequent stoppages for reloading. The six-barrelled mounts planned for *Aquila* were a new development and would not be ready before 1943. Only a few prototypes were manufactured, and *Aquila* was to be equipped in the interim with 20mm twin mounts.

The magazines for the 135mm LA and the 65mm HA guns were located fore and aft of the main machinery spaces (see inboard profile). There was a magazine for aircraft bombs amidships, between MR 3 and MR 4, and a magazine for aircraft torpedoes on the centreline aft, at Hold level. The magazines for the 20mm MG were above the bomb magazine amidships and on the centreline fore and aft, below 'F' deck.

The rangefinder and fire control director for the main guns were superimposed on the forward tower; the rangefinder was mounted at a height of 36m above the waterline. There were two pairs of smaller fire control directors for the 65mm guns, mounted on sponsons on both sides of the hull amidships. It is safe to assume that *Aquila* was to be equipped with a 'Type 1' fire control system previously developed for the main calibre guns fitted in Italian cruisers. The island was equipped with a general fire control device (*apparecchio di punteria generale*, or APG), located in the fourth deckhouse, and the platform between the main FC director and the rangefinder had multiple observation posts fitted with binoculars.

Complement

Aquila was to have a complement of 1,532 officers and men, divided between Navy and Air Force personnel. This would include 65 officers and 1,110 petty officers and seamen of the *Regia Marina*. The *Regia Aeronautica*

A prototype of the 20mm six-barrelled anti-aircraft mounting. (*Storia Militare* Archive)

would contribute 44 officers and 293 ratings and airmen. There were also 30 civilians. The officers' accommodation was located in the after section of the hull, distributed between 'B' to 'E' Decks up to Frame 60. Accommodation for seamen and petty officers was in the central and forward sections of the hull below the hangar deck and forward of machinery spaces, up to Frame 228. Medical facilities were located on 'C' Deck between Frames 170 and 210. They included an operating room, a clinic, an X-ray room, a dental room, an isolation room, a pharmacy and large sick bays for ratings.

The Long Road to the Embarked Air Wing

During the conversion of *Roma* to *Aquila*, the *Regia Marina* focused much of its attention on the composition of the embarked air wing. In early 1942, following talks with the German and the Japanese navies, the Italian Naval Staff envisaged an air wing composed of seven flights each of six aircraft. Five flights, totalling 30 aircraft, would always be in the air while two squadrons would be ready to launch. A further 26 aircraft were to be stowed in the hangar. *Aquila*'s embarked air wing was to total 68 aircraft, a figure higher than initially calculated (66 planes) taking into account the hangar volume and the suspension of some planes from the hangar roof. However, this ambitious objective could have been attained only with the adoption of aircraft with folding wings, something that was to be addressed by the Air Force but never materialised.

Initially the Italian Navy and Air Force had selected two types of aircraft to be embarked in *Aquila*. The Re.2000 was a single-seat fighter suitably modified for launch from the catapults fitted in the *Littorio*-class battleships. For the reconnaissance and attack roles, the choice was to fall on the Fiat G.50bis A/N fighter-bomber, duly modified for naval operations. However, it did not meet expectations in terms of range and payload, and never went into production.

The key issue for commissioning the first Italian aircraft carrier therefore remained the number and type of embarked aircraft. The lack of cooperation between the *Regia Marina* and the *Regia Aeronautica* on this issue proved to be a real stumbling block, both for technical and operational reasons. A clear indication about the attitude of the *Regia Aeronautica* towards aircraft carriers derived from the firm conviction of the Air Force Staff that for all types of naval operations shore-based aircraft were far more effective than those embarked.

This belief lasted at least until late 1941, when the then-head of the *Regia Aeronautica* was replaced by General Rino Corso Fougier, who supported a more proactive cooperation between the two services, including the construction of an aircraft carrier. The first outcome of this was the establishment of an organisation tasked with developing and trialling certain types of embarked aircraft for naval operations. The new body, led by the *Regia Aeronautica*, was code-named 'Organizzazione Roma' (Organisation Rome), abbreviated to OR, a clear reference to the liner currently under conversion as *Aquila*. It was based at the Sant'Egidio air base, close to Perugia. Two runways sized to replicate *Aquila*'s flight deck and fitted with arrester wires were built to test new aircraft, whose manufacture was supervised by the *Regia Aeronautica*. The runways were also fitted with lights for night training and the control station had the same communications equipment that was to be installed in *Aquila*. The air base began its activities in the second half of 1942; in addition to aircraft of Italian origin, it also trialled German aircraft modified for naval operations, whose performance would be assessed by the Italian Air Force.

The *Regia Aeronautica* would own and operate the air wing embarked in *Aquila*, and it subsequently decided that the Re.2000 fighter was not well-suited for this role, choosing the new Re.2001 Falco II instead. Meanwhile, a small batch of Fiat G.50bis for training purposes was being manufactured and sent to Sant'Egidio. After the poor performance of a prototype, the *Regia Marina* and the *Regia Aeronautica* finally agreed that *Aquila*'s embarked air wing would comprise 51 Re.2001/OR modified for their expected naval role. Ten aircraft would be parked on the flight deck, 26 would be stowed in the hangar, and 15 would be suspended from the hangar roof. The Re.2001/OR was to be fitted with a tail-hook, attachments to secure it to the catapult carriage, rein-

A Re.2000 fighter being launched from the battleship *Vittorio Veneto* in early 1942. (Giancarlo Garello collection)

A rare photo taken of a Re.2001/'Falco II' fighter fitted with a tail-hook. (Giancarlo Garello collection)

The only Fiat G.50*bis* A/N fighter-bomber manufactured, fitted with a tail-hook. (Giancarlo Garello collection)

forced structures, two 12.7mm machine guns and other equipment. The aircraft was to be fitted for but not with torpedo launching devices, while other minor modifications were proposed after the first training flights. However, if armed with a torpedo the aircraft could not use the catapults for launch due to interference between the torpedo and the carriage. The take-off run of a torpedo-armed Re.2001/OR would therefore exploit as much as possible the flight deck length. Thus the torpedo-equipped Re 2001/OR would be stowed in the aftermost section of the hangar, while those used as fighters would be stowed in its central and forward sections. In early 1942 the *Regia Aeronautica* had already ordered 100 Re.2001, and later requested that 50 be modified as Re.2001/OR and also used for pilot training. However, in early 1943 only ten Re.2001/OR for training purposes were assembled, while delays in *Aquila*'s conversion led the *Regia Aeronautica* to complete the remaining 40 planes as shore-based night fighters.[10]

Delays and Delusions

Work on *Aquila* was badly affected by material shortages, while other urgent requirements stemming from war operations caused significant delays to the planned schedule. A report dated 15 October 1942[11] and drafted by Admiral Riccardi described the situation involving *Aquila* and another carrier, named *Sparviero* (see below).

Explaining that work on *Aquila* had started some fourteen months earlier, the report stated that she was currently in dry dock for the installation of bulges and armour over the magazines. Machinery installation was progressing as planned, and pier-side tests of each shaft would be carried out after *Aquila* left the dock. The replacement of the 'Capitani Romani'- type propellers with a new model designed for *Aquila* would take place after the static trials. However, the report revealed several technical problems due to a lack of manpower and also modifications to the design resulting from the decision to install German-procured equipment. The scheduled completion of *Aquila* was therefore postponed from October 1942 to July 1943, but even this late date was affected by the availability of embarked aircraft; aircraft with folding wings had yet to be procured.

Based on the lessons learned during the first two years of war in the Mediterranean, in the spring of 1942 the *Regia Marina* decided on the conversion of another liner, *Augustus*, similar to *Roma* but powered by diesel engines, to an aircraft carrier. Since *Augustus* had a hull comparable to *Roma*'s, the Navy Staff decided initially that the new carrier – first named *Falco* (Falcon) and finally *Sparviero* (Sparrowhawk) – would be a twin of *Aquila*. However, there were some important exceptions such as hangar size, bulges and machinery. The hangar would be larger and thus capable of stowing 88 aircraft, of which 40 would be suspended from the hangar roof and a further ten aircraft parked on the flight deck.

An artist's impression of the escort carrier *Sparviero*, which was to be converted from the motor vessel *Augustus*. (INHO)

However, implementation relied heavily on folding-wing aircraft, the availability of which was still deemed feasible in the spring of 1942. The reduction in stability caused by the enlarged hangar would be compensated by bulges slightly larger than *Aquila*'s and likewise filled with reinforced concrete.

As for machinery, Ansaldo proposed to use three diesel engines driving three shafts instead of the original four and a different disposition of the machinery spaces. This solution would allow a major reduction in the volume of the exhaust ducts, thereby increasing hangar volume, but the adoption of three shafts would have drastically changed the lines of the after hull, resulting in a much longer conversion time. This in turn complicated the technical situation, and the inability of the German Navy to transfer catapults and arresting gear because of revisions in the programme of completion for *Graf Zeppelin* forced the *Regia Marina* to opt for a simpler conversion of *Augustus* as an escort carrier rather than a fleet carrier. The *Regia Marina* eventually decided to revisit a conversion proposal drawn up in 1936 for the conversion of *Roma*. *Sparviero* would be obtained by dismantling the upper decks and superstructures of *Augustus* and installing a rectangular flight deck. In the preliminary sketch design dating back to 1936 the flight deck was 155 metres long and 25 metres wide; the forward section of the flight deck featured a 50m x 5m catwalk, supported by pillars and used as the final length of the take-off run (see the artist's impression). There was neither an island nor any other structure on the flight deck; the bridge was to be located beneath the forward end of the flight deck.

A midship section of *Sparviero*[12] shows that the lower part of the hull was to have five decks. Two aircraft lifts were planned. It is not clear how many planes *Sparviero* would have embarked, but eventually the *Regia Marina* planned an air wing comprising 35 Re.2001/OR. Since *Augustus* was commandeered on 1 July 1942 for conversion, it is possible that a final design was ready at that time. There were several differences from the initial proposal, notably a small island located on the starboard side of the flight deck and two German-made compressed-air catapults. Additionally, the flight deck

was to be 180 metres long and *Sparviero*'s overall length would be around 216.5 metres. Bulges would increase the maximum beam to 34m, and full load displacement would be 28,000 tonnes. The original machinery would be retained. Four Savoja/MAN diesel engines would give a total horsepower of 28,000bhp, but maximum speed would remain a modest 18 knots. The exhaust gases would be collected in a cofferdam attached to the hangar roof and then passed to two horizontal ducts that ran along the hull sides.

As far as *Sparviero*'s armament was concerned, the initial design featured six 152/55 single mountings and four 102mm HA guns. However, the armament of the final design would benefit from experience with *Aquila*: it comprised eight 135/45 guns in single mountings, twelve 65/64 in single mountings and twenty-four 20/65mm in four six-barrelled mountings. Horizontal protection included plating with a thickness of 60–80mm, while underwater protection likewise relied on concrete-filled bulges and tight compartmentation.

Photos taken in late 1942 show that by that time *Augustus*' superstructures had been dismantled but no other work had been carried out. During Allied air raids on Genoa carried out in October and November 1942, *Aquila*'s flight deck had suffered splinter damage, and she

A bow view of *Aquila* during her fitting-out at quayside in early 1943. (INHO)

A stern view of *Aquila* in July 1942. (Ansaldo Archive)

was towed to a supposedly safer zone of the Ansaldo shipyard. However, the Allied landings in North Africa in November 1942 drastically changed the naval balance in the Mediterranean, forcing the *Regia Marina* to devote more resources and manpower to the construction of new escort vessels.

A memorandum issued in January 1943[13] stated that *Aquila*'s machinery would be ready for sea trials on 15 June 1943, while trials of other systems were planned on 30 September 1943. Whereas the first deadline could be regarded realistic, the second was over-optimistic because it failed to take into account potential issues arising from building and fitting out an inherently

A photo of *Aquila* taken on 23 August 1943; she is shrouded in camouflage nets. (INHO)

The aircraft carrier *Aquila* during her static machinery trials, carried out at quayside in July 1943. (INHO)

This photo of *Aquila* moored at Sampierdarena basin, Genoa, was taken by the Italian Air Force in the spring of 1943 to assess the effectiveness of camouflage nets. (INHO)

Aquila moored at Genoa in early May 1945. (US Navy)

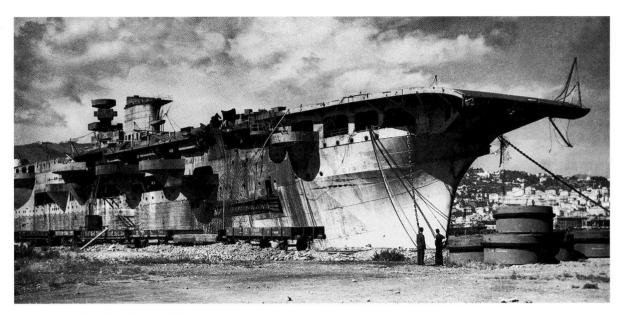

Another view of *Aquila* in Genoa in 1946. (INHO)

complex warship such as a fleet or escort carrier, especially in the fraught political and military situation prevailing in Italy.

In a further memorandum issued in January 1943[14] but related to new naval construction, the Italian Naval Staff urged *Aquila*'s commissioning (planned for the summer of 1943) and *Sparviero*'s conversion (to be completed spring 1944). Meanwhile, war requirements caused further delays to the work on both *Aquila* and *Sparviero*, while Allied air supremacy over a large part of Italy obliged the *Regia Aeronautica* to cancel plans for embarked aircraft. Despite these constraints, between April and early June 1943 *Aquila*'s steam turbines and catapults were successfully tested with the ship moored alongside; this led the *Regia Marina* to plan for sea trials to be carried out three months later. However, on 22 June the Naval Staff abruptly ordered all work on *Aquila* to be halted, and assigned manpower to the construction of escort ships and submarines then badly needed for Italy's war efforts. *Aquila* remained with a reduced crew for maintenance purposes, including 120 workers needed to complete some items of equipment. Since the light anti-aircraft armament had yet to be installed, *Aquila*'s air defence was to be provided by a handful of German Flakvierling Mod 38 guns. In July 1943 and after the Allied landings in Sicily, *Aquila*'s machinery was successfully tested, this being an opportunity for the carrier to steam autonomously to a safer place. By late August 1943, *Aquila*'s readiness status was as follows: hull 99% complete; machinery 98%; fitting out 70%. However, work installing weapons and flight equipment was significantly delayed.

When Italy signed the armistice with the Allies on 8 September 1943, *Aquila* was seized by German troops and looted of any equipment useful for war purposes. The aircraft carrier lay abandoned at the pier side, was slightly damaged by Allied bombings in June 1944 and

finally moved to a supposedly safer place within Genoa's port. When Northern Italy was occupied by German troops and a civil war erupted thereafter, the *Regia Marina* moved to Southern Italy, in the part of the peninsula liberated by Allied forces. In order to prevent the Germans from using *Aquila* to block the entrance of the port, it was decided to sink her at the pier side, but an operation carried out by *Regia Marina* underwater raiders on 19 April 1945 failed. *Aquila* remained at the pier side, listing slightly to starboard. A second attempt by German troops to sink her failed. When a few days later Allied forces entered Genoa, they found *Aquila* still afloat, in the narrow passage between the eastern ancient port and the new western basin of Sampierdarena. The hulk was moved again to the pier side and remained there until 1949. It was later towed to La Spezia, where it remained for almost two years moored to a buoy, and was eventually scrapped in 1952. *Sparviero* was seized and looted by German troops and remained afloat after a failed attempt to sink her; she was scrapped in 1947.

An Evaluation

The unsuccessful conversions of *Roma* and *Augustus* as, respectively, *Aquila* and *Sparviero* ended the most difficult period of the long and troubled history toward the establishment of a mature Italian naval air arm. Whereas *Sparviero* remained essentially a paper design, many have judged *Roma*'s conversion as a worthwhile undertaking, especially if compared with similar activities carried out in Britain and Japan. Apart from the potential value of *Aquila* to Italian fleet operations, the solutions adopted for her hull structure, machinery and underwater protection were compliant with the requirements established by the Navy Staff and would probably have demonstrated their effectiveness. The weakest point of *Aquila*'s design was her aviation facilities and equipment, a shortcoming stemming from concurrent political, technical, operational and industrial factors, the most significant of which was the different agendas of the Navy and the Air Force.

A proper evaluation can be carried out by comparing *Aquila* with similarly-sized foreign aircraft carriers obtained by conversion. This comparison takes into account two parameters, namely *Aquila*'s full load displacement (27,800 tonnes) and an embarked air wing including 40 aircraft (*ie* without counting the projected spare aircraft suspended from the hangar roof). The Japanese carriers *Kaga* and *Akagi* were both converted from warships; *Aquila* was therefore more closely comparable to *Hiyo* and *Junyo*, both conversions of passenger ships. She was also similar in conception and capabilities to the British interwar carriers *Eagle*, *Glorious* and *Courageous*, and would have had a similar role limited to air defence of a task force.

When assessing *Aquila* from an operational point of view, it is safe to assume that even if Italy had entered the war with an operational aircraft carrier, this would not have significantly have changed the outcome. An Italian fleet carrier would probably have prevented the painful losses of Matapan, but the *Regia Marina* was afflicted by much greater problems, above all the absence of cooperation between warships and land-based aircraft and the highly-centralised command structure. The decision to build *Aquila* and *Sparviero* was a rushed decision born out of irrational reactions to the course of the naval war in the Mediterranean. For an Italian aircraft carrier to have been useful and effective, its development should have begun in the late 1920s and early 1930s, not during the Second World War. Even the potential loss of one or two Italian aircraft carriers during the conflict would not have impacted adversely on the experience of naval air cooperation that the mere presence of such a warship necessarily brings to a fleet.

Endnotes:

1   See Michele Cosentino, 'The Quest for an Italian Aircraft Carrier: Design And Achievements', *Warship 2020*, 152–166.

2   The memorandum is dated 2 December 1940 and is cited by G Giorgerini in his book *Da Matapan al Golfo Persico*, September 1989.

3   Enrico Cernuschi, 'La portaerei Impero?', *Storia Militare* no 152, May 2006. *Impero* was the fourth and last unit of the *Littorio* class. She had been launched on 15 November 1939 and was never completed. A model of *Impero* as an aircraft carrier is on display at the newly-established Italian Naval Aviation Museum of Grottaglie, Taranto (see illustration).

4   Italian Navy Historical Office (INHO), General Collection, Folder Aircraft Carrier, 'Progetto di trasformazione del Piroscafo Roma in Nave portaerei' no 18273 dated 6 June 1941.

5   INHO, Collection *Aquila*, Folder 11, Plans, Ufficio Progetti Ansaldo, Genova, 1941–1942.

6   The two cruisers were being built at Genoa, so their machinery was readily available for *Aquila*.

7   INHO, Collection *Aquila*, Folder 1, Letter to Ansaldo, 'Nave P.A. – Varianti', dated 6 December 1941.

8   Michele Cosentino, 'The Bonfiglietti Project: An Aircraft Carrier for the Regia Marina', *Warship 2015*, 44–61. The Pugliese system was used for the newly-built *Littorio*-class battleships. It was also installed in the modernised battleships of the *Cavour* and *Doria* classes. In *Aquila*, the lower part of the double bottom was likewise filled with concrete.

9   Although a front-line warship, *Aquila* was not fitted with a flag bridge because Italian naval tradition demanded that only a battleship play the role of flagship.

10 It appears that only two Re.2001/ORs were completed and tested.

11 INHO, Collection *Aquila*, Folder 4, Memorandum, 'Navi portaerei Aquila e Sparviero', 15 October 1942.

12 Giorgio Giorgerini and Augusto Nani, *Le navi di linea italiane, 1861–1961*, USMM (Rome 1962).

13 INHO, Collection Aquila, Folder 7, Memorandum for Admiral Riccardi, no 90, dated 28 January 1943.

14 INHO, 'La Marina italiana nella 2^ G.M., vol. XXI. Promemoria no 2. Orientamenti circa lo sviluppo delle costruzioni navali', dated 14 January 1943.

THE 'STEALTH' FRIGATES OF THE *LA FAYETTE* CLASS

The light frigates of the *La Fayette* class were the world's first 'stealth' frigates, designed for a minimal electro-magnetic signature, and were hugely influential abroad. **Jean Moulin**, co-author of a seminal French-language book on the class, and **John Jordan** look at the ground-breaking design principles and construction techniques involved.

During the late 1950s the *Marine Nationale* embarked on the construction of a series of 'escort sloops' (*avisos-escorteurs*) of the *Commandant Rivière* class, designed to perform overseas patrol in peacetime and the escort of logistical and mercantile ships in wartime. Nine ships were completed between 1962 and 1970, and they proved remarkably successful in the overseas 'presence' and surveillance/patrol missions. They had a balanced armament, and were powered by four SEMT-Pielstick diesel engines which gave them an impressive endurance of 7,500nm at 16 knots. However, by the 1980s they were becoming due for replacement, and in 1982 the French Naval General Staff requested a study from the Direction des Constructions Navales (DCN) for a ship with good endurance, able to operate either one medium or two light helicopters, with a useful surface attack capability using surface-to-surface missiles (SSMs) and a moderate antisubmarine capability, plus the ability to defend itself against hostile aircraft and missiles.

In 1984 the DCN presented studies for a *frégate légère* of 2,500 tonnes designated FL 25, intended both as a replacement for the *Commandant Rivière* class and for export. A contemporary photo of the model of the FL 25, published in the in-house Navy magazine *Cols Bleus* and reproduced here, shows a scaled-down C 70 type (*Georges Leygues* class – later F 70), armed with a single 100mm, two quad MM 40 Exocet launchers, the Crotale lightweight SAM system, and a hangar for two light helicopters.

The 1987–91 programme envisaged orders for three ships. Replacement for the *Commandant Rivière* class was now becoming urgent, as these ships were due to be decommissioned 1987–94. However, lessons from the Falklands War of 1982 and the Iraq War of 1986–88 suggested the need for a significant reduction in electro-magnetic signature and for resistance to light damage. Missions now included the escort of tankers through the Straits of Hormuz without firing the first shot; electronic countermeasures (ECM) and anti-missile decoys were to be used to deflect enemy missiles.

In 1987 the DCN produced nine sketch designs to fulfil the new staff requirements. Two were selected: one for operations in medium/high-threat zones such as the Straits of Hormuz, the other for patrol and surveillance where the threat was lower. The first, derived from the FL 25, evolved into the *frégate légère à usage général* (FLUG, or 'General-Purpose Frigate' – also referred to as the FL 3000): with a displacement of 3,000 tonnes and a length of 113 metres, it was armed with a single 100mm gun, eight MM 40 SSMs, the Crotale SAM system and two Sagaie chaff dispensers, and could embark a helicopter in the 8/10-tonne class. A novel propulsion system featuring a 'silent' mode in which an electric motor powered by diesels via an alternator was capable of driving the ship at speeds of up to 12 knots in the sub-hunting role was considered.

This proposed 'high-low' mix was further refined by the Marine Nationale. The solution finally adopted was as follows:

- *frégate de surveillance*: cheap, constructed using mercantile standards, for low-threat patrol missions
- *frégate légère*: more sophisticated, built to naval standards, to operate in medium/high-threat zones and possibly join a carrier Task Force.

Six ships of the first type, the 3,000-tonne *Floréal* class, were duly laid down between April 1990 and August 1992 at the private shipyard Chantiers de l'Atlantique, Saint-Nazaire. Due to the simplicity of the design and the adoption of modern, modular construction each ship took an average of only 24 months to complete. Following their entry into service the *Floréal* class

This model of the early FL 25 design, published in *Cols Bleus* during the mid-1980s, shows a scaled-down C 70 (*Georges Leygues* type). Note the significantly smaller funnel and the absence of 'heavyweight' sonars. (DCN)

deployed on a semi-permanent basis to overseas stations, their crews being rotated.

Development of the *frégate légère* was less straightforward, and after further discussion displacement rose to 3,500 tonnes. Even after the adoption of a simpler all-diesel propulsion plant, estimated cost was three times that of the *frégate de surveillance*. The first ship, to be built by DCN Lorient and funded under the 1988 Estimates, was to be followed by two more units of the class within the current 5-year programme.

The design finally adopted was drawn up by an STCAN team of constructors led by Patrick de Leffe, and was radically different to anything which had preceded it. It had superstructures inclined at 10 degrees to enclose all equipment and passageways, and featured the extensive use of composites both to improve stealth and to minimise topweight. The forecastle, with its anchor cable and line handling gear, proved to be a particular design problem, and the solution adopted was inspired by the Danish ocean patrol vessels of the *Thetis* class, the first of which was launched in June 1989. Maintenance and logistics were an important consideration in the design process. The diesel propulsion units could be lifted out in less than 24 hours through openings after dismantling the funnel casing.

The term *Frégate Légère* (FL) was superseded in November 1993 by *Frégate type La Fayette* (FLF). The ships would subsequently be reclassified *frégates de premier rang* in 2008 with a view to establishing a minimum force of fifteen 1st class frigates: six F 70 ASM type (*Georges Leygues* class), two FAA (*frégate antiaérienne*, *Cassard* class), two FDA (*frégate de défense aérienne*, *Forbin* class) and five FLF.

Missions

The principal mission of the FLF as designed was to participate in the resolution of crises outside Europe, in particular in and around the Gulf. It was also to conduct surveillance around French overseas territories, ensure the security of maritime traffic and, when required, form part of the escort of a carrier Task Force. The FLF was intended to operate in moderate/high threat zones, and 'stealth' characteristics and avoidance of (and resistance to) damage were key features. The FLF was designed to operate alone for long periods, often in remote areas, so endurance and ease of maintenance were likewise emphasised. Autonomy was 50 days, and maximum speed limited to that of the *avisos-escorteurs* they replaced, 25 knots. Active stabilisation would allow them to operate a medium helicopter in Sea State 6.

The armament as designed was focused on surface warfare (MM 40 SSMs and helicopter-launched ASMs); the remainder of the armament was to provide self-defence (100mm and Crotale SAM system). Particularly important were the detection, data analysis and electronic warfare capabilities that enabled them to conduct surveillance and to defend themselves in moderate threat zones.

Orders & Construction

Notification of the order for the lead ship, which would be named *La Fayette* in January 1989, was issued to DCAN Lorient on 14 March 1988, with trials scheduled for 1993. The second and third ships, *Surcouf* and *Courbet*, were to have been funded under the 1989 budget but the programme was put back a year. Three further ships, to be named *Aconit*, *Guépratte* and *Ronarc'h*, were ordered in September 1992 as part of the following five-year programme, but in November 1995, faced with a budgetary crisis, the Defence Minister Charles Millon decided to delay construction by two years, and in May 1996 the order for the sixth ship was cancelled.

The first unit, *La Fayette*, was built using conventional construction techniques. From the second ship, *Surcouf*, modular construction was introduced, with assembly of prefabricated and pre-fitted sections. There were eleven blocks for the hull and four for the superstructures (see schematic opposite). The hull and the forward superstructure block, incorporating the bridge and command spaces, was of steel; the after superstructure blocks (Nos 1–3) were of composites.

The construction and fitting out of the blocks was shared between the naval dockyards of Lorient, Cherbourg and Brest, the blocks being transported by the barge *Dino II*. A fully-equipped block weighed up to 300 tonnes. Equipment was modular for rapid repair/maintenance by exchange, and could be manufactured off-site by sub-contractors in their own factories. The eighty modules were installed in the blocks, which were then assembled in the ship hall at Lanester (see photo of *Surcouf*). For the third ship, *Courbet*, the three forward sections were built at Cherbourg, the five after sections at Brest, and the three centre sections and the four superstructure blocks at Lorient.

The propulsion machinery was installed by DCN Indret. DCN Ingénieurie supplied, fitted and tested the armament and electronics, while DCN Ruelle supplied the aviation equipment and the reloading mechanisms for Crotale. Lorient was responsible for assembly, armament and trials. The benefits of modular construction can be seen in the time taken for assembly at Lorient:

- *La Fayette*: 579 days
- *Surcouf*: 365 days
- Courbet: 179 days
- Aconit: 312 days
- Guépratte: 154 days
  [Note that Aconit was affected by the slowing of the programme]

Each of the ships took on average 1,500,000 hours to complete, and unit cost was €240m.

Hull & Superstructures

The hull, which had a total weight of 1,100 tonnes, was divided into eleven blocks, each of which constituted a

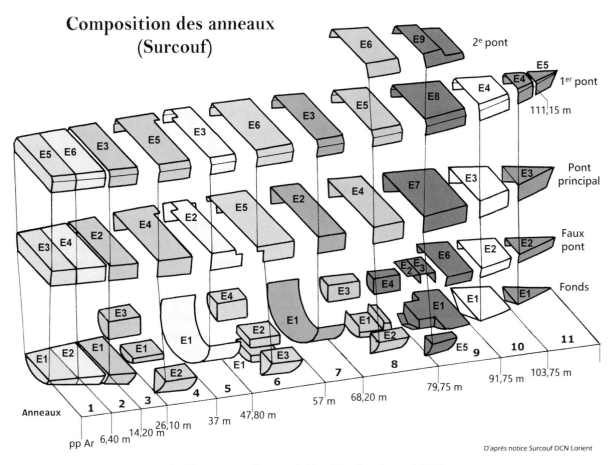

Composition des anneaux (Surcouf)

2e pont

1er pont

111,15 m

Pont principal

Faux pont

Fonds

D'après notice Surcouf DCN Lorient

Schematic showing the prefabricated hull sections for the second ship of the class, *Surcouf*. (DCN)

The prefabricated hull sections for *Surcouf* being assembled in the Lanester ship hall at Lorient naval dockyard. (DCN)

Les blocs de superstructures

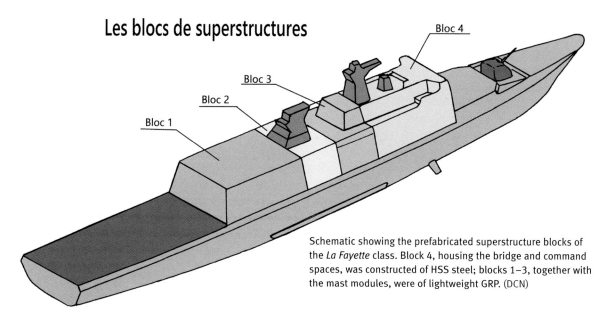

Schematic showing the prefabricated superstructure blocks of the *La Fayette* class. Block 4, housing the bridge and command spaces, was constructed of HSS steel; blocks 1–3, together with the mast modules, were of lightweight GRP. (DCN)

Principes d'assemblage

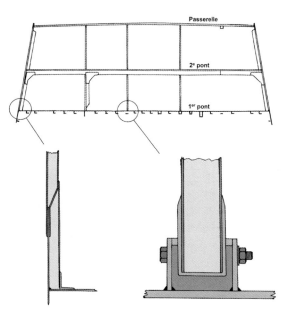

Schematic showing the connections between the steel of the hull and the GRP superstructures. (DCN)

watertight compartment. The hull and the forward superstructure block were constructed of HSS steel (E355FP), and there was 10mm Mars 190 protective plating over the operations centre, the bridge and the magazines. The hull was given tight compartmentation (*c*400 compartments) to ensure the vessel's survival in the event of flooding, and there was a transverse void compartment amidships to enable the forward and after ends of the ship to be isolated in the event of fire or flooding (see inboard profile). A central passageway connected the forward and after decks at the level of the main deck (*pont principal*); it was routed around the modular structure of the forward turret, but was otherwise straight. The platform deck below was divided by watertight bulkheads, so movement fore and aft had to be at main deck level. Above the main deck, the 1st deck was at the level of the helicopter hangar and platform. Forward of the hangar there was a central passageway and two lateral passageways. The lateral passageways would also be a feature of the 'Horizon'-class frigates: they were pressurised, provided protection for vital spaces – there was protective plating over their outer sides – and were used for cabling; they facilitated the movement of the damage control teams and the evacuation of smoke. Forward of the bridge, space was reserved

Building Dates

No	Name	Callsign	Estimates	Laid down	Launched	Trials	Commissioned	In service
F710	*La Fayette*	YE	1988	15 Dec 1990	13 Jun 1992	15 Apr 1904	15 Dec 1995	22 Mar 1996
F711	*Surcouf*	SF	1989	3 Jul 1992	3 Jul 1993	26 May 1994	10 Jan 1997	7 Feb 1997
F712	*Courbet*	CO	1989	15 Sep 1993	12 Mar 1994	25 Jul 1995	4 Nov 1996	1 Apr 1997
F713	*Aconit*	AT	1992	1 Aug 1996	8 Jun 1997	28 Mar 1998	21 Oct 1998	3 Jun 1999
F714	*Guépratte*	GT	1992	1 Oct 1998	3 Mar 1999	16 Nov 2000	30 Jun 2001	27 Oct 2001
F715	*Ronarc'h*	–	1992	[Cancelled 1996]				

Notes:

First three ordered 14 Mar 1988; last three ordered 23 Sep 1992.
Aconit originally to have been named *Jauréguiberry*.

Surcouf (2008)

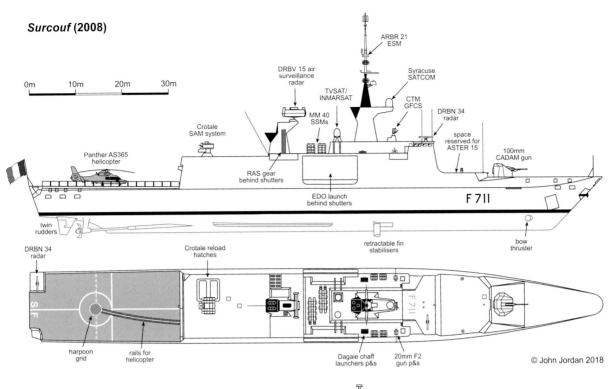

La Fayette Class: Inboard Profile

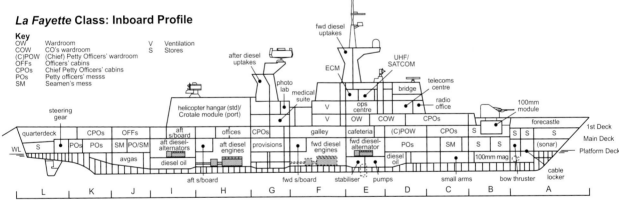

© John Jordan 2018

Characteristics (*Surcouf* as completed)

Displacement:	3,200 tonnes normal, 3,600 tonnes full load

Dimensions:

Length	124.10m oa, 115.00m pp
Beam	13.6m wl, 15.43m max
Draught	4.80m mean

Propulsion:

Engines	4 SEMT-Pielstick 12 PA6 V 280 STC diesels on two shafts
Horsepower	21,000CV (15,400kW)
Speed	25 knots max
Endurance	7,000nm at 15kts, 9,000nm at 12kts

Electrical generation:

Diesel-alternators	3 UD 30 RVR V12 diesels with Jeumont- Schneider alternators
Generator power	2,340kW

Armament:

Anti-surface	8 x MM 40 Exocet SSMs, 1 x 100mm CADAM
Anti-air	Crotale CN 2 SAM, [1 x 100mm], 2 x 20mm
Helicopter	AS 365SA Panther

Electronics:

Surveillance radars	DRBV 15C surface/air, 2 DRBN 34 nav/helo control
Fire control	CTM with Castor 2J radar
ESM/ECM	ARBR 21, ARBG 1, 2 Dagaie Mk 2 chaff dispensers

for two 8-cell vertical launch modules for the Aster 15 local area defence surface-to-air missile under development when the ships were being built. The second batch of three ships was to have been so fitted from completion; however, budgetary constraints resulted in the abandonment of this proposal, and the space was ultimately employed for additional berthing.

The masts, after superstructure blocks and the cover over the forecastle were of glass-reinforced plastic (*composite verre-résine*, or CVR in French parlance). The preference for GRP over aluminium was influenced by the Royal Navy's experience with the Type 21 frigates in the Falklands War. The advantages of GRP, which had previously been used only in French minehunters, were as follows:

– weight was 55 per cent that of steel
– more flexible when hull deformed by swell
– better thermal isolation
– better suited to stealth measures such as radar-absorbent paint
– in the event of fire slow to combust, and poor thermal conductivity
– smoother finish than welded steel
– good thermal isolation.

However, there were drawbacks:

– higher cost
– the 'metallic culture' of the shipyards, which were accustomed to working with steel
– the difficulty of securing it to other structures
– increase in thickness of the GRP panels, which implied a reduction in internal volume
– precautions needed for electro-magnetic protection
– poor resistance to splinters.

The GRP 'sandwich' comprised an outer skin of fibreglass and resin over a balsa wood core. The GRP sheets

In order to reduce the electro-magnetic signature of the FLF to a minimum, all apertures in the bow and the stern, including the anchor bays, are covered by hinged doors. The photo shows the starboard anchor door of *La Fayette* in 1995. When lost in heavy seas it proved costly to replace. (Y Violette)

for the outer walls of the superstructures had thin metallic sheets glued to them; these were then welded to the steel plating of the upper hull, resulting in completely smooth external joints. The internal partition bulkheads were bolted into steel 'U'-shaped brackets secured to the 1st deck (see accompanying graphic).

The configuration of the hull and superstructures was conceived to minimise the radar signature: outer walls were angled at 10 degrees, and protrusions and cavities kept to minimum. There were only two scuttles (for the CO's cabin), and the masts incorporated the diesel uptakes. Radar-absorbent paint was employed, and the resulting radar signature of the ships (*surface équivalente radar*, or SER) was claimed to be comparable to that of a trawler. Other 'discretion' measures included cooling for the engine exhausts to reduce infrared signature, and minimising the underwater noise signature by mounting all machinery on isolating rafts and fitting the Prairie/Masker system, which generates air bubbles around the hull and propellers.

The forecastle is covered with a GRP roof with a surface area of $150m^2$, supported on pillars; this had to be reinforced after the first firing trials with the 100mm gun. The quarterdeck, with the line handling equipment, is covered by the helicopter deck, which is of steel. Both the forecastle and the quarterdeck can be closed off by watertight panels, and there is a large hinged door over the starboard anchor recess which fits flush with the hull (see photo). The quarterdeck has an inclined ramp for a boat offset to port of the centreline and an access door in the stern. It was originally envisaged that this space would be re-worked to fit a towed sonar, although due to budgetary constraints this was never fitted.

For stabilisation of the hull there is a *système de tranquillisation automatique du flotteur* (STAF), comprising twin rudders angled outboard and retractable fins, both constructed of GRP. This makes it possible to operate a helicopter in Sea State 6, and also preserves the stealthy features of the hull and superstructure design by keeping angles of heel below 10 degrees.

Machinery

The propulsion system finally adopted was a simple two-shaft CODAD (COmbined Diesel And Diesel) arrangement. The four SEMT-Pielstick 12 PA6 V 280 STC 12-cylinder turbo-charged diesels, each with a nominal rating of 5,220CV, are paired, with each pair driving one of the two shafts via single reduction gearing and hydraulic couplings at a maximum 231rpm; the variable-pitch, 5-bladed propellers have a diameter of 3.5 metres. Each propulsion unit is located in its own engine room (F & H); these are separated by auxiliary machinery rooms (G) and the transverse void space (see drawing). The engine rooms are unusually spacious with easy access for maintenance, and acoustic discretion is ensured by mounting the diesels on isolating rafts.

The CODAD arrangement makes possible the following modes of operation:

Machinery Spaces

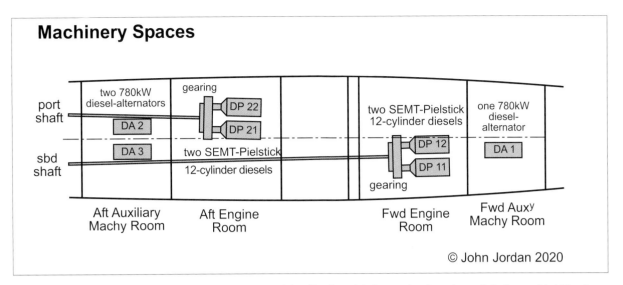

Schematic of the machinery layout. Note that the main propulsion diesels and their associated gearing and shafts are at hold level, whereas the diesel alternators are mounted on the platform deck above.

- four diesels together (21,000CV max, 23.5–25 knots)
- two diesels on two shafts (15 knots max)
- one diesel on one shaft (the other decoupled and streamed).

There is also a bow thruster rated at 250kW to enable the ship to manoeuvre in confined spaces.

Electricity is supplied by three diesel alternators (DA) located on the platform deck; each comprises a UD 30 RVR V12 diesel plus a Jeumont-Schneider 780kW alternator. A single unit is located forward of the main machinery spaces in compartment E, and two aft in compartment I (see drawing). Each of the two conventional evaporators can supply 24 tonnes of fresh water

One of the five-bladed, variable-pitch propellers of *La Fayette*. Note that the starboard-side rudder is angled outboard; the twin rudders form an important element in the stabilisation system. (Y Violette)

per day and a modern osmosis unit supplies 10 tonnes per day. There is sufficient tank stowage for 60 tonnes of fresh water.

The machinery controls, electrical switchboards for 440V/60Hz and 115V/400Hz circuits, and damage control panels are grouped together in a single compartment designated the *PC Machine, Electricité et Sécurité* (MES). The propulsion machinery is controlled and monitored remotely from the *PC Machine*, which also houses the controls for the stabilisation system, comprising the two rudders, two retractable fins and a list correction system (*Co-Gîte*) which manages the transfer of fuel oil between tanks.

The damage control centre handles leaks, firefighting and the movement of casualties. It is equipped with surveillance panels to show detection of leaks and fire, together with the consoles for the electrically-driven pumps, fire collectors and counterflooding. The FLF is classified as *citadelle permanente*, with the entire interior of ship able to be closed down in an NBC environment. Air entering the ship can be controlled via a filtration and air conditioning system, which also provides dehumidification.

Armament

MM 40 Exocet

The French Navy's first surface-to-surface missile, the MM 38 Exocet, was installed in the A/S frigate *Tourville* in 1974; its maximum range of 42km corresponded to the radar horizon of a medium-sized warship. Its longer-range successor MM 40, which entered service in 1981, is likewise a 'fire-and-forget' sea skimmer using solid fuel for propulsion, but features folding wings which enable it to be housed in a more compact cylindrical fibreglass tube. After launch the missile descends to a preset cruise altitude of 3–15m (depending on sea state), using a radar altimeter to ensure stabilisation. Guidance is inertial until the missile is 12–15km from the target; the missile then switches to active radar homing. Detonation of the 165kg warhead is on contact or is triggered by the radar altimeter when the missile passes over the ship. Initial target position is determined either by the ship's surface surveillance radar or, for over-the-horizon (OTH) targeting, by relay using the ship's heli-

copter or another friendly unit. Prefiring data inputs include target range and bearing, own-ship heading and speed, and vertical reference.

In the FLF two quadruple ramps are fitted between the bridge structure and the first funnel; they are angled out at 80 degrees from the ship's axis and the missiles are launched at a fixed angle of elevation of +15°. The stock of MM 40 missiles is rotated between ships, and generally only two missiles are embarked for peacetime deployments. Initially the missile was the Block 2 (range 65–75km), but this has recently superseded by the Block 3 SSM, which has an improved seeker and turbojet propulsion for a range of 180km.

100mm DP Gun

The 100mm Mle 68 CADAM TR (*CADence de tir AMéliorée, Technologie Rénovée*) mounted on the forecastle of the FLF is capable of close-in defence against aircraft, and fire against surface targets and shore. The gun is entirely automatic; it has a maximum theoretical range of 17km at an elevation of 40 degrees, but effective range is 12km against surface targets and 6km against aircraft. Ammunition includes the OEA F1 high-explosive shell and the OPF F4 high-explosive pre-fragmented shell; both weigh 13.5kg and employ contact/proximity fuzing. The OEA F1 has a conventional 1.1kg bursting charge, while the OPF F4 contains 1,350 high-density steel balls each weighing 1 gramme. Magazine capacity is 600 rounds.

The outer walls of the gunhouse are faceted to disperse radar waves, and in the FLF the lower edge of turret is located just below the level of the forecastle cover to conceal the sharp edge. The turret is unmanned and is controlled from the Operations Centre (see below).

Crotale

Crotale was initially developed for the French Air Force as a short-range missile system to defend air bases; it was designed and built by Matra and Thomson/CSF. The development contract for the navalised variant was placed on 1 Feb 1974, and following testing on the missile trials ship *Ile d'Oléron*, the first production launcher was installed on board the frigate *Georges Leygues* in 1978. Crotale would subsequently be fitted in the frigates of the F 67 and F 70 types.

Missile Characteristics

	MM 40 Exocet	Crotale CN 2
Dimensions:		
length	5.80m	2.34m
diameter	0.35m	0.16m
span	1.13m	0.54m
Weight:		
missile	850kg	95kg
warhead	165kg	13kg
Performance:		
speed	Mach 1	Mach 3.5
range	70km	13km

Gun Characteristics

	100mm CADAM	20mm F2
Barrel length	55 cal	90 cal
Weight of mounting	21t	332kg
Weight of round	23.5kg	0.12kg
Muzzle velocity	870m/s	1,050m/sec
Rate of fire	78rpm	720rpm
Elevation	-15°/+80°	-15°/+65°
Training speed	40°/sec	[manual]
Elevation speed	29°/sec	[manual]
Range	12km surface	2km AA
	6km AA	[effective]

La Fayette running trials of her propulsion machinery in October/November 1993. The gunhouse is faceted and its lower edge is below the GRP cover for the forecastle to minimise radar reflections. (DCN)

The 8-cell missile launcher is on the centreline atop the helicopter hangar, with the reloading module occupying the port side of the hangar itself. There are eight missiles in the launcher, and two complete sets of reloads are carried in the module, the launcher being trained to starboard and reloaded via two large rectan-

Rear view of the eight-cell Crotale launcher on *Surcouf*. (R Reboul)

The 100mm turret is embarked during during the fitting out of *Guépratte* at Lorient on 28 February 2000. (DCN Lorient)

Surcouf, the second ship of the class, at Nice on 15 June 2005. She is unmodified since her completion in 1997. (J Pradignac)

gular doors. Reloading takes 30 minutes. The launcher incorporates an on-mount Doppler monopulse tracking radar with a range of 20,000m. Guidance is line of sight, using the radar to track the transponder on the missile.

In early versions of Crotale the fire control radar found it difficult to detect low-flying missiles because of surface clutter, particularly in rough seas. The EDIR (*Ecartométrie Différentielle InfraRouge*) variant, developed specifically for naval use, used an infrared camera to detect the flare from the missile rocket motor and was capable of intercepting sea-skimming missiles manoeuvring at 5g down to three metres above the surface; range was 13,000m. The more advanced CN 2 uses the VT 1 hypervelocity missile; range is as for the earlier model, but the maximum speed of the missile has been increased from Mach 2.3 to Mach 3.5.

La Fayette was fitted on completion with Crotale EDIR; later ships have the CN 2 variant, and *La Fayette* was retro-fitted with this in 1998.

20mm guns

The frigates of the *La Fayette* class were fitted from completion with two 20mm F2 guns in the bridge wings. The F2 is a navalised version of the GIAT M 693 manufactured for the French Army, and weighs 470kg including ammunition; it is fed by two 160-round magazines on the mounting. Rate of fire is 720rpm and maximum range 10,000m (2,000m effective). The gun has a three-man crew: the gun commander designates the target, the gunlayer is seated on the mounting, and there is a loading number.

Machine guns

Overseas deployment to counter piracy and drug-running led to a requirement for a light machine gun capable of engaging small boats. From 2007 the frigates were equipped with four Browning M2 12.7mm automatic MG derived from the 0.5in US model adopted by the *Marine Nationale* in 1943. The model fitted in the FLF is the air-cooled M2 HB (Heavy Barrel) on an *Affût Léger Démontable Portable* (ALDP, or 'Light Portable Mounting'). Two are fitted forward of the 100mm gun and two on the helicopter deck. Reloading is via a belt of 100 cartridges in an on-mount box. Weight is 210kg and manning as for the 20mm F2 gun.

When first completed two 7.62mm AANF-1 guns were mounted in the after bridge wings; these had a rate of fire of 900rpm and a range of 800m. These were replaced from 2015 by the Dillon M134 7.62mm, a six-barrelled gatling-type gun developed by General Electric that entered service in 1963 and was widely used in Vietnam. The gun is capable of 2,000–6,000rpm and has a range of 1000m; it weighs 39kg.

Electronic countermeasures (ECM)

Two AMBL 1C Dagaie Mk 2 chaff dispensers are located in the bridge wings; they fire rockets carrying metallic

One of the two Dagaie chaff dispensers on board *Guépratte*.
(J Moulin)

The modular fire control system (CTM) of Guépratte, with the
antenna for the Castor IIJ gun fire control radar on the left.
(J Moulin)

The antenna for the DRBV 15C air/surface surveillance radar on
Guépratte. (J Moulin)

and infrared chaff out to a range of 750m. Fittings have
also been installed forward of the bridge and on the after
funnel for the installation of a BSM 1/2 jammer.

Antisubmarine warfare

As completed the FLFs were not fitted for ASW; there
was neither a sonar nor antisubmarine weaponry,
although there was provision for the embarkation of an
AN/SLQ 25A Nixie noise-maker to decoy torpedoes.

Electronics

The DRBV 15C is a combined air/surface surveillance
radar with associated IFF developed by Thomson-CSF.
The lattice antenna of early variants has been superseded
by a solid antenna. DRBV 15 is an 'S'-Band Doppler
pulse radar featuring frequency agility and pulse
compression. It can detect an aircraft out to 100km and
a missile flying at medium altitude at 50km, and has
good low-altitude performance against sea-skimmers.

The DRBV 15C radar supplies target designation data
for the 100mm gun and for Crotale. It cues a modular
fire control system (*Conduite de Tir Modulaire*, or CTM)
incorporating a Castor IIJ radar, a TV 33 camera and an
IR scartometer. The CTM is complemented by two TDS
90 optical sensors (*Dispositifs Légers de Désignation
d'Objectifs*, or DLDO), located atop the bridge.

For navigation and helicopter control two DRBN 34

radars (Racal Decca 20 V 90) were fitted in these ships on
completion: one to starboard atop the bridge, the other to
port at the after end of the helicopter deck. DRBN 34 was
replaced in 1913–14 by the Italian Consilium, which is
designated DRBN 35 in French service.

For early warning of hostile threats there is a radar
detection system. *La Fayettte* was fitted on completion
with ARBR 17; this was superseded in 1998 by the more
advanced ARBR 21A (DR 3000), which was fitted from
completion in the other units. For interception of transmissions the early ships of the class were fitted with
ARBG 1A SAIGON (*Système Automatisé d'Interception
et de GONiométrie des émissions VHF/UHF*); the last
two ships have ARBG 2A MAIGRET (*Matériel
Automatique d'Interception et de Goniométrie des
Radiocommunications en Exploitation Tactique*).

Combat System

The Operations Centre (*Central Opérations*, or CO)
receives and processes all data and is the command centre
of the ship. It houses the consoles for the surface and air
pictures, for ESM and ECM, for the 100mm gun and
Crotale – the latter under the command of the Target
Designation Officer (initially designated *officier de désignation d'objectifs* or ODO, subsequently *officier d'au-
todéfense* or OAD), and the plot and console for the
Chief of Watch (Operations).

Tactical data is processed and displayed by the STI
(*Système de Traitement de l'Information* – ex-SENIT 7)
based on the Thomson-CSF TAVITAC 2000 system,
employing two computers and five colour consoles.
Operating in real time, it evaluates and displays the
tactical situation, analyses aircraft and missile threats,
and operates air defence systems including weapons and
electronic countermeasures; it can handle 400 tracks
simultaneously. Tavitac 2000 uses ruggedised civilian-
standard equipment rather than dedicated military equivalents; this has resulted in substantial cost savings.

The command system initially embarked was
OPSMER/SEAO (*Système Embarqué d'Aide aux
Opérations*), replaced from 2010 by SIC 21 (*Système
d'Information et de Commandement*) from
Thomson/DCN.

The foremast of *Surcouf* in 1995, shortly after her completion. The ARBR 21 ESM suite can be seen above the main yard, and the direction-finding loop at the masthead is ARBG 1A. (Y Violette)

An impressive bow view of *Aconit* at Toulon on 16 January 2012. The DF loop is the later ARBG 2A and the navigation/helicopter control radar is DRBN 35 (Consilium). Note the CTM modular fire control system atop the bridge. (J Pradignac)

Aviation

The FLF was designed from the outset to be able to operate a medium helicopter and to land a heavyweight helicopter such as the Super Frelon (13 tonnes). The flight deck measures 30m x 15m and has a surface area of 450m². The hangar, which is offset to starboard in order to accommodate the Crotale reload module to port, is 10 metres long and has an internal floor area of 150m², with maintenance facilities and stowage for spares.

In the centre of the flight deck is a 1.8m-diameter grid for a harpoon device, the helicopter is secured and moved to and from the hangar using the SAMAHE 10 system (*Systéme Automatique de MAnutention pour Hélicoptère Embarqué*); the control position is to port, abaft the Crotale reloading module. The TR 5 fuel replenishment system has an aviation fuel tank with a capacity of 80m³. In addition to the DRBN 34/35 helicopter control radar atop the hangar, a TACAN SRN 6 homing beacon is fitted, and there is accommodation for a flight crew comprising two officers and nine petty officers.

It was initially envisaged that the FLF would embark the NH 90 multi-role helicopter, but development of the

latter was protracted and in its final configuration the French Navy's NH-90 Caïman has proved difficult to accommodate in the hangar. The ships normally embark an AS 365 SA Panther. The Panther is a navalised version of the AS 365 N2 Dauphin fitted with a harpoon securing device and folding rotors. It was originally equipped with an ORB 32 surface search radar, a Titus communications link, a 172kg winch and a 4,560kW searchlight. Its principal mission is to relay over-the-horizon (OTH) guidance data for the MM 40 missiles; it can also be used for reconnaissance, for ship identification and for rescue; in the latter role it has space for six passengers. Operational range from the ship is 200nm.

The first fifteen Panthers were delivered 1993–98, entering service with 35F (Hyères) in March 1994 and with 36F (also Hyères) in September 1995; a further unit was delivered in 1992 for trials of anti-tank missiles and was converted to the same standard as the others in 2007. All sixteen Panthers are currently serving with 36F. In 2014 there were eight detachments: five for the nine frigates based at Toulon plus three for the *frégates de surveillance* based at Réunion (*Floréal*, *Nivôse*) and

The helicopter platform and the open hangar, offset to starboard to accommodate the Crotale reloading module. On landing, the helicopter is secured by its onboard harpoon mechanism to the circular grid in the foreground, then moved to the hangar using the SAMAHE system. The windows of the helicopter control room can be seen to port of the hangar door. (R Reboul)

Martinique (*Ventôse*). One was taken out of service in 2015 for the training of maintenance personnel.

The Panthers had their avionics and data link modernised to Standard 2 in 2007–14. The winch can be disembarked to compensate for the lack of power in hot climates; this enables the helicopter to establish the surface situation within a 100nm radius of the ship in 90 minutes. However, the modernisation was limited by cost constraints: the original ORB 32 search radar and turbines had to be retained, and ESM capabilities are lacking. It is envisaged that the Panther will be fitted to fire the ANL (*Anti-Navire Léger*) missile being developed jointly with the UK; this was tested successfully in June 2017, and will replace the French AS 15TT and the RN's Sea Skua.

Boats and Other Equipment

Three boats are normally carried, all of which are stowed, together with their handling gear, behind metallic shutters or hinged doors.

When the ships were first completed there was an ETN 90 service boat (*Embarcation de Transport en Nombre*) in a large central bay to port and an EDL 700

An EDO semi-rigid inflatable in the starboard-side bay of *Courbet* in 2010. The bay is closed by metallic shutters, seen here in the raised position. (R Reboul)

(*Embarcation de Drome Légère*) to starboard; the latter was replaced in 2001–02 by a semi-rigid inflatable designated EDO (*Embarcation de Drome Opérationnelle*),

A stern view of *Courbet* at Toulon on 16 January 1912. The large door offset to port conceals the ramp for the 20-seat EFRC Hurricane boat used by the commando detachment. Note the Browning M2 12.7mm automatic MG mounted on either side of the helicopter deck. (J Pradignac)

and this was in turn superseded by an improved model (EDO NG) from 2011.

Either of these boats can be replaced by a semi-rigid inflatable for commandos: the 20-seat ETRACO (*Embarcation de Transport RApide pour COmmandos*), or the more recent 32-seat ECUME (*Embarcation Commando à Usage Multiple Embarquable*).

The third boat is the EFRC Hurricane RIB, which is stowed on an inclined ramp and launched via a hinged door in the stern.

The FLF was equipped from completion with two 2.8-tonne anchors: one was located in the bow to permit future installation of a bow sonar, the second to starboard. When the ship is underway the starboard anchor well is covered by a hinged door which fits flush with the hull.

The replenishment at sea (RAS) gear is located on either side of the mainmast, and is normally concealed behind sliding steel shutters to ensure stealth.

Complement

A high degree of automation was adopted for these ships with a view to reducing the size of the complement. The initial design had accommodation for a crew of 180. In 1997 the breakdown was as follows: 12 officers, 68 petty officers and 56 ratings, together with 12 aircrew and, when the mission required it, 12–25 commandos. These figures were revised in 1999 to 15 officers, 85 petty officers and 53 ratings – a reflection of the need for higher skill levels. *Guépratte* was the first of the class to be built with accommodation for women, and in 2014 there were 25 women in a ship's crew of 140 (not taking into account aircrew or commandos).

The central passageway on the main deck is used for all movement on board, and the main services are at this level; the superstructure decks are normally out of bounds, and are not even fitted with guard rails. The CO's quarters are on the 2nd deck close to the Operations Centre, and the officers' cabins on the 1st deck and at the after end of the main deck forward of the hangar; the wardroom is on the 1st deck. The senior petty officers are in 2/4-berth cabins on the 1st, main and platform decks, with a wardroom on the main deck, while the junior POs are in four messes: two forward and two aft of the machinery spaces on the platform deck. Seamen have 20-man berthing on the platform deck, and there are three messes for women (other than officers) at the after end of the same deck.

The personnel are currently organised as seven branches (*services*):

– *Pont* (= 'deck')
– *Flotteur* (= 'hull')
– SIC (*Système d'Information et de Commandement*)
– LAS (*Lutte Anti-Surface*)
– Art (*Artillerie*)
– *Commissariat*
– *Machine* (*propulsion*)
– *Electricité*

An overhead of *Guépratte* taken shortly after her completion. Note that the Crotale launcher is in the process of being reloaded, a procedure which takes around 30 minutes. (Marine Nationale)

The latter two branches form a single *compagnie*.

In a major departure from earlier practice, a single galley prepares meals to be distributed to the cafeteria (seamen), and to the messes and dining rooms for the petty officers and the CO/officers. The FLF, which was designed for extended deployments overseas, has sufficient stowage for 50 days' provisions. The Supply Officer (*Commissaire*) is responsible not only for supplies but also for publicity, working conditions and all legal aspects; he/she has a team of seventeen.

There is a well-equipped sick bay with an operating room and X-ray facilities, although a Medical Officer is embarked only for particular missions.

Modifications & Modernisations

Stealth considerations have prevented any important modification to the ships' silhouette; the only significant modifications since completion have been the fitting of the four 12.7mm MG (see above), changes to the paint on the diesel uptakes and an upgrade to the ESM suite. ARBR 21 replaced ARBR 17 in *La Fayette* during her first refit (IPER), February–June 1998. The latter ship also underwent changes to the shape of the uptakes to

The lead ship of the class, *La Fayette*, at Toulon in 2009. The two starboard-side Browning M2 12.7mm MG can be seen forward of the 100mm gun and on the helicopter deck aft. (J Pradignac)

The fourth unit of the class, *Aconit*, at Toulon on 10 June 2013. (J Pradignac)

improve stealth, and received the improved CN 2 variant of Crotale; the other four units had these modifications from completion.

More recent updates include:

2007	4 x 12.7mm
2010–11	VIGISCAN IR on foremast
2011	EDO replaced by EDO NG
2012	Planned to replace MM 40 Block 2 by Block 3 and install six Dillon 7.62mm M 134
2013–14	DRBN 35 Consilium in place DRBN 34
2014	*Surcouf/Aconit* to receive new jammers at the base of the bridge.

The FLFs as completed had provision for two 8-cell Sylver vertical launch modules for the Aster 15 local area defence missile; had they been fitted, the associated Arabel tracking/FC radar would also have been required. The proposal was eventually abandoned when the Navy failed to secure the necessary funding. Space was also reserved for bow and towed sonars, and the anchors disposed accordingly. The towed sonar space was subsequently converted into a store for commandos. Despite this, the stealth provisions incorporated into the FLF made it inherently well-suited to ASW operations, and it was still envisaged that the necessary sensors and weaponry would be fitted in a mid-life refit.

These and further proposals for a radical modernisation of the ships were essentially killed off by the advent of the FREMM frigate, which was designed for a complement of 100 compared with 150 for the FLF. A new programme drawn up in 2015 envisaged the embarkation of a hull sonar to counter the diesel submarines entering service with secondary powers. However, the final contract specified only the renovation of structures and the installation of electronic/data systems comparable to that of the carrier *Charles de Gaulle* to ensure operational compatibility.

The FLFs are now due to be replaced by a new design designated the *frégate de défense et d'intervention* (FDI – formerly the *frégate de taille intermédiaire* or FTI). In May 2015 it was stated that this programme would be brought forward, with the first of class to commission in 2023. Plans to upgrade the FLFs were thrown into doubt by this development. A report dated March 2017 envisaged an order for five FTIs and a modernisation of three of the five FLFs involving an upgrade to the combat system and refurbishment of the hull and machinery. The modernised ships would then serve until 2030. The proposed modernisation was to include:

- replacement of Crotale by two Sadral point-defence missile launchers, each with six Mistral missiles, the launchers being removed from decommissioned frigates of the F 70 type.
- replacement of the MM 40 Block 2 SSMs by Block 3
- installation of a Thales KingKlip Mk II hull sonar
- upgrading of the electronics outfit and combat data system
- installation of a torpedo countermeasures system.

As a temporary measure, in December 2017 *Surcouf* was fitted with the Thales Blue Watcher sonar. The latter is designed for small ships (3.5 frequency at 6.1kHz), whereas the KingKlip sonar earmarked for the full modernisation is intended for medium units (5.25 frequency at 8kHz, 25km range).

The following dates for the planned half-life refits

A model of the *frégate de taille intermédiaire* (FTI – now FDI) on display at the 2016 Euronaval Exhibition. The ships will be armed with an OTO Melara 76mm Super Rapid gun, eight MM 40 Exocet Block 3 missiles, and two 8-cell launchers for the Aster 15 local area defence missile; they will embark a single NH 90 Caïman helicopter and/or an unmanned drone helicopter. Note the integrated mast assembly, which is designed to be installed as a single module and which incorporates the operations room at its base. The first of class, *Amiral Ronarc'h*, was laid down at Lorient Naval Dockyard on 24 October 2019. (DCN)

(*rénovation à mi-vie*, or RMV), which are to be carried out by the Naval Group at Toulon, were posted on the website *Mer et Marine* on 17 February 2020:

– *Courbet*: from Autumn 2020
– *La Fayette*: 1922
– *Aconit*: 1923

Evaluation

The revolutionary design of the FLF had a considerable impact abroad and resulted in a number of export orders for modified ships. In August 1991 a contract was signed by Thomson-CSF and the China Shipbuilding Corporation of Taiwan for the construction of six similar units of the *Kangding* class, to be constructed at Lorient. Built from 1993 to 1998, the ships had the same hull and propulsion system as *La Fayette* but modified superstructures and a completely different weapon/sensor outfit.

Similar to the French units, but slightly larger and with Crotale replaced by Aster and its Arabel tracking radar, were the three ships of the Saudi *Al Riyadh* class ordered under the Sawari 2 programme and built at Lorient between 1999 and 2004.

Finally, in March 2000, an order was placed by Singapore for six slightly smaller ships of the *Formidable* class. Only the first was built at Lorient 2002–07, the remaining five being built at Singapore; the last two were completed in early 2009. Like the Saudi units, the Singaporean ships embarked the Aster missile system, and the 100mm gun was replaced by the OTO 76mm Super Rapid gun; the MM 40 Exocet SSMs were replaced by the US Harpoon.

All of these ships have been built according to the same principles, with steel hulls and extensive use of GRP for the superstructures. Movement is internal, with the result that the ship operates more like a submarine than a conventional surface ship. The need to preserve stealth characteristics has limited the ability to fit new systems; the forward 12.7mm MG embarked in the *La Fayette* class since 2007 have had to be located atop the GRP cover for the forecastle, which places the guncrew in a precarious position due to the absence of guard rails. There have been unconfirmed reports that the ships are prone to man-overboard emergencies. The covered forecastle has also meant that water shipped via the hawsepipes has been difficult to clear, and specially-designed fittings had to be developed to prevent water ingress. The arrangement of the internal spaces is also somewhat cramped, with relatively narrow passageways and some awkwardly shaped compartments.

In fairness the British Type 23 frigate, which was likewise designed to take into account the lessons of the Falklands War, began with 'clean' superstructures angled inboard, and has since had so many new items of equipment 'bolted on' externally that its stealth features have been compromised. The lesson to be taken from this is that for ships which will need to be regularly upgraded during their service lives, the key to design is to maximise hull size and internal volume to give the necessary flexibility for later changes, as it is difficult to predict the dimensions and weights of systems in the early stages of development. The RN Type 45 destroyer is a recent example of this trend.

Arguably more serious has been the failure to take advantage of the possibilities for upgrades that were built into the design. There can be little doubt that the military capabilities of the *La Fayette* class would have benefited from the installation of the more capable Aster missile system and the fitting of bow and towed sonars, for which space was reserved from the outset. Note, however, that for the Saudi frigates fitted with Aster hull length was increased by 8 metres and beam by 1.8 metres. It should also be noted that a suite of underwater sensors would need to be complemented by anti-submarine weaponry, which was completely lacking in the FLFs on completion.

It was initially envisaged that A/S homing torpedoes would have been delivered by the NH 90 helicopter, but hangar dimensions proved insufficient for the latter in its final configuration. The obvious solution would have been to suppress Crotale and its space-intensive reloading module and to extend the hangar, but as the hangar was constructed of GRP this would have been expensive, particularly when added to the cost of Aster and its associated Arabel fire control radar and of the two sonars. Also, facilities which have proved useful in the unaccompanied surveillance and patrol missions, such as the additional accommodation which has replaced the Aster modular launch system, and the commando boat and stores currently accommodated in the stern, would have been lost. The decision to restrict modernisation to an upgrade of the combat system and a relatively small, inexpensive bow sonar is therefore understandable.

Other features of the ships have been more successful. They have proved surprisingly robust and reliable, and easy to maintain when deployed overseas; in 2015 they achieved a remarkable 90 per cent availability. The stabilisation system works well, and the FLFs have proved to be excellent sea-boats, with a moderate roll even in heavy seas. Most importantly, they have performed extremely well in the peacetime missions for which they were primarily designed. They have deployed to West Africa and (on numerous occasions) to the Indian Ocean, and have regularly participated in NATO and other international exercises. All five ships of the class were rotated in sequence off Libya during the military intervention of March–October 2011 (Operation 'Harmattan'), and since their completion they have been prominent in the monitoring and suppression of drug-running and piracy in the Gulf and off the Horn of Africa.

Principal source:
Patrick Maurand & Jean Moulin, *Les Frégates Furtives Type 'La Fayette'*, Editions Lela Presse (Le Vigen, 2018).

FROM GREYHOUNDS TO SHEEPDOGS

Home Fleet Destroyers at the Turning Point of the Battle of the Atlantic, March–May 1943

This new study by **Michael Whitby** highlights the anti-submarine operations of the Fleet destroyers of the 'O' and 'P' classes in support of the transatlantic convoys during the spring of 1943.

At a particularly bleak moment in *The Cruel Sea*, Nicholas Monsarrat's classic novel of the Battle of the Atlantic, a beleaguered convoy escorted by the fictitious corvette HMS *Compass Rose* is reinforced by two Royal Navy *Laforey*-class Fleet destroyers. 'The two destroyers joined punctually at six o'clock,' Monsarrat related, 'coming up from the south-east to meet the convoy, advancing swiftly towards it, each with an enormous creaming bow-wave. They both exhibited, to a special degree, that dramatic quality which was the pride of all destroyers: they were lean, fast, enormously powerful – nearer to light cruisers than destroyers – and clearly worth about three of any normal escort.'[1]

Prior to the spring of 1943, North Atlantic trade convoys typically were protected by a close escort of one or two aged destroyers and four or five corvettes. Only infrequently were they reinforced by modern Fleet destroyers – in one notable instance in December 1942, HM Ships *Milne* and *Meteor* bolstered the escort of the battered ONS-154. Admiralty policy was 'to allocate new destroyers for service with the Fleet where air attack or offensive action with enemy surface units is probable. Replacement or reinforcement of ocean escorts have therefore, to be made from older destroyers released from the Fleet ...'[2] This policy was departed from only under dire circumstances. Such was the situation in the spring of 1943 when more than a dozen Fleet destroyers joined Western Approaches Command to alleviate the crisis in the North Atlantic. Deployed in support groups or as the screen for escort carriers, their role in turning the Battle of the Atlantic – in combination with enhanced air cover and better-trained close escort groups – is acknowledged in most historical accounts. Nevertheless, the details of their experience have largely escaped attention; thus, their precise impact remains unexplained. This study of the destroyers that formed the support group EG-3 aims to shed light on their role in the decisive events of the spring of 1943.

The Crisis in the North Atlantic

The situation in the North Atlantic was grim. Germany's *U-bootwaffe* was swarming around the convoy routes, exacting horrendous losses. In mid-March more than 35 U-boats ravaged the eastbound convoys HX-229 and SC-122, overwhelming the escort to sink 22 merchant vessels against the loss of only one of their own. Such losses were unsustainable. The Admiralty's Vice Chief of the Naval Staff, Vice Admiral Sir Henry Moore, currently in Washington discussing convoy policy with his American and Canadian allies, was well-positioned to comment on the crisis.

> If we attempt to run these large convoys with an escort of six or seven vessels without the possibility of reinforcement we shall suffer quite unacceptable losses.
>
> The only alternative is to provide support groups and the minimum number which could meet the situation is two on each side of the Atlantic. The provision of support groups is in our opinion essential not only for the protection of convoys but also because they will give us a means of killing the U-boats thereby gaining the upper hand ... Without these support groups our losses may well force us to take drastic emergency steps to increase the escorts. This could only be done by reducing the number of convoys which would be disastrous from the imports point of view.[3]

The idea of employing support groups of fast destroyers to reinforce close escorts was not new; there were just never enough suitable ships available. The chronic shortage of destroyers persisted into March 1943, and the Commander-in-Chief Western Approaches balked when reports from the conference held in Washington suggested he might have to form additional close escort groups, which could mean foregoing support groups. Admiral Sir Max Horton warned: 'Such sacrifice would be viewed by me with the greatest concern since I consider the supporting groups are absolutely essential to the successful prosecution of the Battle of the Atlantic during the coming months.' But it was not simply a matter of finding the necessary destroyers; Horton thought the crisis demanded trained, experienced groups 'to obtain maximum killing efficiency' – they had to hit the ground running.[4] Destroyers serving with the Home

Fleet met this standard; however, heavy commitments had prevented their release. That changed in mid-March 1943 when enhanced *Kriegsmarine* capital ship strength in Norway caused the Admiralty to suspend the Russian convoys. Since they absorbed most of the Home Fleet's destroyer strength, the interruption enabled Fleet destroyers to be loaned to Western Approaches. *Offa*, *Orwell*, *Obedient*, *Onslaught*, *Oribi* and *Impulsive* formed the Third Escort Group (EG-3), while other destroyers worked alongside escort carriers in EG-4 and EG-5 – frigates and sloops from Western Approaches formed EG-1 and EG-2.[5] In a letter to a colleague, Horton betrayed his anticipation of better days ahead:

> This job has been pretty sombre up to date, because one hadn't the means to do those very simple things for which numbers are essential, and which could quash the menace definitely in a reasonable time; but in the past few days things are much brighter and we are to be reinforced, and I really have hopes now that we can turn from the defensive to another and better role – killing them.[6]

Support Group EG-3 is Formed

Beyond numbers, EG-3 met Admiral Horton's plea for experience. In particular, EG-3's senior officer (SO EG-3) was almost uniquely qualified to lead a support group. A respected 'destroyer man', Captain James 'Bez' McCoy had commanded no fewer than eight destroyers before

taking over *Offa* and the 17th Destroyer Flotilla on 12 March 1943. Beyond that wide experience, which included being senior destroyer officer at the Second Battle of Narvik, McCoy knew the Battle of the Atlantic intimately, having served as Captain (Destroyers) at Western Approaches from July 1941 to August 1942, then as captain of HMS *Hecla*, the escort depot ship based in Iceland. Given those qualifications, and the fact McCoy was appointed to command the 17th Flotilla at the same time the destroyers joined Western Approaches, it seems obvious that Horton hand-picked him for the job.

EG-3's other captains were also seasoned veterans. Reflecting RN practice, each had previously commanded a 'Hunt'-class or older destroyer before taking over a modern Fleet destroyer. All were also 'Salt Horses'; non-specialists who spent the vast majority of their career at sea, and who were typically talented ship-handlers, a plus given the atrocious conditions they would soon encounter. Finally, each had significant experience in convoy warfare, gained in the North Atlantic, along the UK's east coast or on the Mediterranean and Russian runs. This combination of destroyer and fighting experience gave them a huge amount of professional confidence, and they were complemented by wardroom teams and senior ratings that featured a mix of professionals and volunteers who had been similarly tested.

Ships and equipment also mattered. EG-3's destroyers were of recent vintage: the 'O's and 'P's commissioned in 1941–42, *Impulsive* in 1938. However, that did not mean

HMS *Offa* off the Orkneys, probably in the autumn of 1942 before she was fitted with centimetric radar. McCoy's destroyer was the only one to participate in all of EG-3's support operations. (Conrad Waters collection)

Captain James McCoy, centre, with King George VI on the bridge of HMS *Onslow* in August 1943. The seasoned judgement of the veteran destroyer officer was a key factor in the success of EG-3 and support group operations overall. (IWM A-18565)

that they were fitted with the latest fighting equipment. It is often assumed that newly-commissioned Fleet destroyers received the best of everything, but EG-3 exposes that fallacy.[7] In particular, they were ill-equipped in terms of the specialised sensors critical to antisubmarine warfare: specifically Types 271 or 272 centimetric RDF, or radar, which could detect the conning tower of a U-boat running on the surface, and High Frequency Direction Finding systems (HF/DF or 'Huff-Duff'), which enabled escorts to pinpoint enemy radio transmissions. This was because the Admiralty chose not to fit such equipment, especially valuable RDF sets, on an interim basis to ships that would not remain permanently on escort duties.[8] *Offa* was the only ship fitted with centimetric radar – she had Type 272 whereas the others had the far less effective Types 286 and 291 metric sets – and none had HF/DF. In terms of other equipment, all had variants of Type 128A asdic, as good as any contemporary destroyer set, and before joining Western Approaches they shipped new, more powerful Minol depth charges. Their greatest attribute was their speed. Capable of more than 30 knots, Fleet destroyers were faster than most escorts, which enabled them to quickly chase down U-boat contacts and move to threatened convoys. Speed, however, came at the cost of endurance, and Fleet destroyers were relatively short-legged, a drawback that would rear its head in the weeks to come.

With the situation critical at sea, EG-3 received only a measure of specialised training before joining the convoy war. After formulating the original support group orders with Admiral Horton at Western Approaches headquarters in Liverpool, McCoy took an abbreviated version of the invaluable escort commander's course at the Tactical Unit. Then, after *Offa* joined the rest of EG-3 at Moville, Northern Ireland, on 22 March, the commanding officers undertook a crash course at the Tactical School in Londonderry before carrying out a night exercise at sea under the expert eye of Western Approaches' training commander, Captain Joe Baker-Creswell. Plans for further exercises had to be abandoned on 23 March, when EG-3 received orders to put to sea to support SC-123. It was the first support group to deploy.

Convoys SC-123 and HX-230

The 50 ships of the slow eastbound convoy SC-123 had sailed from New York City on 14 March. Before EG-3 joined, from 21–26 March the close escort was bolstered

101

Admiral Sir Max Horton with Captain Joe Baker-Cresswell. Horton's insistence that support groups be comprised of seasoned units was a key factor in their ability to have an immediate impact on the Battle of the Atlantic. As Training Commander at Western Approaches, Baker-Cresswell oversaw the final fine-tuning support groups received before joining the convoy war. (IWM A-17821)

by the escort carrier USS *Bogue* and two destroyers. On 26 March the Admiralty reported two U-boats in the vicinity, but no attacks developed, probably due to *Bogue*'s air coverage. Unhappily, the ideal scenario of a convoy being reinforced by both a support and a carrier group was frustrated when rough seas prevented *Bogue* from refuelling her destroyers and she was forced to leave the convoy that afternoon.

When EG-3 joined the convoy 240nm south-east of Cape Farewell late the next afternoon, the six destroyers deployed in accordance with the instructions conceived by Horton and McCoy.[9] By day, 'supporting groups will sweep and search the vicinity of convoy, their own SO acting on the requirements of SO of [Close] Escort Group.' By night, they 'will take up night disposition on extended screen except that in the case of destroyers not fitted with RDF Type 271 they should be placed on the close screen and ships of Escorting Groups thus released should join the extended screen.'[10] Admiral Horton also promulgated additional instructions with a psychological bent: 'The small number of close escorts for large convoys is liable to act adversely on merchant ship morale. SO of Support Groups should therefore bear in mind advisability of showing their Group to Convoys

when convenient opportunities occur.'[11] McCoy accordingly led EG-3 through the columns of merchant ships when he met SC-123, but any psychological boost proved short-lived. After just 16 hours with the convoy, Western Approaches signalled McCoy: 'If SC-123 is not actively threatened proceed to join HX-230 after dark.'[12] Here was the advantage of fast, independent support groups: if one convoy was not threatened the group could move to another that was. Horton left discretion of when to move to the senior officer on the spot, but since SC-123 had not been attacked and no U-boats appeared to be shadowing, McCoy complied immediately. Unfortunately, as so often happens in the volatile North Atlantic, severe weather intervened. Battered by a Force 8 gale, whose violent seas damaged all six destroyers, it took EG-3 until daybreak on 29 March to cover the 240nm to HX-230.

The Admiralty deemed the threat to HX-230 to be serious, identifying eleven U-boats in its vicinity. Although the close escort had driven off four U-boats, the SOE, Commander Edward Bayldon in HMS *Hurricane*, thought two were still shadowing. Regrettably, coordination between Bayldon and McCoy never seemed to gel. When *Offa* pulled alongside *Hurricane* it took four hours for the two officers to exchange information by loud

The cruel sea: the frigate HMCS *Swansea* battles the North Atlantic in the autumn of 1943. High seas regularly hampered support group operations, disrupting refuelling, frustrating anti-submarine efforts, impeding movements between convoys, and battering ships and men. (Department of National Defence, DHH GM-1441)

hailer and transfer line. When the process was finally completed, Bayldon positioned two of EG-3's destroyers at visibility distance 40 degrees on each bow of the convoy, and sent the remainder astern as a strike force. When *Hurricane* picked up two HF/DF transmissions about 25nm astern of the convoy, Bayldon asked McCoy to sweep down the bearings:

> As these two U-boats appeared to be the only ones in contact with the convoy, I considered it desirable that both should be kept down if this reduced the chances of either one being sunk. This was reported to Escort Group 3. He concurred but I do not know what dispositions he made.

Apparently unaware of McCoy's subsequent actions, Bayldon later reported:

> I do not wish to imply that the Third Escort Group was not handled efficiently, on the contrary, the Senior Officer did everything to make things easy for me, but I am certain that had either he or I have had a free hand and control of escort, support force and aircraft, we could have achieved more definite results on the 29th of March.[13]

This is perplexing. As an experienced SOE, Bayldon knew from Western Approaches Convoy Instructions that he retained complete control of a convoy's defence when reinforcements joined; the cautious tone of his criticism suggests he was perhaps unsettled by McCoy's reputation and seniority. Whatever the reason, word of his dissatisfaction reached McCoy – probably through Horton – and when later acknowledging smooth co-ordination with another SOE, McCoy noted with evident sarcasm 'no difficulties were experienced in connection with the so-called "vexed question" of command.'[14] Despite this *contretemps*, U-boats were prevented from attacking HX-230. Bayldon concluded that sweeps by EG-3, *Hurricane* and the frigate HMS *Kale*, combined with shore-based air cover, were 'the last straws which caused the U-boats to break off the attack.'[15] Unable to refuel in persistent heavy seas, McCoy left the convoy at dawn on 30 March; all six of EG-3's destroyers would require docking to repair the hammering inflicted by the North Atlantic.

Evaluation and Recommendations

Since EG-3's deployment was the first test of the new support groups, it engendered significant comment. Overall there was satisfaction, the Western Approaches war diary noting: 'the fact that attacks on convoys ceased on arrival of Support Groups is a most encouraging sign.'[16] McCoy went further, concluding that 'such a system offers every possibility of rendering the enemy's attack on trade in the North Atlantic extremely difficult, if not abortive.' Nonetheless, there was still much to sort out, and McCoy detailed three requirements to improve the system:

Even with the Home Fleet's contribution to Western Approaches, there were never enough Fleet destroyers for the North Atlantic. After repairing storm damage suffered during EG-3's initial support operation, HMS *Onslaught* shifted to the support group EG-4 to work with the escort carrier HMS *Archer*. (Author's collection)

(i) the provision of an oiler with the floating hose method of oiling

(ii) a standard form of passing information to the Support Group when it joins the convoy

(iii) the retention of the same ships in each group so that a team spirit can be worked up, knowledge of each other and each other's methods can be furthered, maximum efficiency obtained, and a considerable volume of signal traffic made unnecessary.

With regard to refuelling deficiencies, McCoy explained:

After many years in command of destroyers I am only too aware of the special strain imposed on Commanding Officers in oiling by the trough method and knowing the constant strain imposed on them by the necessarily continuous action of the Support Group I can see that the result of using any other method but the floating hose for oiling will result in self-inflicted damage to the ships.

A censored view of HMS *Obedient* in July 1943. *Obedient* left EG-3 after suffering damage from heavy seas on the group's first deployment. (IWM A-17864)

McCoy also recommended that additional oilers be deployed so that 'the Senior Officer will know that on almost all occasions he will be able to obtain oil from the next convoy to which he may hurriedly be ordered.'[17] As it was, underway replenishment remained a challenge, and the combination of a shortage of escort oilers, poor equipment and training, and rough seas plagued support groups throughout the spring.

Regarding the flow of information, McCoy proposed that the lengthy process endured by *Offa* and *Hurricane* could be remedied by standardised forms that could be easily disseminated among all escorts. McCoy also sought the communications discipline typical of Fleet manoeuvres. He thought the 'endless signalling' amongst escorts could be alleviated by the use of 'spare code-words' to explain basic tactical manoeuvres. He also considered it essential 'for R/T call signs to be allocated to the Senior Officer, Support Group, to the entire group, to each division of the group and to each ship', a practice utilised on the Russian convoys. McCoy admitted this came with risks, and that: 'In due course by reason of this standardisation, the enemy will become aware of such co-operation.' 'This', he insisted, 'will just be too bad. I am convinced the smoother working of Command of the combined force will outweigh any advantage the enemy may obtain from it.' Finally, McCoy warned that R/T was overused, particularly when homing a support group onto a convoy. 'The whole value of the Support Group system lies in the unpleasant surprise which its

arrival with the convoy gives to the enemy', and those chances would dissipate if the enemy monitored an increase in R/T traffic. Relying upon his enormous experience, McCoy had illuminated important issues, which he thought could be remedied through well-tried procedures associated with fleet operations.

McCoy's final recommendation that support groups be kept intact to ensure cohesiveness may actually have been tendered in protest, since he was already aware that EG-3 was being broken up. Six destroyers had to be found to screen two escort carriers about to join Western Approaches, which forced changes to EG-3. Upon learning the news, *Onslaught*'s CO, Commander WH Selby, echoed McCoy's perspective:

> The hopes which were raised of being able to work as a team were rudely shattered when HMS *Obedient* and HMS *Orwell* were ordered to return to Scapa. Although it is obvious, I would like to stress the necessity for keeping the same ships together if good results are expected in the highly specialized work of specialized submarine hunting.[18]

The complaints were to no avail, and it was a reduced EG-3 that next went to sea. Only *Offa* and *Impulsive* remained from the original group, joined by the 'P'-class destroyers *Panther* and *Penn*. Through service with both the Home and Mediterranean fleets, the newcomers satisfied Horton's criteria for experience and they quickly became solid members of McCoy's team.

Oiling at sea put a strain on sailors and often impeded the ability of destroyers to fulfil their support duties. Here, sailors undertake the delicate floating hose method. (IWM A-7863)

Convoy HX-233

After reviewing the lessons of EG-3's initial operation with Admiral Horton in Liverpool, McCoy took *Offa* to join *Impulsive*, *Panther* and *Penn* at Moville for two days of exercises under Baker-Cresswell. *Offa* and *Impulsive* had just completed repairs to storm damage while the two 'P'-class were fresh from post-refit work-ups. Unhappily, while in dock none of the destroyers' systems had been upgraded. The Admiralty considered fitting centimetric radar to *Panther* and *Penn*, but it would have extended their refit another 30 days and they were needed at sea. This left *Offa* as the only ship with that equipment, and the entire group still lacked HF/DF. No matter, in the early afternoon of 13 April they sailed to reinforce HX-233. With 35 U-boats estimated to be in its vicinity, 'some doubt was felt for the safety of this convoy'; therefore, besides reinforcing it with EG-3, it was routed farther south than usual, even though this would reduce coverage from shore-based aircraft.[19] What transpired was a text-book example of the impact of a support group on a threatened convoy.

HX-233's close escort was the mixed American/British/Canadian group A-3, led by the US Coast Guard cutter *Spencer*, and filled out by a second cutter, the destroyer HMCS *Skeena* and five corvettes. When EG-3 rendezvoused with the convoy in the early hours of 17 April, reflecting instruction received at the Western Approaches Tactical Unit, McCoy approached from astern where it was thought U-boats would most likely be lurking. The tactic paid dividends when *Offa* sighted a submarine, which immediately submerged. The destroyers carried out a concerted attack, and although

While EG-3 filled gaps on the other side of HX-233, USCG *Spencer* launched the attack that resulted in the destruction of *U-175* on 17 April 1943. The image demonstrates how close the attack was to the convoy. (NARA 26-G-1517)

they heard the sound of 'hammering', which indicated they might have inflicted damage, they lost contact. They had, nonetheless, thwarted an attack on the convoy. Shortly afterwards, *Panther* sighted another U-boat about 6nm away. Charging aggressively at 28 knots, she forced the U-boat down, but when she slowed to initiate her attack run at about 800 yards, Lt-Cdr Viscount Jocelyn elected to classify the contact rather than carrying on directly into the attack. *Panther* overran the target before depth charges could be fired, enabling the U-boat to escape. Lesson learned: Jocelyn reported that next time he would not waste time on classification when the contact was so obviously a U-boat.

HMS *Penn* coming alongside HMS *Rodney* during the passage of convoy WS19 to Freetown in June 1942. EG-3's 'P'-class destroyers were seasoned veterans of Fleet operations in support of convoys in the Atlantic and Mediterranean. (Conrad Waters collection)

Just before EG-3 engaged these submarines, *U-628* torpedoed the steamer *Fort Rampart* and, shortly afterwards, *Spencer* detected another U-boat lurking beneath HX-233. As *Spencer* engaged this target, ultimately destroying *U-175*, the SOE directed EG-3 to abandon their various hunts and assume position ahead of the convoy to cover his absence. This was an important consideration. As emphasised in the Atlantic Convoy Instructions, the primary objective was the safe and timely arrival of the convoy; killing U-boats was secondary. Adhering to that maxim, on several occasions over the spring, EG-3 aborted hunts to return to a convoy, which undoubtedly cost kills. To return to HX-233, after learning that *Fort Rampart* remained afloat astern of the convoy, McCoy sent *Panther* and *Penn* to investigate the wreck thinking it might attract U-boats. Again, his instincts proved correct, and the two destroyers attacked a faint contact, badly shaking *U-628*. Now far astern, they broke off the hunt.

After darkness fell, EG-3 ran down several HF/DF intercepts by *Skeena*. For five hours the four destroyers searched for the enemy while dropping random 'scare' charges. Since *Offa* made no contacts with her Type 272

RDF, McCoy assumed they had succeeded in forcing the U-boats down. He was right: a *Kriegsmarine* analysis lamented 'four boats in succession attempted to attack but each in turn was pursued by the escorts, and suffered continuous depth charge attacks.'[20] This gave credence to British official historian Stephen Roskill's subsequent appreciation that '*Offa*'s ubiquitous support group was key to the defence of HX-233.'[21] As it was, anxiety over the convoy eased and, in the afternoon of 18 April, Western Approaches directed EG-3 to reinforce SC-126, some 200nm to the southwest. A similar situation developed there: U-boats were in contact but air cover and EG-3's sweeps foiled attacks. After two days supporting SC-126, Western Approaches ordered McCoy to proceed west to St John's, Newfoundland providing cover to ONS-3 and ON-178 *en route*. After an uneventful passage, EG-3 put into 'Newfy John' on 25 April.

Review and Report

McCoy again struck a positive note in his post-deployment report, observing: 'Experience has shown that the provision of Support Groups is not only justified but is in

Ravaged by depth charges and shell-fire, *U-175* wallows on the surface before sinking. The loss of so many U-boats during the spring of 1943 marked the turning point in the North Atlantic. (NARA 26-G-1512).

fact acting as a very great deterrent to the attacks of the enemy.' Nonetheless, deficiencies lingered. In particular, 'the lack of an HF/DF set, not only in the leader but in the group, very seriously affected its efficiency.'[22] EG-3 was forced to rely upon HF/DF information from escorts, which hampered their own response. Had they had been able to process their own HF/DF information on the night of 17/18 April, for example, they might have been able to pin-point the exact location of the U-boats. Regarding the deficiency in centimetric radar, *Panther*'s Lt-Cdr Viscount Jocelyn complained:

> Throughout this operation, I have felt at a great disadvantage in not having the services of RDF Type 271 or 272. Type 291 is good but not good enough, and the restrictions on its use mean that operators do not get enough practice and are inefficient when the restrictions are lifted. I understand that the replacement of searchlights by Type 272 has priority, and I most sincerely hope that it is a high one as I do not consider that the functions of a supporting destroyer at night can be carried out most satisfactorily without the equipment.[23]

Offa finally obtained a cherished HF/DF set at St John's, but radar remained a weakness.

McCoy also thought that more aggressive routing would take a bigger bite out of the U-boats. 'Bad weather', he observed, 'operates more favourably for the enemy than it does for us. It is therefore imperative that convoys be routed, in so far as is humanly possible, to pass through areas of fine weather:

(i) so that ships of the support groups can refuel expeditiously
(ii) so that the support groups can move at very high speed to counter any threat by the enemy
(iii) to facilitate the use of carrier borne aircraft.

When supporting HX-233 and SC-126, McCoy thought since 'the weather conditions were good I felt, due to the fact that I could proceed at high speed [on sweeps] and refuel without delay, that I was completely on top of the enemy.' McCoy wanted to take the fight to the enemy and thought routing convoys through calmer conditions would bring more opportunities to tackle the U-boats. However, Admiral Horton was unwilling to take that risk since routing convoys farther south would reduce shore-based air cover. Moreover, the situation in the North Atlantic had improved in the final two weeks of April 1943, striking a 'hopeful note' at Western Approaches. More U-boats were on patrol than during the previous half-month; however, 'the losses of ships in convoy

Although HMS *Oribi*'s pendant number has been censored, the lantern for the Type 271 RDF she received later in the war is clearly visible in the former searchlight position astern, while her Type 291 antenna is atop the foremast. The lack of Type 271 during the battle for ONS-5 hampered *Oribi*'s ability to hunt U-boats in the fog that descended on the convoy. (John Maber collection, courtesy of Richard Osborne)

Despite lacking both HF/DF and centimetric RDF, HMS *Panther* proved a valuable member of EG-3, particularly in support of HX-233 and SC-131. During the campaign, Lieutenant-Commander Viscount Jocelyn, an experienced destroyer officer who took over *Panther* in October 1941, contributed insightful commentary on support group operations. (US Navy, courtesy of Richard Osborne)

showed a significant decrease.' The reasons seemed clear: 'It appears that the greater efficiency of Escort Groups, the increased assistance from Support Groups and the additional strength of air cover provided, have not only resulted in driving off the packs but have prevented them from assuming the offensive.'[24] To reinforce this trend, Western Approaches positioned support groups at either end of the air gap, with EG-3 the first based at the western terminus in Newfoundland.

The Battle for Convoy ONS-5

Long after the war, Stephen Roskill fixed the battle for ONS-5 amongst the loftiest annals of British naval history. 'The seven-day battle fought against thirty U-boats', he reflected, 'is marked only by latitude and longitude and has no name by which it will be remembered; but it was, in its own way, as decisive as Quiberon Bay or the Nile.'[25] Unlike those epic encounters, the battle for ONS-5 was fought in stages over several days. The first, from 28 April to 1 May, featured clashes between the close escort and some eight U-boats. Then, after the convoy endured a harsh three-day gale, the second confrontation occurred between 4 and 6 May when as many as 40 U-boats stalked ONS-5. Fighting alongside an escort depleted by fuel shortages, two of EG-3's destroyers proved indispensable in this second, most decisive part of the battle.

The newly assigned HMS *Oribi* was the first of EG-3's destroyers to reinforce ONS-5, joining from SC-127 at midnight on 1 May south-east of Cape Farewell as it turned on its southerly course for Halifax. U-boats had sunk only one merchantman to that point but things promised to get worse. Although there was a temporary blackout with special intelligence, the Admiralty estimated that some 60 U-boats were in the North Atlantic, the bulk of which were congregated around ONS-5. After seizing an opportunity to top up with fuel, *Oribi*, now fitted with HF/DF, chased down two contacts the first evening, both of which proved negative. *Offa*, *Impulsive*, *Panther* and *Penn* met ONS-5 in the evening of 2 May after enduring a four-day search for the convoy that exasperated McCoy and sucked precious fuel from their bunkers. The reinforcements were welcomed by ONS-5's

SOE, Commander Peter Gretton in HMS *Duncan*, but the day after EG-3 joined, he was forced to detach to St John's when high seas prevented refuelling. The rough seas also frustrated antisubmarine capability. McCoy complained 'any form of A/S operations was out of the question. Weather conditions caused complete quenching and endangered A/S domes, and indeed it was impossible to proceed at sufficient speed to drop Minol depth charges at any but the deepest settings.'[26] Each destroyer sustained weather damage and *Oribi*'s gyro was knocked out, impeding her ability to make the intricate manoeuvres critical in antisubmarine warfare. As the gale lingered, *Impulsive*, *Panther* and *Penn* were also unable to refuel, forcing them to abandon the convoy over the night of 3/4 May, leaving only *Offa* and *Oribi* in support of a close escort reduced to the old 'declassed-Fleet' destroyer HMS *Vidette*, the frigate HMS *Tay*, whose asdic had been knocked out by high seas, and four corvettes.

It would require a much more lengthy study to give full value to *Offa*'s and *Oribi*'s contribution to the defence of ONS-5. For four days they were deluged with contact reports as U-boats swarmed around the convoy, sinking twelve merchant ships against the loss of seven of their own (six by the escort, one by a Catalina aircraft). Records are imprecise due to the volume of concurrent activity, but *Offa* and *Oribi* conducted at least 30 sweeps into the flanks of ONS-5 in response to HF/DF and radar contacts, or supposition as to where U-boats might gather.[27] The 5th and 6th of May were particularly arduous. After a night of numerous sweeps, on the morning of the 5th, *Oribi*'s commanding officer, Lt-Cdr John Ingram, noted: 'enemy W/T activity was frequent and numerous HF/DF bearings gained.' At 1010, Lt-Cdr Robert Sherwood RNR in the frigate HMS *Tay*, who had taken over as SOE, directed Ingram to sweep down a HF/DF bearing to a distance of 12nm. That bore fruit and between 1057 and 1110 *Oribi* sighted three U-boats running on the surface at about 6nm range. All three dived when they realised they were being overhauled by the destroyer – something a corvette or frigate could not have accomplished – and *Oribi* carried out a series of attacks over the next two hours. Ingram wanted to force the U-boats down, not necessarily destroy them. After one attack he reported: 'Object of impeding submarine

One of EG-4's 'enthusiastic desperados.' While supporting ONS-4, on 25 April 1943 HMS *Pathfinder* finished off *U-203* after it was damaged by a Fairey Swordfish aircraft from HMS *Biter*. Note that although *Pathfinder* is not fitted with centimetric radar in this March 1943 image, she has a HF/DF antenna on her pole mast aft. (Author's collection)

achieved. No other results expected from this attack.'[28] Realising they were in for a long fight, he also conserved depth charges, dropping only two in this instance. Ingram broke off the hunt when McCoy ordered him back to the convoy, but he achieved his goal and none of these U-boats returned to attack ONS-5.[29] After seizing an opportunity to refuel in gentler seas, that evening *Oribi* and *Offa* teamed up against another U-boat detected by asdic. After five attacks, McCoy broke off the hunt when 'Heavy W/T activity indicated that the convoy was threatened with annihilation and I considered it imperative to return to it before dark.' Although they did not destroy *U-266*, they caused enough damage to force her to withdraw.

During the early hours of 6 May *Offa* and *Oribi* continued their incessant sweeps as they worked with the close escort to thwart further attacks. Besides keeping U-boats down, they had two close-quarters scraps with the enemy. Dense fog typical of the Northwest Atlantic in spring now blanketed the area, which particularly impeded *Oribi* with her lack of centimetric radar. While steaming through the murk at 22 knots to assist the corvette *Sunflower*, at 0252 *Oribi*'s asdic operator reported 'Echo bearing Green 30 – close.' Ingram recalled:

> An instantaneous decision had to be taken whether this might be the *Sunflower*, of whose position I was doubtful, or a submarine. However, as I had no RDF contacts in that sector, I swung the ship to starboard and it was with great relief that I saw a submarine slide out of the fog about one cable on the starboard bow steering from right to left. The conning tower was obscured by the forecastle, and the stem must have hit just abaft of this. The inclination of the submarine was approximately 090 left on impact, the force of the collision slewed her round to port and she passed down the port side, heeled over with her bows and conning tower out of the water.[30]

With *Oribi* making 22 knots and the U-boat only 200 yards away this was more a controlled collision than a deliberate ramming. The impact stove in the destroyer's bow, flooded her forepeak and rendered her asdic ineffec-

tive – she ultimately had to spend two months in dock. Admiral Horton complimented Ingram for his 'great skill in ship-handling', and the 'difficult and instant decision which resulted in the ramming and final destruction of an enemy submarine.'[31] It was not as it seemed, however; *U-125* actually stole away in the fog – with centimetric RDF, *Oribi* could have tracked her escape – but, unable to dive, the submarine fell victim to *Snowflake*'s guns a short time after.

Offa's encounter proved less gratifying. Chasing down an RDF contact abaft the convoy's beam, at 0300 the destroyer sighted a U-boat through the fog 100 yards on the starboard bow. While *Offa* sprayed the submarine with Oerlikon fire, McCoy put the wheel hard to starboard to ram but the U-boat dived. He 'could still see the hull of the U-boat underwater as it disappeared aft very rapidly', but his order to fire depth charges miscarried when bungled drill prevented a full pattern being dropped, thus 'certain destruction was not obtained.'[32] Learning of more attacks on the convoy, a frustrated McCoy broke off the hunt; although badly shaken, *U-223* escaped with only slight damage.

That lowered the curtain on EG-3's involvement with ONS-5. Running low on fuel, concerned about the damaged *Oribi*'s vulnerability and with EG-1's sloops arriving as reinforcements, at daybreak on 6 May McCoy departed for St John's. ONS-5 suffered no further losses, but EG-1 and the close escort destroyed two more U-boats. Accounting for the decisive reverse, U-boat command acknowledged the aggression of EG-3 and the close escort: 'in the fog the escorts succeeded in attacking 15 boats with depth charges, and many found themselves suddenly and unexpectedly under fire from destroyers using radar. On the morning of the 6th the pursuit had to be abandoned.'[33]

ONS-5 demonstrated that convoys with proper support could be fought through U-boat packs. Admiral Horton saluted EG-3's 'inestimable assistance in driving off very heavy attacks by U-boats', and he saluted McCoy for conducting operations 'in a very able manner.'[34] For his part, McCoy recognised that they were gaining the upper hand, and repeated his plea to route convoys through better conditions:

It has become quite clear that in good weather conditions escorts which are fitted with RDF Type 271 or 272, and which are handled with determination, will always defeat the U-boat at night or in fog.

It therefore follows that to make a successful attack U-boats must do so submerged. It is probable, though, that such attacks will only be made by day.

The proper use of support groups by day will keep the U-boats down and, therefore, evasive turns by even the slowest convoy will be effective, because in order to keep in touch the U-boats must surface in order to attain the necessary speed, thereby offering themselves for attack. Good weather is of course essential for the success of these tactics.

Horton was still not prepared to expose convoys; however, he acknowledged that if the situation improved, as it would, routing convoys 'so as to invite attack by a small number of U-boats deserves consideration.' In his report, *Panther*'s Lt-Cdr Viscount Jocelyn focused on the endurance issues that had forced several destroyers to abandon the convoy. He proposed that instead of relatively short-legged Fleet destroyers, 'Sloops and Frigates (who are not constantly faced with a fuel problem) ought to make up support groups, and that destroyers should always form part of a definite [close] escort group.' Horton and McCoy vehemently disagreed, the latter countering that: 'The whole point of the Support Groups is that they shall have the speed to move quickly from one convoy to another and, when in company with a convoy, move at very high speed to put the U-Boats down.'[35] McCoy was correct under the circumstances then at play; nevertheless, as the U-boat threat dissipated in the months ahead, British support groups would typically comprise frigates and sloops, releasing modern destroyers to their more traditional duties.

The Last Days of EG-3

After a short lay-over in St John's, EG-3, now reduced to *Offa*, *Panther* and *Penn*, dodged snow squalls and ice fields to reinforce ONS-7. Numerous U-boats remained in the North Atlantic, and early on the morning of 17 May *U-657* torpedoed a merchant ship, only to be destroyed in turn by the close escort. After that, no other U-boats were reported in the convoy's vicinity. Meeting ONS-7 at 0240Z 18 May, EG-3 did not stay much longer than it took to refuel. *Penn* participated in one hunt that came up empty, but otherwise EG-3's activity focused on oiling – Lt-Cdr Jocelyn complained: 'after a long voyage in Sub-Arctic waters and through ice with no heating coils, [the fuel] had the appearance and consistency of axle grease.'[36] EG-3 left the convoy in the forenoon of the 19th, entering St John's the next day.

After a quick turnaround, on 22 May EG-3 embarked upon its final operation. Since the ships were on return passage to the UK, they remained with the eastbound SC-131 for five days, the longest they supported any convoy. SC-131 was routed farther south through calm weather, causing Lt-Cdr Jocelyn to observe it was 'a great relief' to

encounter conditions 'which did not interfere with the capabilities of a destroyer.' Consequently, 'throughout I felt the situation was well in hand and unlikely to turn against us.'[37] An incident on the evening of 25 May contributed to that confidence. *Offa* was sweeping astern of the convoy when a U-boat was sighted on the surface 8nm away. *U-558* crash dived, but when *Offa* approached to within 3nm she broached, betraying her position. McCoy called in *Panther* for support and the two destroyers worked over a contact that varied from 'very weak' to 'firm.' After one attack produced oil, they eventually lost contact and, now 38nm astern of SC-131, McCoy returned to the convoy. 'No evidence of damage was obtained,' he reported, 'but since the U-boat sighted was the first to report the convoy and no attack developed on the convoy (although slight W/T activity continued for 12 hours), it is possible he may have been dissuaded from further activity.'[38] That assessment was on the mark: *U-558* withdrew due to damage sustained in the attacks. The next day *Offa* chased down two more HF/DF bearings. No U-boats were sighted, but the sweeps kept them down and SC-131 continued unmolested.

Based on this experience, Lt-Cdr Jocelyn echoed McCoy's comments after ONS-5 and recommended a more aggressive strategy. 'During the period 14th April–21st May (ie 38 days),' he observed, '*Panther* was in harbour for 11 days, at sea for 27, and only had 6½ in company with any convoy, which seems too little.' In contrast, EG-3 accompanied SC-131 for five days. Jocelyn proposed support groups adopt a schedule that would enable them to remain with convoys longer or even support all convoys. Rather than moving between threatened convoys, he recommended: 'if 7 Support Groups in the North Atlantic can be formed, then 5 or 6 can be operating and 1 or 2 boiler cleaning, and that the former operated as above are enough to give support to all convoys.' Jocelyn recognised that the policy of diverting convoys around known U-boat patrol lines might extend passages, but repeating McCoy's appeal for more aggressive routing, he thought diversion was no longer necessary. A better method to counter U-boat concentrations 'is to sink or damage U-boats at a rate which keeps the total number at sea down to a reasonable quantity and that Escorts plus Air and Surface Support are now competent to achieve this ...' There were not yet enough suitable ships to field seven support groups, but Jocelyn's recommendation reflects the confidence accrued by EG-3.

Unbeknown to Jocelyn, the pendulum had already swung in the Allies favour, and U-boats were being destroyed at the rate he sought. On 26 May, while *Offa*, *Panther* and *Penn* ploughed eastward with SC-131, Horton signalled his command: 'The tide of the battle has checked if not turned, and the enemy is showing signs of strain in the face of heavy attacks by our sea and air forces.'[39] Confronted by unsustainable losses the *Kriegsmarine* withdrew from the North Atlantic, relieving the crisis that had sparked the demand for the Fleet destroyers. On 27 May, Horton ordered McCoy, whom he addressed by his former title of 'Captain D17' rather than 'SO EG-3', to rejoin the Home Fleet at Scapa Flow. McCoy's final report to Horton as SO EG-3 reflected the pride the group had taken in its task:

> It was with great disappointment that I received the signal informing me that the Third Support Group was to be abandoned in the United Kingdom. The ships of the group had only just got to know each other well enough to start working together as a team, and thus were becoming really efficient as an A/S Striking Force. We all thoroughly enjoyed our time in the Western Approaches and left your command with a real sense of regret.

Conclusions

Support groups were just one factor in the turn-around in the North Atlantic. Contemporary analysts and historians agree, almost universally, that the combination of support groups, increased shore-based air cover, better trained escorts and the introduction of escort carriers were responsible for the victory over the *U-bootwaffe*.

What did EG-3 contribute during its two months in the North Atlantic? A postwar Admiralty study calculated that support groups not only added strength to convoy defence but were responsible for a 25 per cent increase in the number of U-boats destroyed by the close escort.[40] As the experience of EG-3 demonstrated, Fleet destroyers' speed enabled them to harass shadowers, enabling close escorts to maintain their defensive positions around convoys. Sweeps may not seem important – and the fact that EG-3 aborted many hunts cost kills – but they

deterred U-boats: Stephen Roskill's reference to '*Offa*'s ubiquitous support group' is telling. EG-3's skill and experience, guided by McCoy's seasoned judgement, was a significant factor in this success, enabling the ships to tackle challenging circumstances with confidence. Moreover, as the first support group to deploy and with McCoy playing a role in formulating their tactics and analysing their initial operations, there was a pioneering aspect to their contribution. Admiral Horton alluded to this in a June 1943 letter to a colleague: 'The Support Groups inaugurated the change, when we got reinforcements from the Home Fleet late in March – then came our own Support Groups and the escort carriers (very well trained too) – then new weapons and increased support from Coastal Command.'[41] It was a combined effort to be sure, but EG-3 was in the vanguard.

This study opened with Nicholas Monserrat. The novelist based *The Cruel Sea* on his personal experiences as a wartime escort captain so he may well have witnessed the arrival of reinforcements in the form of Fleet destroyers. Another writer did for certain. Lt-Cdr Evelyn Chavasse, captain of the antiquated 'four-stacker' HMS *Broadway*, was SOE of HX-237 in May 1943 when it was reinforced by the 'enthusiastic desperadoes in *Opportune*, *Obdurate* and *Pathfinder*'. Chavasse observed that the Fleet destroyers 'were modern ships compared with old rattle-traps like the *Broadway* and our aging V and W class destroyers.' 'They did not have, I think,' he reflected, 'the long experience we had in anti-submarine tactics, but, my word, they had speed and above all they had dash.' Praising their contribution, Chavasse reckoned: 'Whether as Support Groups or in

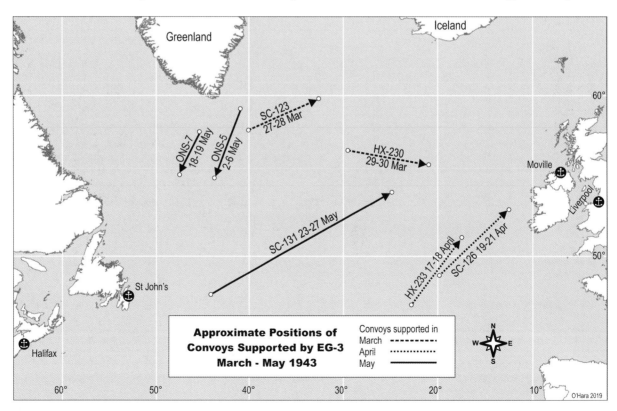

the close escort they were invaluable.' Coupled with Roskill's 'ubiquitous', Chavasse's 'invaluable' well describes the role of Fleet destroyers such as those of EG-3 in helping to turn the tide in the North Atlantic.[42]

Acknowledgements:

The author wishes to thank Malcolm Llewellyn-Jones for sharing documents and ideas in the initial stages of the study, and Norm Jolin, Gil Lauzon and Vincent P O'Hara for reviewing the manuscript; Vince also produced the accompanying map. Richard Osborne of World Ship Society and Conrad Waters generously shared images.

Sources:

This study is based upon convoy records held at the UK National Archives (TNA), the US National Archives and Records Administration (NARA), and the Directorate of History and Heritage (DHH) in Ottawa, Canada. David Syrett (ed), *The Battle of the Atlantic and Signals Intelligence: U-boat Situations and Trends, 1941–1945*, Ashgate Publishing (Aldershot, 1998) provides the Admiralty's U-boat assessments, while the Admiralty's *The U-Boat War in the Atlantic: Volume II, January 1942–May 1943* (1952) gives the U-boat perspective. As always, www.naval-history.net was a treasure trove of information. The most useful secondary sources are Stephen Roskill's *The War at Sea*; Clay Blair, *Hitler's U-boat War: The Hunted, 1942–1945*, Random House (New York, 1998); Michael Gannon, *Black May: The Epic Story of the Allies' Defeat of the German U-boats in May 1943*, Harper Collins (New York, 1998); and WA Haskell, *Shadows on the Horizon: The Battle of Convoy HX-233*, Chatham Publishing (London, 1998).

Endnotes:

1   Nicholas Monsarrat, *The Cruel Sea*, Cassell (London, 1951), 190.

2   British Deputy Prime Minister to Canadian Prime Minister, 23 February 1943, Admiralty War Diary, NARA.

3   VCNS to Admiralty, 17 March 1943, Admiralty War Diary.

4   C-in-C Western Approaches to Admiralty, 5 and 13 March 1943, Admiralty War Diary.

5   Support Groups were designated 'Escort Groups' to avoid the need to introduce new terminology into naval cyphers.

6   Horton to Rear Admiral RB Darke, 23 March 1943, quoted in Rear Admiral WS Chalmers, *Max Horton and the Western Approaches*, Hodder and Stoughton (London, 1954), 188.

7   Clay Blair, for example, incorrectly claimed EG-3's destroyers 'All had new radar and several had Huff Duff.' Clay Blair, *Hitler's U-Boat War: The Hunted, 1942–1945*, Random House (New York, 1998), 282.

8   Admiralty to Western Approaches, 10 February 1943, Admiralty War Diary.

9   Unless otherwise indicated, all times are local.

10 C-in-C Western Approaches, 24 March 1943, Admiralty War Diary.

11 C-in-C Western Approaches, 29 March 1943, Admiralty War Diary.

12 C-in-C Western Approaches to EG-3, 0837Z/28 March 1943, Directorate of History and Heritage (DHH), 2000/15.

13 *Hurricane*, 'Report of Proceedings', 1 April 1943; and *Offa*, 'Report of Proceedings from 13th April 1943 to 25th April 1943', 26 April 1943, TNA, ADM 199/353.

14 'Report of Proceedings of the Third Escort Group for the period 29th April to 8th May, 1943', 9 May 1943, ADM 199/353.

15 *Hurricane*, 'Report of Proceedings', 1 April 1943.

16 Western Approaches, General Survey of Events, 15–31 March 1943, TNA, ADM 199/631.

17 'Report of Proceedings', Third Escort Group 22 March 1943–1 April 1943.

18 *Onslaught*, 'Report of Proceedings', 31 March 1943, TNA, ADM 199/579.

19 Flag Officer Newfoundland, 'Monthly Report – April 1943', 19 May 1943, DHH, 81/520/1000-5-00.

20 Admiralty, *The U-Boat War in the Atlantic: Volume II, January 1942–May 1943* (1952), 103.

21 SW Roskill, *The War at Sea, Volume II: The Period of Balance* HMSO (London, 1956), 372.

22 SO EG-3, 'Report of Proceedings 13–25 April 1943', 27 April 1943, TNA, ADM 199/575.

23 *Panther*, 'Report of Proceedings', 24 April 1943, TNA, ADM 199/575. Type 291 could be monitored, therefore its use was restricted unless the enemy was in confirmed contact.

24 Western Approaches, General Survey of Events, 16–30 April 1943, TNA, ADM 199/631.

25 Quoted in Peter Gretton, *Convoy Escort Commander*, Cassell (London, 1964), 147.

26 *Offa*, 'Report of Proceedings from 29th April, 1943 to 8th May', 1943, 8 May 1943, TNA, ADM 199/353.

27 The analysis of the sweeps is derived from the after-action reports of *Offa* and *Oribi*.

28 *Oribi*, 'Report of Attack on U-boat', 6 May 1943, Library and Archives Canada (LAC), RG 24, Volume 11329, File 8280-ONS 5.

29 In *Black May: The Epic Story of the Allies' Defeat of the German U-Boats in May 1943*, Harper Collins (New York, 1998), Michael Gannon calculates that *Oribi* forced down four U-boats.

30 *Oribi*, 'Report of Proceedings', 8 May 1943.

31 'HM Ships *Oribi* and *Offa* in Attacks on Enemy Submarines While on Convoy Escort Duty: Awards', TNA, ADM 1/14459.

32 *Offa*, 'Report of Proceedings from 29th April, 1943 to 8th May, 1943'.

33 Admiralty, *The U-Boat War in the Atlantic: Volume II*, 105.

34 C-in-C Western Approaches minute, 14 June 1943, TNA, ADM 199/353.

35 SO EG-3, 'Reports of Proceedings of the Third Escort Group for the period 29th April to 8th May, 1943'; C-in-C Western Approaches minute; and *Panther*, 'Report of Proceedings', 6 May 1943.

36 *Panther*, 'Report of Proceedings – May 14th to May 27th', 26 May 1943, TNA, ADM 199/353; and SO B-5, 'Report of Proceedings', 22 May 1943, LAC, RG 24, Volume 11329, file 8280-ONS 7.

37 *Panther*, 'Report of Proceedings – May 14th to May 27th'.

38 *Offa*, 'Report of Proceedings – May 14th to May 29th', TNA, ADM 199/353.

39 C-in-C Western Approaches, 26 May 1943, Admiralty War Diary.

40 Admiralty, *Defeat of the Enemy Attack on Shipping: A Study of Policy and Operations Vol I* (1957), 96.

41 Horton to Rear Admiral RB Darke, 15 June 1943, quoted in Chalmers, *Max Horton and the Western Approaches*, 200.

42 EH Chavasse ms, 'Business in Great Waters: War Memories of a Semi-Sailor', 86-87, DHH, 88/181.

THE BATTLESHIP *CARNOT*

Generally considered to be one of the worst examples of the notorious *Flotte d'échantillons* that emerged from the 1890 Naval Programme, the French battleship *Carnot* had an undistinguished career and was stricken before the outbreak of the Great War. **Philippe Caresse** looks at the rationale for her construction and her subsequent years of active service with the *Marine Nationale*.

By general agreement the least successful of the battleships of the *Flotte d'Échantillons*, *Carnot* was designed by engineer Saglio, the director of the French Navy's propulsion establishment at Indret, and built at Toulon naval dockyard.

The lengthy construction time of *Carnot* would be heavily criticised, and proposals by the renowned naval architect Louis-Emile Bertin for the incorporation of a cellular layer to improve the protection of the lower hull were rejected. The dimensions of the main gun turrets were the subject of heated discussions between the builder and the *Conseil des travaux*. The original plans for a conning tower located around the military foremast were rejected by the Navy's chief constructor, the *Directeur du matériel*, who wanted the mast moved aft to free up this key strategic space, and major modifications would be necessary before *Carnot* was commissioned for trials. The problems experienced during the fitting-out process were such that, on 22 June 1896, a ministerial telegram requested that the commission compile a dossier of the issues raised, notably concerning the ship's stability. It quickly became apparent that there was a requirement for a dramatic reduction in topweight. However, despite numerous modifications, *Carnot* achieved notoriety as a failed and obsolescent ship of limited military value even before her entry into service.

Characteristics

For the ship's general characteristics see the accompanying table.

The armour belt, of homogeneous steel, was supplied by Marrel, the contract being signed on 30 September 1891. It was 2 metres high (0.5m above the waterline) and 450mm thick amidships, reducing to 350mm at the bow and 240mm at the stern. The belt was tapered beneath the waterline, the thickness of the lower edge declining from 250mm amidships to 140mm at the ends. Above the main belt was a light belt (*cuirasse mince*) of 120mm steel plates.

The protective deck was ordered on 19 August 1891 from Châtillon & Commentry. The main deck was of 70mm iron plating on a double layer of 10mm steel, with a 20mm splinter deck beneath over the machinery spaces and magazines.

The conning tower, which weighed 65 tonnes, had 230mm steel plating on the face and walls and 200mm

General Characteristics	
Length oa	117.00m
Length wl	115.16m
Beam	22.50m
Depth of keel	7.90m
Draught	7.62m fwd, 8.31m aft
Freeboard forward	7.10m
Normal displacement	12,029 tonnes
Full load displacement	12,247 tonnes
Wetted area of underside	3,070.52m²
Displacement per cm at wl	19.10 tonnes
Complement	32 officers, 622 men

on the communications tube. The bridge was located directly forward of the conning tower. The total weight of protection was 4,055 tonnes.

As designed there was a sharp 90-degree angle between the forward end of the armour belt and the upper section of the bow, but this was eliminated in January 1901 by adding a piece of steel. At the same time bilge keels were fitted to improve stability.

Armament

The main armament, which was mounted exclusively in enclosed turrets, comprised two 305mm 45-calibre guns Mle 1887 fore and aft and two 274.4mm 45-calibre Mle 1887 guns on the beam in the standard 'lozenge' arrangement.

Both the 305mm and the 274.4mm turrets were constructed by the Hauts Fourneaux, Forges et Aciéries de la Marine et des Chemins de Fer Company, and the hydraulic systems by the Société des Batignolles. The turrets themselves were protected by plates of homogeneous nickel steel 370mm thick, with 70mm on the roofs and the armoured hood, and 320mm on the barbettes. Each of the 305mm turrets weighed 93.70 tonnes, each of the 274.4mm turrets 79.10 tonnes.

The 305mm gun fired a steel shell of 340kg or a cast iron shell of 292kg; maximum range with the CI shell at an elevation of 10 degrees was 12,400m with a combat charge. The COs who served on *Carnot* were unified in their criticism of the firing cycle, which was exceptionally slow: 90 seconds per round using the ready-use rounds stowed in the turret, and 3 minutes when using the hoists for replenishment. The 274.4mm gun fired a steel shell

Carnot: Profile

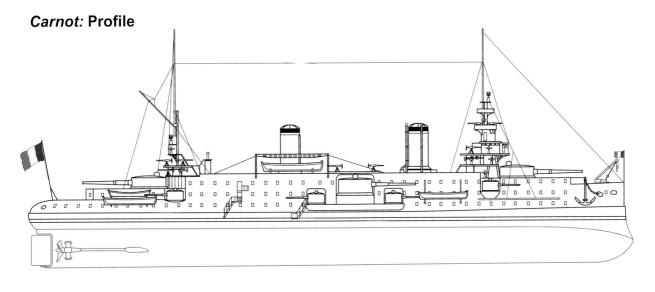

METRES

weighing 255kg or a cast iron shell of 216kg. Rate of fire was better: one round per minute. In December 1897/January 1898 external sights were fitted to the main guns, and from December 1901 they were capable of night firing.

The 138.6mm turrets had 100mm armour plating on the face and sides and 20mm on the roofs. The total weight of protection for the guns was 1,343.45 tonnes.

The submerged torpedo tubes were located between Frames 47 and 48, and were 2.03m below the waterline. They were fixed at an angle of 90 degrees to the longitudinal axis of the ship. The Mle 1892 torpedo had a length of 5.05m. Total weight was 501kg including the 75kg warhead. Compressed air was provided by two Elwel pumps with a 750-litre capacity.

Propulsion Machinery

Steam was supplied by 24 Lagrafel & d'Allest-pattern boilers built by Indret operating at a pressure of 15kg/cm^2 grouped in three boiler rooms. The two forward boiler rooms were served by a broad funnel of rectangular cross-section, the after boiler room by a slimmer funnel of elliptical cross-section. The funnels were replaced in 1908 under a contract with Ropars placed on 13 July 1907.

The boilers supplied steam for two vertical triple-expansion (VTE) engines built by Indret on the plans of *Ingénieur de la Marine* Garnier. Assembly of the engines on board was supervised by *Chef monteur* Richoux. The engines were installed side by side in separate compartments designated *Machine bâbord* and *Machine tribord* (Port/Starboard Engine Room). They powered two bronze 3-bladed propellers that turned inwards. At normal load the immersion of the tips of the propellers was 2.35m.

The surface area of the single non-balanced rudder

Armament

Two 305mm 45-cal Mle 1887 BL guns in two single turrets (100 rounds)

Two 274.4mm 45-cal Mle 1887 BL guns in two single turrets (106 rounds)

Eight 138.6/45 Mle 1888–91 QF guns in single turrets (1,844 rounds)

Four 65mm 50-cal Mle 1891 QF guns in open single mountings (1,644 rounds)

Sixteen 47mm 40-cal Mle 1885 QF guns in open single mountings (8,448 rounds)

Eight 37mm 20-cal Mle 1885 revolver cannon (4,200 rounds)

Two 450mm submerged torpedo tubes (6 Mle 1892 torpedoes)

Calibre	Shell weight	Muzzle velocity	Firing cycle
305mm	292kg CI	815m/s	0.6rpm
	340kg APC	780m/s	0.6rpm
274.4mm	216kg CI	815m/s	1rpm
	255kg APC	780m/s	1rpm
138.6mm	35kg	730m/s	4rpm
65mm	4kg	715m/s	8rpm
47mm	1.5kg	650m/s	8–15rpm
37mm	0.5kg	435m/s	20–25rpm

* Cast Iron (CI) shell

Calibre	Angle of elevation	Range
305mm	+10°/-5°	12,400m
274.4mm	+10°/-5°	11,700m
138.6mm	+15°/-5°	9,400m
65mm	+20°/-15°	5,400m
47mm	+20°/-20°	4,000m
37mm	+25°/-30°	2,000m

Propulsion Machinery

Boilers	24 Lagrafel & d'Allest (Indret)
Engines	two Indret VTE
Propellers	two 3-bladed 5.3m bronze
Rudder	non-balanced, 23.40m^2
Power on trials	16,344CV
Maximum speed	17.86 knots
Coal	702 tonnes
Endurance	5,250nm at 8.7kts (25 days)
	811nm at 17.9kts (1.9 days)

proved inadequate; and in January 1901 it was enlarged by 30cm aft and 10cm forward. At 12 knots the turning circle was about 868m; at 10 knots the angle of heel was 1 degree.

For the evacuation of water from the bilges there were longitudinal drains on either side of the ship, linked by a transverse pipe and connected by valves to the various compartments. Four Thirion pumps with a capacity of 600 tonnes per hour and a Stone hand pump served these installations. For the compartments in the double bottom there was a collector on each side of the ship served by less powerful (20t/hr) Thirion pumps. *Carnot*'s officers considered these arrangements inadequate; in particular an alternative means of pumping out the boiler rooms was considered necessary.

Twelve ventilators were provided for the boiler rooms: six vertical each rated at 40,000m^3/h, and six inclined each rated at 54,000m^3/h. Electrical power for the ship was supplied by four Sautter & Harlé 83V dynamos each rated at 600A.

Equipment

Between March and May 1903 Germain hydraulic order transmission (see John Spencer, '*Conduite du tir:* The Birth of Centralised Fire Control', *Warship 2010*, 167–69) was installed. This would later be complemented by 2-metre optical rangefinders purchased from Barr & Stroud and installed atop the bridge between November and December 1909. For torpedo fire control there were 'large tables' of the *aviso torpilleur* type located on either side of the conning tower.

The six searchlight projectors were equipped with 60cm Mangin mirrors fitted with mixed horizontal 45A lamps from Sautter & Harlé. One was mounted on the upper platform of the foremast and a second on the after side of the mainmast (*ligne haute*); both were on fixed pedestals with local control. Of the four lower projectors (*ligne basse*), the two in the admiral's gallery aft were hand-worked; those on either side of the bow had remote electrical controls.

The boat complement (see accompanying table) included two 10-metre steam pinnaces that could be equipped with the Desdouits torpedo-launching apparatus. The 11-metre pulling pinnace and the two 10-metre cutters could be fitted with a 37mm QF gun. The larger boats were lifted and moved by a traversing

Equipment

Searchlights	six 60cm 45A projectors
Boats	two 10-metre steam pinnaces
	one White 7.6-metre steam launch
	one 11-metre pulling pinnace
	two 10-metre pulling cutters
	one 9-metre pulling cutter
	two 8.5-metre whalers
	two 5-metre dinghies
	one 3.5-metre punt
	two 5.6-metre Berthon canvas boats
Anchors	two 13.03-tonne bower anchors
	one 2.39-tonne kedge anchor

gantry mounted around the second funnel. There were also hinged davits above the wing 274mm turrets, between the bridge structure and the first funnel, and at the ends of the ship to handle the boats when moored in an anchorage. Two 2-cylinder steam winches rated at 3,000kg were provided to handle the boats.

The 60cm anchor chains on either side of the ship had 24 links; the smaller anchors had four 30mm links. The main capstan, which was rated at 40,000kg, was a Bossière 2-cylinder steam model.

Complement

The designed complement was 32 officers plus 622 petty officers and ratings. Following a Ministerial directive dated 6 December 1897, the installations for embarkation of an admiral and his staff were suppressed and converted to provide additional accommodation for the crew. An admiral would be embarked on *Carnot* only in exceptional circumstances and for a very short period.

The ship had sufficient stores capacity for 60 days at sea with a crew of 700 officers and men; one complete ration was calculated at 1.65kg. Drinking water sufficient for 13 days was provided in cast iron tanks with a

Commanding Officers

CV Pissère	17 Mar 1896 – 17 Mar 1898
CV Ravel	17 Mar 1898 – 17 Mar 1900
CV Campion	17 Mar 1900 – 25 Jun 1902
CV Perrin	25 Jun 1902 – 29 May 1904
CV Imhoff	29 May 1904 – 20 Oct 1904
CV Lecler	20 Oct 1904 – 10 Feb 1906
CV Motet	10 Feb 1906 – 8 Sep 1908
CV Pradier	8 Sep 1908 – 5 Oct 1909
CV Journet	5 Oct 1909 – 5 Apr 1911
CV Jochaud du Plessis	5 Apr 1911 – 15 Feb 1912
CV Le Nepveou de Carfort	15 Feb 1912 – 28 Sept 1912
CV Raffier-Duffour	28 Sept 1912 – 6 May 1913
CV Lesquivit	6 May 1913 – 27 Dec 1913
CV Fatou	27 Dec 1913 – 7 May 1914

Notes:

CV	*Capitaine de Vaisseau*	Captain

Carnot at her moorings in 1899; the ship's boats are in the 'anchorage position'. (DR)

A much later image of *Carnot* steaming off the coast of Provence. She has exchanged her original black and buff paint scheme for a uniform blue-grey livery. Note the two Barr & Stroud rangefinders atop the navigation bridge; these were fitted in December 1909. (Marius Bar)

total capacity of 37.7 tonnes. There were also 23,625 litres of wine and 1,875 litres of spirits, 1,220kg of salted provisions, 10,200kg of meat, an unknown quantity of fresh vegetables, 18,130kg of flour and 7,390kg of biscuit. Coffee, sugar, cheese and rice accounted for a further 1,950kg. The galleys were fired by wood, of which 20 cubic metres (*stères*) were embarked.

Completion

The order for *Carnot* was placed with Toulon Naval Dockyard. Construction was supervised by naval engineers Henry, l'Homme, Maugas, Schwartz and Janet. The battleship was laid down as the *Lazare Carnot* (a prominent politician during the revolutionary wars known as 'The Organiser of Victory') on 10 September 1890, and the first rivet was put into place by President Sadi Carnot. Work on the hull progressed extremely slowly, and the launch ceremony took place almost four years after the keel laying.

A dramatic event was to change the name of the ship. On 24 June 1894, during a visit to Lyon, President Sadi Carnot was assassinated by an Italian anarchist. On 7 July Navy Minister Félix Faure decided to honour both the 'Organiser of Victory' and the former president by changing the name of the ship to simply *Carnot*.

The launch took place as planned at about 1100 on 12 July 1894. A large crowd gathered to witness the launch, which was presided over by naval engineer Maugas in the presence of the Maritime Prefect of the 3rd Region. As a

Carnot is launched at Toulon Naval Dockyard on 12 July 1894. The armour belt would be fitted once the ship was afloat. Note the pronounced tumblehome of the sides. (Service Historique de la Marine)

sign of respect for the late president, no decoration was applied to the hull, which still lacked superstructures (see photo); the national flags were at half-mast and furled.

Following her launch *Carnot* was towed by the tugs *Lagoubran* and *Hercule* to Post No 2 in the Mourillon Basin where she would be fitted out. Completion was delayed by the late delivery of the turrets, which led to the disciplining of at least three engineers attached to the port of Toulon. The delay was almost certainly compounded by a decision to modify the military masts to secure a reduction in topweight of 128 tonnes.

The first (static) machinery trial took place on 27 December 1895. The ship then embarked on her official sea trials off the Hyères islands, to the east of Toulon, in December 1896, with further sea trials in January, March and April of the following year.

The battleship was commissioned on 25 June 1897. She was attached, together with the battleships *Brennus* and *Jauréguiberry*, to the 1st Division of the Mediterranean Squadron, which was commanded by Vice Admiral Jules de Cuverville. *Carnot* then took part in all the sorties of the squadron. Criticisms of the ship were not slow to emerge. With 85 large ports per side it proved impossible to work in a cellular layer above the armour belt. During trials with the wing turrets, which were undertaken with the turrets trained at 15 degrees to the ship's axis, the six scuttles for the lower battery were damaged. More dramatically, tests revealed that if one of the two engine rooms was flooded, the ship would quickly capsize.

Here is an extract from the report by *Carnot*'s first commanding officer, Captain Pissère:

> At speeds of up to 14 knots the ship handles well; she holds a more stable course than *Dévastation*, which I commanded for two years. Above 14 knots she tends to yaw; according to the official trials reports *Carnot* does not handle as well as *Charles Martel* and *Jauréguiberry*. She has a pronounced roll, which is hardly surprising given her high GM (1.15m), but pitches even more than she rolls. Unfortunately it has to be acknowledged that handling is the worst aspect of the ship's performance. Even in the most favourable circumstances the turning circle is in excess of 800 metres. With a cross-wind and a strong breeze it is difficult to maintain a steady course even if the propellers are turning at different revolutions.
>
> Other defects include the proximity of the 138.6mm turrets to the waterline, the inadequate height and thickness of the upper belt, the absence of any means of clearing water from the armoured battery, the defective positioning of the underwater torpedo tubes, and the layout of the midship magazines.

This was a harsh judgement on the new French battleship, at a time when the Royal Navy had recently commissioned the eight battleships of the *Royal Sovereign* class followed by the nine *Majestics*, all of which were well-armed and well protected, with excellent sea-keeping characteristics.

Carnot being fitted out at Post No 2 of the Mourillon Basin. The military masts would be considerably lightened before she underwent speed trials in December 1896. (Marius Bar)

A close-up of the bridge structure of *Carnot* in 1899. The forward port-side 138.6mm gun can barely made out against the black hull. (DR)

Early Years of Service

In September the examinations for the school of higher studies of the Navy took place on board *Carnot*, anchored for the occasion at Golfe-Juan, under the direction of the First Officer, CF Davin. For the remainder of the year there were torpedo firing and gunnery trials between Villefranche and Golfe-Juan.

At midnight on 9 February 1898 *Carnot* departed Toulon to serve as a target for the Light Squadron, which cast off at around 0400. She would then undertake a simulated engagement with her half-sister *Charles Martel*, dropping anchor at Villefranche, in company with the battleship *Formidable*, the following day.

During the night of 14/15 April, between 0100 and 0300, the 1st Division, comprising the battleships *Brennus* (VA Humann), *Carnot*, *Magenta* and *Formidable*, the 2nd Division with *Charles Martel* (CA Dieulouard), *Neptune* and *Marceau*, an armoured cruiser, four 2nd class cruisers and five torpedo boats, set sail to practise long-distance signalling in the waters off the Ile du Levant, before proceeding at 11 knots to Villefranche. There they joined the British battleship *Ramillies* (VA Sir John Ommanney Hopkins, C-in-C Mediterranean Fleet) and the cruiser *Surprise*, which were already moored in the anchorage. After the customary exchanges of courtesies the admirals went

Left: *Carnot* is docked for maintenance at Brest in 1900. (DR)

Below: *Carnot* at anchor at Golfe-Juan. (DR)

ashore, where they were received by the President of the Republic and Queen Victoria. The following day the President embarked in *Brennus* to observe manoeuvres that took place 30 miles off Cape Ferrat. The return voyage to Toulon was particularly testing due to a storm that battered the squadrons.

Following their return, during the Fashoda Incident, the ships of the Mediterranean Squadron made few

Two fine views of *Carnot* at anchor in Brest roads in 1901. (DR)

sorties; they remained at Toulon awaiting orders to sail.

The year 1899 began with a visit by the Squadron to Villefranche, on the occasion of the carnival. However 16 February saw the sudden death from a seizure of President Félix Faure, putting an abrupt end to the festivities.

Shortly afterwards, in order to further the good relations that were being established with Italy, Vice Admiral Fournier was instructed by the government to take the Squadron to Cagliari to greet the Italian king and queen. Fournier flew his flag in *Brennus*, which visited the capital of Sardinia on Sunday 9 April in company with the battleships *Carnot*, *Charles Martel*, *Bouvet* and *Masséna*, the armoured cruisers *Latouche-Tréville*, *Pothuau* and *Chanzy*, and the 1st class cruisers *Du Chayla*, *Galilée* and *Linois*. On the 12th there was a combined review of the French and Italian fleets; at 1000 on 18 April Fournier's squadron left for Toulon.

Left: A close-up of the military foremast and bridge structure of *Carnot* in 1903. The admiral's bridge can be seen above the navigation bridge. The QF guns are as follows: a 65mm Mle 1891 gun in the bridge wings with a 47mm Hotchkiss Mle 1885 gun above; 37mm QF and revolver cannon scattered around the bridge decks; and 47mm Mle 1885 guns on the lower platform of the military foremast. (DR)

Below: *Carnot* moored to a buoy in Toulon roads following the modification to her bow. (Author's collection)

In July, the Squadron made visits to Marseille, Sète and Port-Vendres before heading for Barcelona, where it dropped anchor on the 16th. It returned to Toulon via Port Mahon (Minorca). During August there were visits to Les Salins d'Hyères and Golfe-Juan. From 28 to 30 July gunnery practice took place off the Ile du Levant, the

Carnot at Spithead in August 1905. (Photo Hopkins)

Exercise for the evacuation of the wounded on board *Carnot* on 14 November 1905. (DR)

target being the decommissioned sloop *Amiral Perseval*.

At 1230 on 11 October, Admiral Fournier sailed for the Levant in the presence of Navy Minister De Lanessan and the Maritime Prefect. The Squadron spent a week at Piraeus, the port of Athens, then left on 28 October for Beirut. On the 31st, when the Squadron was 18 miles from its destination, *Charles Martel* had to leave the line when a member of her crew fell overboard. *Carnot* launched a lifebelt, and the sailor was picked up by a boat from *Jauréguiberry*. At 1515 the Squadron moored a mile from the Lebanese capital. The French ships left their moorings on 11 November and headed for Jaffa, where they remained until the 27th. Following this, *Brennus* led the formation to Smyrna (Asia Minor), where it remained until 5 December, followed by visits to Sevastopol and Salonika (13 December). The Mediterranean Squadron returned to Toulon at 0800 on 21 December, ending a cruise that was regarded as a complete success.

On 4 January 1900, at about 1600, *Carnot* and *Masséna* departed Toulon and headed for Brest to join the Northern Squadron; they would be replaced in the Mediterranean Squadron by the new battleships *Gaulois* and *Charlemagne*. They visited Tangier on 8 January, made the transit to Brest at a speed of 12 knots, and *Carnot* moored to a buoy in the outer roads on the 14th. The battleships *Masséna*, *Amiral Baudin* and *Carnot* would make up the 1st Division. *Carnot* was docked for maintenance at Le Salou on 31 January, emerging on 21 February. On 6 March *Carnot*, accompanied by *Masséna*

and the torpedo boats *Mangini* and *Aquilon*, conducted gunnery exercises, then torpedo firings. The exercises continued all week, with the ships anchoring overnight in Le Fret Bay. Similar sorties were made by the Squadron during the months that followed. In April *Carnot* lost an anchor when mooring in Morgat Bay. For the annual naval manoeuvres the Squadron departed Cherbourg on 10 July, and was reviewed on the 19th by President Emile Loubet, who was embarked for the occasion in the sloop *Elan*.

In October, when *Masséna* was in dock and *Formidable* was undergoing maintenance in the dockyard, *Carnot* flew the flag of Vice Admiral Ménard and *Amiral Baudin* that of Rear Admiral Touchard.

On 5 December *Carnot* was placed in 2nd category reserve and entered the dockyard ten days later for the fitting of bilge keels, the enlargement of her rudder, and the installation of a servo-motor for the latter, together with a modification to her bow (see above). During these two months of immobilisation her place in the 1st Division would be taken by the old battleship *Courbet*.

Recommissioned on 18 March 1901, *Carnot* ran her 6-hour trials on the 27th and her turning trials two days later, followed by gunnery trials on 25 April. She subsequently made visits to Cherbourg and Quiberon.

With the start of the grand manoeuvres, in which the Northern Squadron, representing the enemy, was given the task of penetrating into the Mediterranean, *Carnot* and the fleet departed Brest on 22 June and headed for Spain. Commanded by Vice Admiral Ménard, the group

Carnot in the River Penfeld at Brest. (DR)

Carnot at sea following the removal of the admiral's bridge. (DR)

visited Vigo from the 24–26 June, Lagos from 28–29 June, and finally Tangier from 1–3 July. Following the completion of the exercise, which took place mainly in Corsican waters, a port call was made to Ajaccio before the ships entered Toulon on 11 July. At the conclusion of the manoeuvres the combined Squadrons would visit Port-de-Bouc, Aigues-Mortes, les Salins, Cavalaire, Saint-Tropez, La Ciotat, Ajaccio (again) and finally Toulon, the base of the 3rd Maritime Region, on the 30th.

On 18 August *Carnot* was attached to the Reserve Division of the Mediterranean Squadron. For the remainder of the year she participated in various sorties, visiting Ajaccio from 10 to 16 December. She was then docked from 14 March to 20 April 1902 for removal of the bilge keels, which appear to have been incorrectly fitted at Brest. *Carnot* would be at Villefranche from 14 to 29 May, would conduct gunnery exercises off Golfe-Juan, and would then visit La Ciotat on 10 June followed by Port-de-Bouc. *Carnot* would again have the opportunity to leave the coasts of Provence when she replaced *Charlemagne* in the 1st Division for a visit to Cartagena, Spain. In July she would participate briefly in the grand manoeuvres directed by Admiral Gervais, who flew his flag in *Bouvet*.

From 2 August 1902 to 10 April 1903, *Carnot* made only five training sorties to Les Salins, followed by visits to Marseille and la Ciotat between 10 and 17 April. She again visited Cartagena from 23 to 27 June, and was at Porquerolles at the end of the year.

On 22 February 1904, *Carnot* and *Jauréguiberry* rejoined the 1st Division of the Northern Squadron. The latter ship left Toulon for Brest on 25 March, while

Carnot departed on 14 May, dropping anchor at Brest on the 21st. The 1st Division now comprised *Masséna* (VA Caillard), *Carnot* and *Jauréguiberry*, and the 2nd Division the coast defence ships *Bouvines* (CA Leygue), *Amiral Tréhouart* and *Henri IV*. In June, visits were made to Cherbourg, Saint-Vaast and Boulogne, and in July to Les Sables d'Olonne, La Pallice, the Iles d'Oléron, and Ile d'Aix. On 2 August the ships were at Royan, then Quiberon and La Pallice, returning to Brest on 6 September.

Carnot was docked from 4 November to have new bilge keels fitted. On 10 March 1905 the ship was equipped with wireless transmission (W/T). After various routine activities in company with the division, *Carnot* and the other ships of the Squadron arrived at Spithead on 6 August to take part in a great naval review that took place on the 9th: the Royal Yacht *Victoria and Albert*, with King Edward VII on board, filed past the assembled French and British fleets, an event that symbolised the new rapprochement between the two nations.

The French ships sailed for Brest on 14 August, and on 24 October *Carnot* was docked for maintenance. During the docking it became apparent that the starboard bilge keel was damaged over a length of six metres, and that there was similar damage to the port bilge keel, apparently caused by the anchor chains scraping the hull. The keels were not repaired but were shortened by six metres at their forward ends. At the same time the admiral's bridge above the navigation bridge was removed.

Carnot was refloated on 15 December and remained in refit until 13 March 1906. The Northern Squadron sortied on 20 June to take part in the grand manoeuvres scheduled to take place in the Mediterranean. Visits were made to Mers el-Kebir on 25 June, then to Algiers from 3 to 6 July. When the exercises concluded, there were visits to Bougie from 9 to 10 July and Bizerte until the 23rd. Manoeuvres then resumed in the waters off Les Salins d'Hyères.

Carnot departed for Marseille on 2–3 August, where she conducted firing practice for her light guns and manoeuvres with submarines. She would be at Grau-du-Roi from 10 to 12 August, joining up with the combined Squadrons in Tangier. On the 16th, the ships headed for Royan, dropping anchor on the 21st. The Northern Squadron would be at Ile d'Aix on the 25th, La Pallice on the 26th and Les Sables d'Olonne on the 27th. On the 28th the ships returned to Brest, and *Carnot* would be docked from 3 to 17 November. The battleship's activities would then be few and far between, with a single sortie to Le Fret 12–13 December for gunnery exercises.

The Fateful Year of 1907

From 1 January 1907 the two divisions based at Brest were to form the 2nd Division of the Mediterranean Squadron. In consequence, at midday on 7 February, *Carnot* and her consorts departed Brest for Toulon, where they dropped anchor on the 13th. Following the loss of the battleship *Iéna* (see the author's article in *Warship 2007*) *Carnot* replaced her in the 2nd Division until 1 July. During the months of March, April and May the battleship made several sorties around the coasts of Provence for gunnery exercises and torpedo firings. She

was in Missiessy Dock No 3 from 14 to 26 June, and visited Mers el-Kebir, Philippeville and Les Salins from 30 June to 15 July.

There were numerous gunnery exercises in the second half of August. It was during these exercises that major problems developed with the forward 305mm turret. On 2 September *Carnot* entered the dockyard for investigations, and engineer Davaux made an alarming discovery. The turret support structure was deformed, a split had appeared in the pivot, and the reinforcements around the base of the turret showed abnormal signs of wear. Further investigations revealed similar problems with the after turret. In order to effect repairs it was necessary to disembark the armour plating of the turret, the guns and mountings, the turntable, and finally the pivot itself. It was estimated that repairs would take nine months and cost 109,000 francs. More seriously, it was suspected that the turrets for the 274.4mm guns would be found to have the same defects.

Carnot returned to Toulon roads on 7 October and was placed in special reserve on 8 December. The repairs were carried out by the Paul Boursier & Jolidon Company and were completed in May 1909. The battleship began sea trials 26 May, but there were serious problems with the HP cylinder of the starboard engine, and the ship had to return to the dockyard. *Carnot* was finally recommissioned on 5 October, joining the 2nd Division of the 2nd Battle Squadron under the command of Rear Admiral Berryer, who flew his flag in *Bouvet*.

The formation then sailed for Quiberon on 23 February 1910, calling at Bizerte, Algiers, Mers el-Kebir, Tangier, Cadiz, Lisbon, Vigo, La Pallice and Ile

The 274.4mm wing turret flanked by single turrets for two of the eight 138.6mm QF guns. (DR)

Carnot and the submarine *Bonite* in Toulon roads. (DR)

d'Aix. *Carnot* and *Bouvet* left the Quiberon anchorage on 13 March for Cherbourg, where they were to be docked for maintenance. On 9 May the Squadron returned to Toulon via Mers el-Kebir, arriving 30 June. In late June there were gunnery exercises at Golf-Juan and the 2nd Squadron returned to Brest. *Carnot* was docked at Cherbourg on 25 September for a short refit, returning to Brest on 11 February 1911.

With the entry into service of the battleships of the *Danton* class in 1911, the 2nd Battle Squadron became the 3rd Squadron, and sailed for the Mediterranean on 12 August to take part in the grand manoeuvres. *Carnot*

then took part in the Great Naval Review of 4 September in Les Vignettes roads, where the fleet was inspected by President Armand Fallières.

In the early morning of 25 September the battleship *Liberté* blew up in the anchorage at Toulon and the boats of all the ships present took part in the rescue of the survivors. On 4 October the 3rd Squadron returned to Brest via Algiers and Mers el-Kebir. Durant the voyage firing practice against targets was conducted and, as anticipated, the starboard wing 274.4mm turret showed signs of structural damage. *Carnot* was duly placed in normal reserve at Cherbourg 15 February 1912.

Carnot after 28 January 1908, with her hull and upperworks repainted in a uniform blue-grey. (DR)

Carnot in Brest roads in 1912. The two rings on the second funnel mark her out as the second ship of the Second Division. (Musée Maritime)

The ships of the 2nd Squadron moored at Villefranche in December 1912. In the background, from left to right: the battleships *Saint Louis*, *Bouvet* and *Carnot*. In the foreground: *Jauréguiberry* and *Gaulois*. (Musée Maritime)

Jauréguiberry (foreground) and *Carnot* at Cherbourg. Note the uniform blue-grey livery; during this period these two ships were serving with the Northern Squadron. (DR)

Following repairs similar to those carried out on the 305mm turrets, the ship was recommissioned on 22 April but was rapidly returned to reserve status.

On 28 September *Carnot* joined the 2nd Division of the 3rd Squadron, and this formation sailed for Toulon on 16 October. The voyage was broken by port calls at Lagos (19–21 October), Mers el-Kebir (22–23 October) and Algiers (24 October to 5 November). After a brief visit to Les Salins d'Hyères the ships arrived at their destination on the 9th. There were exercises off the coasts of Var and Alpes-Maritimes, followed by visits to Villefranche, Golfe-Juan, Saint-Tropez and Bizerte from 6 to 21 January 1913.

Carnot was by now obsolescent and her guns were barely serviceable; it was time for the ship to be definitively decommissioned. Her farewell tour took in the customary towns of Villefranche, Golfe-Juan and Les Salins, then *Carnot* made her solitary passage to Brest, departing 25 April and dropping anchor on 2 May. She was docked, then placed in normal reserve from 7 May.

Carnot was decommissioned on 1 April 1914. She was attached to the administration of the port of Brest, and her crew was disembarked the following day and allocated to the 2nd Depot. She would become a floating prison, and in 1915 her guns were disembarked. The 305mm guns would be re-bored at the Navy's artillery establishment at Ruelle and converted into 370mm howitzers Mle 1915. In 1916 they were assigned to the ALVF (*Artillerie lourde sur voie ferrée* – railway-mounted heavy guns) and baptised *Staïfia* and *Surcouf*. *Carnot*'s

boilers and engines were likewise recycled by *Constructions Navales*.

On 30 October 1919 the ship was stricken and put up for sale for scrapping. On 8 April 1920 *Carnot* was bought by the Dutch Franck Rijsdijk Ship Breaking Company for the sum of 1,100,000 francs. Taken in tow for Rotterdam in June of the same year, the old battleship was severely battered by a storm; she capsized and sank.

Editor's Note:

This article was translated from the French by the Editor. The official plans of *Carnot* have not survived; the profile has been adapted from artwork published in secondary sources, allied to a careful study of photographs.

Carnot at Cherbourg at the end of her career. (DR)

HMY *VICTORIA AND ALBERT* (III)

In 1897 the great naval architect Sir William White was called upon to design a new Royal yacht. The yacht faced disaster while building, but survived; the DNC's career was over. **Ian Sturton** has used the three volumes of the Ship's Cover, the twenty-six accompanying draughts and the Ship's Book to shed fresh light on the drama.

The decision to build a new Royal yacht was announced on 1 January 1897, Diamond Jubilee year. The existing yacht, the veteran wooden side-wheeler *Victoria and Albert* (II), dating from 1855 and dwarfed by more recent European rivals such as the German *Hohenzollern* (II) and Russian *Shtandart*, could no longer be considered a fitting symbol of the premier maritime power (see Table 1); worn-out boilers and decayed timbers made her increasingly difficult and expensive to keep in service. Her interior and fittings had been designed by Prince Albert; the Queen's dearest wish (as an exact replica could not be constructed) was for the Royal apartments in the new ship to copy those in the old as closely as possible.

Pembroke Dockyard (1814-1926) was the immediate choice for building the new yacht. It had constructed the Queen's earlier Royal yachts *Victoria and Albert* (I), *Victoria and Albert* (II), *Alberta* and *Osborne* (the tiny *Elfin* came from Chatham) but had in recent years specialised in battleships and large cruisers – big, heavy ships.

At the Admiralty, Sir William H White, Director of Naval Construction (DNC) and Assistant Controller since 1885, had overall responsibility for the design. White entrusted his principal lieutenant, Chief Constructor WE Smith, with immediate supervision of the Royal yacht design and calculations; detailed work was assigned to WJ Luke and PL Pethick, two experi-

enced Assistant Constructors. A constructor not involved in the design process, Chief Constructor HE Deadman, dealt with detailed working drawings for hull and fittings. In the dockyard, the ship was in the personal charge of Chief Constructor H Cock.

For exceptional steadiness and ease of motion, she was designed for a moderate metacentric height (GM) of 2ft at deep load and 9in in the light condition. An excessive metacentric height would provide exceptional stability and resistance to underwater damage, but give a quick roll, making a ship 'stiff' and uncomfortable and, if a warship, a poor gun platform.

The Ship's Cover lists particulars of the torpedo depot ship *Vulcan* alongside those of the new yacht, suggesting she was considered as a possible model for hull and weight distribution. An *Engineering* article on the *Shtandart* with plans is bound into the Cover, together with a note that drawings of the ship had been returned to the Russian Embassy, but there is no direct suggestion of a Russian design connection.

The proposed design stage ('As Proposed') was reached in May 1897, when a detailed description and set of plans (signed by White and dated 5 May) were drawn up. After modifications and further deliberation the Board of Admiralty approved the final design ('As Designed') in October 1897. The final design of an ordinary warship could not be changed significantly without

Table 1: **Characteristics of Several Royal Yachts, May 1897**

	Hohenzollern (II)	*Shtandart*	*V & A* (II)	'New Yacht'
Date	1893	1896	1855	1899
Length	380ft 6in	370ft	300ft	380ft
Breadth	45ft 11in	50ft 6in	40ft 3in	50ft
Draught (mean)	16ft 3in	20ft 7in	16ft 3in	18ft
Displacement	4,120t	5,480t	2,470t	c4,600t
IHP	9,000–10,000	15,000	2,400	11,000
Speed, knots	21.0–21.5kts	21.25kts	17kts (trials)[1]	20kts
Armament	8 – 5cm QF[2]	8 – 47mm QF	2 – 3pdr[3]	2 –3 pdr[3]

Notes:

[1] 2,980ihp required for this speed.

[2] Later two 52mm QF; planned wartime armament 3 – 10.5cm QF, 12 – 5cm.

[3] Bronze MLR saluting guns.

The paddle-driven HMY *Victoria and Albert* (II) photographed in 1876. She carried the new King and other royalty from Cowes to Portsmouth, following *Alberta*, on 1 February 1901; her final royal service was to take the King to and from Flushing (Vlissingen), on 25 February and 3 March 1901 respectively. Decommissioned on 23 July 1901, she was paid off into Dockyard Reserve on 3 December 1901 and broken up in 1904. (Naval History & Heritage Command, NH 110444/45)

The German paddle yacht *Hohenzollern* (I) (sold 1912), was replaced in service by *Hohenzollern* (II), commissioned in 1894 and shown here; this image was used in her portrayal on German colonial postage stamps. She was famed for taking Kaiser Wilhelm II on European cruises and interventions – the secret Treaty of Björkö was signed on board by the Kaiser and Tsar Nicholas II on 24 July 1905. *Hohenzollern* (II) was broken up in 1923. (NHHC, NH 46815)

The Russian Imperial yacht *Shtandart* at Yalta. Built at Copenhagen, she was badly damaged on an uncharted rock in the Gulf of Finland, 20 August (OS)/2 September (NS) 1907. Laid up from 1914, she was converted to a minelayer between 1932 and 1936 as *Marti*. She served as a minelayer and troopship during the Second World War, suffering heavy damage and being grounded in the autumn of 1941. *Marti* was employed for subsidiary duties during the remainder of the war, her guns being used on land. She continued in this role after the war, being renamed *Oka*; she later served as a target for missiles, and was broken up in the early 1960s.

HMY *Alberta* (launched 3 Oct 1863, commissioned 30 Nov 1863, 370t, 1,000ihp = 14kts) conveyed the Queen between the mainland and Cowes and made short excursions with Royalty on board. She carried the Queen's catafalque from Cowes to Portsmouth on 1 February 1901. She was decommissioned on 30 March 1912 and broken up in 1913. (*Navy & Army Illustrated*, Vol 4, 51)

HMY *Osborne* (II) (launched 19 Dec 1870, commissioned 12 Jun 1874, 1,850t, 3,000ihp = 15kts) was almost entirely used by the Prince and Princess of Wales. In the Prince's 1875–76 tour of India she acted as tender to the much larger modified troopship *Serapis*, which carried the Royal party for most of the voyage. *Osborne* was paid off on 6 May 1908 and sold on 31 July for breaking up. (*Navy & Army Illustrated*, Vol 4, 51)

HMY *Elfin* (launched Chatham 8 Feb 1849, commissioned 1 May 1849, 98t, 40nhp = 12kts), tender to *Victoria and Albert* (II), was used on routine duties between Cowes, Southampton and Portsmouth and nicknamed the 'milk boat.' Paid off on 19 March 1901, she was dismantled in 1901. (*Navy & Army Illustrated*, Vol 7, 517)

Board approval. However, the new Royal yacht was no ordinary warship.

The Court and the Constructors

From the outset, Royalty was closely involved in the design process in relation to Royal accommodation, laid down in detail by the Queen on 2 March 1897, and comfort, as well as to wider issues. The Cover contains many letters from Rear Admiral JRT (later Admiral Sir John) Fullerton, Commanding Officer of the *Victoria and Albert* (II),[1] to the Controller, Sir John Fisher,[2] to White and to his staff. An example from early in the design process, dated 4 February 1897, reads:

Dear Sir William ... Queen *most* interested in the new yacht *but* [underlined twice] wished her sitting room to be in the same position as the present one i.e. opposite the bedroom and that the bedroom should be slightly larger. She said two or three feet. She won't *hear* [underlined three times] of the kitchen being aft! Nothing I could say would convince her that there would not be a smell of cooking,

Photographs of (left) Rear Admiral JRT Fullerton, Commanding Officer of *Victoria & Albert* (II), and (right) Sir William White. (*Navy & Army Illustrated*).

so that must be altered. I expect we shall be obliged to change the position of the large dining room altogether. The Queen also asks where her dining room is to be. She always uses the room right aft which we call the Breakfast room. We must manage some dining room for her not too far from her sitting room I think.

A note from Fullerton to Fisher is attached: 'Please remember P(rince) of Wales wishes accommodation for a band.' Fullerton to White on 3 March:

> The Plans have been very much in my mind and I have come to the conclusion that the "Maids and Valets" and "Nursery" must change places. I am *sure* [underlined three times] that it would never answer for the Ladies–in–Waiting to pass through Servants quarters. The childrens quarters must be cut down if the present servants cabins do not give the accommodation I mentioned in my paper. If possible there should be one large cabin with two beds in it and a nurses cabin adjoining and opening into it *one* side and two cabins and a nurses the other. This will do. I will not interfere with the Suites rooms. It will be an advantage in other ways too – as the dinner will go outside Royal apartments altogether. Also you must be prepared for an alteration in position of the *Sick Bay*,[3] that is a "fad" of the Royal people and will have to be attended to.

Fullerton to Deadman 15 March (White being absent): 'The following alterations in the accommodation will be

necessary. The Queen wishes the Munshi[4] to be with the suite aft and a cabin must be provided for him the *same size* as the others. The sick bay will have to be shifted from under the turtle deck. This I thought was very probable from the first ...'. Deadman to Fullerton (17 March): '... as regards placing the Munshi in the after suite it appears impossible to get this without either removing one of the persons already accommodated there or by curtailing the extent of the accommodation afforded them ...'. The constructors' patience was tested. Fullerton suggested to Deadman (6 April) that certain cabins amidships be abolished, making it possible to put the Royal kitchen there. In response, Deadman observed (8 April) that in the new position it would be 'immediately under Her Majesty's apartments, a feature which it was understood on a former occasion was greatly objected to by Her Majesty.' He asked for 'any further notes as to the arrangement of the lower deck as the drawings are now being rapidly advanced and it is very desirable to provide at first for all known requirements so as to avoid subsequent alterations.' Fullerton replied gently: 'It is, I am afraid impossible to avoid alterations in the plans. Besides the Queen they have to be submitted to the Prince of Wales and the Duke of York.'

The 'As Proposed' drawings and a model were shown to the Queen at Windsor by Fisher, White and Fullerton, 14 May 1897. The letters continued. Fullerton to White (17 May, paraphrased): 'The Prince of Wales writes to suggest yacht be fitted with *bilge keels* – I replied that you

Victoria & Albert:
Profile as Proposed
May 1897

Victoria & Albert:
Profile as Designed
October 1897

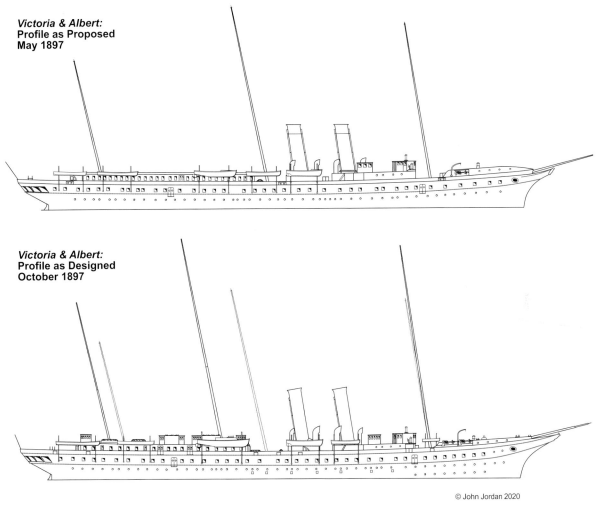

© John Jordan 2020

Presentation of the 'As Proposed' and 'As Designed' profiles on the same page and to the same scale emphasises the similarities and differences between the designs. A careful study of the drawings, bulkheads, compartmentation etc, indicates that the funnels in each are in similar positions, although steam pipes differ, and that the position of the *base* of the foremasts in each is similarly located in the compartmentation. The 'As Designed' profile shows the mast positions before (short, grey) and after (tall, black) the intervention of the Prince of Wales (see Table 9).

have already arranged for them'. Fullerton to White (4 June): 'I cannot understand how … [you will manage] … the ventilation of the engine room if the top of the pavilion is to be a promenade deck. There is a growing feeling that the Queen will suffer from the noise if the engines [are] immediately below her but I understand and have said that it is guaranteed there will be nothing felt – I am right I hope?'

Structure

The new yacht was a graceful steel-hulled vessel, sheathed and coppered, with a clipper bow, two funnels, three masts and an elliptical stern. The hull was constructed on the double bottom system, with an internal vertical keel 2ft 9in deep and Z-section frames 4ft apart. The double bottom was carried out to the bilges. Watertight longitudinal bulkheads formed inner walls to coal bunkers in boiler rooms and to oil and water tanks in engine rooms.

Eight transverse bulkheads were carried to the upper deck. The bilge keels were 2ft 6in deep and extended for about three-quarters the length of the ship. The shell plating was 35lb at the keel plate thinning to 20lb, and 30lb in contact with anchors; plates were double-riveted and butt-jointed, with inside straps quadruple-riveted. The shell plating was covered by teak planking 4in thick to the turn of the bilges, reducing to 3in above the waterline, then 2.5in to the bulwarks; the teak was in turn coppered to above the waterline. The hull was to be elaborately decorated with carved and gilded decorations, some by Frederick Hellyer, who had executed carved work on the *Victoria and Albert* (I) of 1844. Rigols, mahogany mouldings carved and gilded to represent two 15in cable-laid ropes encircling the hull above and below the square ports, were the work of S Trevanan. Decorations were to be completed and fixed in place for the launch, after which they would be removed during fitting out to prevent damage.

Description

General characteristics 'As Designed' are listed in Table 2, with a breakdown of estimated weights in Table 3(a). Supplementary information on dimensions was of particular importance to port officials responsible for securing the yacht alongside. Data sent to the Harbour Commissioners at Kingstown (Dún Laoghaire), to the Belfast Harbourmaster and to Cherbourg for the French authorities are listed in Table 4.

Victoria & Albert: Sections as Designed May 1897

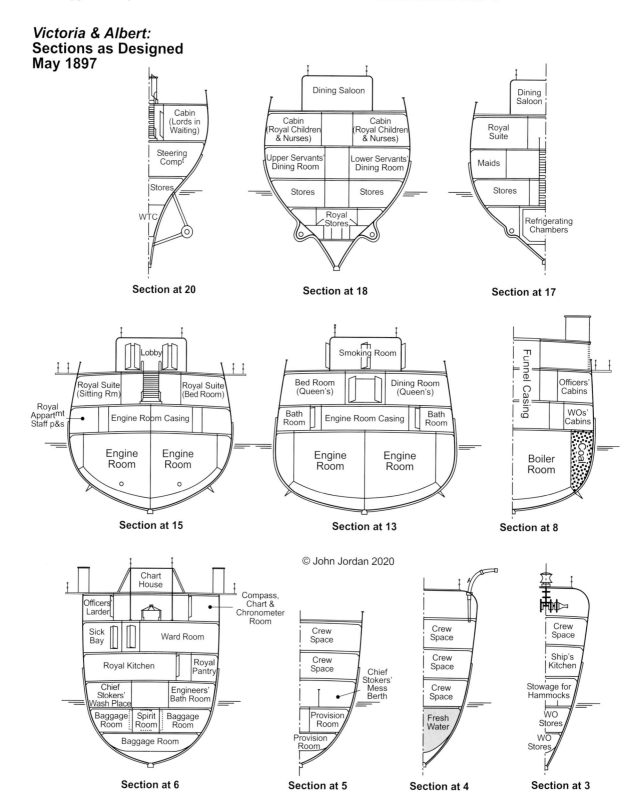

© John Jordan 2020

Table 2: **General Characteristics of *Victoria and Albert* (III)**

	As designed	Actual	List
Displacement, normal	4,700t (350t coal)		
Displacement, deep load	5,000 (650t coal)		5,500t
Length (pp)	380ft	380ft 1in	380ft
Beam (extreme)	50ft	50ft 3.5in	50ft
Breadth (moulded)	49ft 5in	49ft 7in	
Draught, fwd, 350t coal	17ft		
Draught, aft, 350t coal	19ft		
Draught, mean, 350t coal	18ft		
Draught, mean, 650t coal	18ft 10in	19ft 7.5in[1]	
Draught, extreme			20ft 6in
Depth, moulded	37ft		
Tons per inch immersion	31t		
Area of midships section	760 sq ft		
Freeboard at fore perp	24ft 9in		
Freeboard, minimum	19ft		
Freeboard at aft perp	21ft		
Gross tonnage (Suez Canal)	4,387.38t		
Register tonnage (Suez Canal)	2,319.72t		

Note:

[1] 650t coal and fully equipped.

Armament

From the beginning the new ship was in no respect intended as or fitted for service as an auxiliary cruiser in wartime; when asked in Parliament, GJ Goschen, First Lord 1895–1900, stated: '… after mature consideration, the comfort of the inhabitants of the yacht would be interfered with by constructing it as a cruiser to a much greater degree than would be the value of the cruiser.'[5]

Two bronze 6pdr RML saluting guns ('howitzers') on the foredeck were only taken aboard for ceremonial occasions; they came from the *Victoria and Albert* (II), as did the four antique bronze ornamental swivel guns dating from 1734 on the poop.[6]

Machinery and Performance

Speed was another matter for debate; *Victoria and Albert* (II) had in her day been very fast. The Queen was informed that White proposed to 'break the record' in yachts with a maximum speed of 23 knots, but '[the Prince of Wales] seemed to think it would be a mistake to devote too large a space to machinery. He said comfort and

Table 4: **Additional Dimensional Data**

Length: 444ft (length of hull), 460ft 7in (oa) actual, being 380ft 1in pp with 53ft for bow and bowsprit added forward and 27ft 6in for stern overhang added aft), 477ft 7in (extreme length including ensign staff), breadth over boats 68ft, entrance port 125ft from after end of hull, width of port about 4ft; the draught with 650t coal and fully equipped was stated as 20ft 3in (September 1900).

The North Sea Canal bridges were 42m (138ft) above water level; the maximum mast height allowed in the Canal was 40m (131ft).

Table 3: **Estimated Weights (Tons)**

(a) As Designed and approved by the Board (22 October 1897).
(b) At deep displacement if ship completed as proposed before the accident, with GM = 3in.

Item	(a)	(b)
Hull	2,550	–
Hull recorded (inc sheathing)	–	3,209
Hull to complete	–	250
Missing weight	–	72
Removed at Pembroke	–	71
Boat hoists	–	20
Wood and copper sheathing	350	–
Sub-total for above: hull	2,900	3,622
Machinery, boilers, condensers etc	960	954
Reserve feed water	50	48
Engineer's stores	20	35
Coal at normal displacement	350	–
Coal at deep displacement	–	650
Sub-total for above: machinery	1,380	1,687
Water	70	70
Provisions, spirits etc	15	15
Stores and slops	10	10
Officers, men (320) and effects	40	44
Masts, yards, spars, rigging, blocks, awnings	50	61
Cables and anchors	60	55
Boats	25	27
Warrant Officer's stores for __ months and armament	25	25
Deck coverings	–	10
Royal party, baggage and stores	50	50
State furniture (estimated)	–	20
Sub-total for above: general equipment	345	387
Board Margin	75	–
Total	4,700	5,696

accommodation were the chief points with a certain speed of course – and thought a regular speed of 17 or 18 knots with power to go 20 if necessary would be sufficient.' White accordingly fixed the desired continuous sea speed as 18 knots with 7,500ihp and 11,000ihp maximum for eight hours (Table 5). The estimated rate of coal consumption and weights of engines, boilers, shafts and propellers were obtained from the Engineer-in-Chief (Table 6).

The machinery took up 174 feet of the ship's length (engine room 58ft, boiler space 90ft, passage in between 26ft). The two sets of vertical four-cylinder triple-expansion engines by Humphrys, Tennant had cylinders of 26.5in (HP), 44.5in (IP) and 53in (2 x LP) diameter by 39in stroke. Steam from eighteen Belleville water-tube boilers with economisers was generated at 300psi, reduced to 250psi at the engines. The engines were side by side, separated by a watertight longitudinal bulkhead, the boilers in two compartments one ahead of the other and separated by a watertight transverse bulkhead.

The forward boiler room had nine boilers, six each with eight generator and seven economiser elements and three with ten generator and eight economiser elements, while the after boiler room had nine boilers, each with ten generator and eight economiser elements. The generator and economiser tubes were of 4in and 2.75in diameter respectively. Heating surfaces totalled 26,000 sq ft (2.35 sq ft per ihp) and grate surfaces 840 sq ft (13ihp per sq ft). The forward boiler room supplied the starboard engine, the aft room the port engine, with connections in the engine rooms. The coal bunkers, in addition to those already mentioned, were placed across the ship forward of the forward boiler room and aft of the after boiler room. The designed maximum coal stowage of about 750 tons was reduced to 714 tons during construction.[7] The openings to and from the engine rooms were necessarily governed to a great extent by the requirements of the Royal apartments. To keep the vessel cool and to deaden sound, all boiler and engine-room casings were

Table 5: Performance Data

Estimated speeds (*) for specified powers, and estimated power (**) for specified speed:
(a) as designed.
(b) as estimated before sea trials.

Horsepower	Speeds in knots	
	(a)	(b)
11,000ihp (natural draught)	20.0*	19.5*
7,500ihp (continuous at sea)	18.0*	17.5*
5,000ihp	16*	15.5*
1,150ihp**	10	

Table 6: Estimated Coal Consumption and Endurance

Speed	ihp	Daily consumption[1]	Endurance with: (a) 350t coal	(b) 650t coal
17–18kts	7,500	168t	850nm	1,600nm
16kts	5,000	115t	1,150nm	2,150nm
10kts	1,250	36t	2,350nm	4,300nm

Estimated coal consumption: 2.0–2.1lb per ihp per hour.

Note:
[1] Includes 8t for auxiliary and cooking purposes.

Table 7: Trials Data

	48hr at 5,000ihp	48hr at 7,500ihp	8hr at full power	Measured mile
Draught fwd	18ft 2in	18ft 2.5in	18ft 2in	18ft 2in
Draught aft	20ft 1in	19ft 11.5in	20ft 1in	20ft 1in
Steam press at boilers	254psi	272psi	306psi	306psi[1]
Vacuum	26.9in	25.25in	25.25in	25.25in
Shaft revolutions	110.9rpm	128.4rpm	147.4rpm	148.1rpm
Horsepower, mean	5,142ihp	7,649ihp	11,298ihp[2]	11,500ihp
Speed over 23.2nm course	16.32kts	18.47kts	Not seen	20.53kts[3]
Coal consumption	1.94t[4]	1.87t[4]	Not taken	–

Source: *Engineering* and the Cover.

Notes:
[1] Average steam pressure at engines 247psi.
[2] 5,620ihp starboard engine, 5,678ihp port engine.
[3] Speed over measured mile.
[4] For 40 hours.

Victoria & Albert: State Deck

As Proposed May 1897

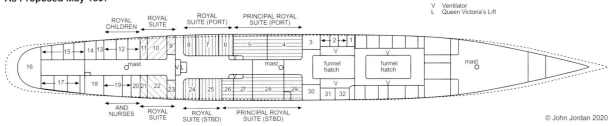

© John Jordan 2020

State Deck As Proposed May 1897.
The State Apartments are numbered – see Key below.

KEY
1   Mess Berth for Indian Servants
2   Indian Servant
3   Cabin for Royal Personage
4   The Queen's Drawing Room
5   The Queen's Dining Room
6   Dressing Room
7   Bedroom
8   Bath Room &c
9   Cabin for Dresser
10 Bedroom
11 Bath Room &c
12 Cabins
13 Bath Room &c for Lords in Waiting
14 Munshi
15 Cabins for Lords in Waiting
16 Dining Room for Lords and Ladies in Waiting
17 Cabins for Ladies in Waiting
18 Sitting Room for Ladies in Waiting
19 Cabins
20 Bath Room &c
21 Bath Room &c
22 Bed Room
23 Cabin for Dresser
24 Cabin for Dresser
25 Sitting Room
26 The Queen's Bath Room
27 The Queen's Dressing Room
28 The Queen's Bedroom
29 The Queen's Wardrobe Room
30 Cabin for Royal Personage
31 Courier
32 Page

lined outside with 1.5in teak, the spaces between planking and casing having two layers of silicate cotton (mineral wool) with an air space in between. Similar arrangements were adopted in the ceilings, walls and floors of the Royal apartments. The three-bladed gunmetal propellers were of diameter 13ft 3in, pitch 17ft 6in (port) and 17ft 7in (starboard), and had expanded surfaces of 12 sq ft. The balanced rudder had an area of 170 sq ft and a maximum breadth of 16.1ft. Trials data are listed in Table 7.

Auxiliary Machinery

The steering engines were in the main engine rooms, acting on the rudder through shafting; this indirect arrangement was considered less intrusive and noisy than the alternative of running steam pipes through the Royal apartments to a steam steering engine fitted aft. Three dynamos providing electricity at 80 volts had a total capacity of 1,800 amps, more than sufficient for the 1,200 amps needed for refrigeration, general lighting and extra heating in the Royal apartments. The refrigerating capacity was 3,500 cu ft including lagging, with internal (useful) space 2,400 cu ft at a temperature of 15°F under tropical conditions.

In March 1897, White directed that an artificial system of heating in addition to the stoves in the principal cabins be provided. The choice was between hot water and steam. 'Baths must be provided with hot and cold

service.' Smith and the Engineer-in-Chief agreed on steam heating as being much lighter although water chests would be needed at intervals to keep the condensed steam. The necessary fittings and appliances would be inserted in the machinery spaces.

Mast, Rig and Boats

The original 'As Proposed' draughts signed by White in May 1897 included three short, raked pole masts without spars or rigging. Between May and October/November 1897, however, numerous changes were suggested: mast heights were greatly increased, a yard added to the foremast and gaffs to all three masts. The purpose of the added sail power was not discussed – in good weather its use would have been picturesque but a tedious exercise for the crew, in bad weather its use for steadying the yacht ('steadying effect') positively dangerous. The original draughts do not show the heavy rig but it is discussed in the Cover (see Tables 8 and 9).

Photos show the gaffs lowered; one photo seen shows the yacht in dock with the yard raised. After the accident Controller decided upon a reversion to the original design 'As proposed', *ie* short, fixed masts without rig or sails. They were too high to pass under the North Sea (Kiel) Canal bridges and were criticised aesthetically as being too short and too thin ('a dreadful eyesore and very adversely commented on by all who see her'). White's technical solution to the Canal bridges conundrum is

Table 8: Masting Measurements
Heights of masts above waterline.

	9 May 1897	Aug – Sep 1898	As built	26 Jan 1900 – 30 Sep 1901
Foremast	144ft	153ft 2in	174ft	134ft
Mainmast	156ft 10in	170ft 6in	187ft	143ft
Mizzenmast	112ft 5in	130ft	151ft	113ft

Notes:

Mast heights may include top ornaments and vanes.
In Aug/Sep 1898 the bowsprit length, excluding housing, was 43ft, the bowsprit housing 13ft, and the fore, main and mizzen gaffs 45ft, 50ft and 36ft respectively. In January 1899 the foremast yard was 75ft, the length including two 3ft yardarms.

Table 9: Positions of Masts

	As on building drawings	As sent to dockyard
Foremast	70ft	70ft
Mainmast	201ft 6in	236ft 9in
Mizzenmast	314ft 6in	326ft 0in

Notes:

Positions measured along waterline length from fore perpendicular to centre of mast; the set of figures sent to the dockyard was as arranged with the Prince of Wales: rake of foremast and mainmast reduced, rake of mizzenmast increased.

Table 10: Complement in July 1901

Executive & Navigating Branch	8 officers, 119 WO & men	127
Engineer Branch	4 officers, 135 WO & men	139
Artificers Branch	1 officer, 12 WO & men	13
Medical Branch	1 officer, two men	3
Accountant Branch	2 officers, four men	6
Miscellaneous		8
Domestic		8
Marines		51
Total Complement		355

contained in an urbane memorandum comparing the merits of telescopic masts and fidded masts (masts with striking topmasts); the former were selected. Davits for thirteen boats were fitted finally, seven to port and six to starboard; the boats were carried swung outboard, for greater steadiness at sea and less obstruction of deck spaces – and more attractive views for passengers.

Accommodation

The complement, at first provisionally 300, was estimated in October 1897 at 320, in January 1898 as 330, the Marine band in addition. Excepting the engineering officers, almost all the 230 crew of *Victoria and Albert* (II) transferred to the new yacht. The additional men drafted did not include defaulters. The total complement on completion was 355 (Table 10). Provision was made for a Royal party of 50.

The vessel had five decks, one more than in *Victoria and Albert* (II): the upper, state (or main), lower, orlop and platform decks, carried on bulb beams and laid with Douglas fir, treated to be non-flammable. The state deck was 10ft high. In the final plans the after part of the state deck was occupied by the Royal apartments, which had a central passage mostly 14ft wide. The 180ft Royal pavilion on the upper deck was 10ft 6in high; the aft 65ft was occupied by the state dining room, then, centrally, a reception room was opposite the entrances at the companion ladders either side of the ship, and at the forward end a smoking room. The roof of the Royal pavilion served as an extra promenade deck. A 6ft by 4ft lift ('electric hoist') between state and upper deck was for special use of the Queen.[8] The Queen's apartments were more spacious than in the earlier yacht (Table 11). The large bridge deck and a forecastle, later removed, were also on the upper deck.

The seamen's mess decks were forward on the state and lower decks; their accommodation was described as cramped. The length of seamen's lockers for the provisional complement of 300 was 720ft, *ie* 2.4ft per man. An estimated 320 billets for 'hanging up' (hammocks) could be arranged on the men's decks (110 on state deck, 120 lower deck, 70 orlop deck and 20 under forecastle). The wardroom and officers' cabins were farther aft on the state deck, before the Royal apartments were reached, with the warrant officers' and lower servants' cabins one deck lower. The accommodation for the upper servants was on the lower deck aft.

The maids, valets and dressers had the status of upper servants and all would dine together. There would be 22 upper and 14 lower servants. Fullerton (no date): 'There is no reason why the lower servants should not have cabins with two or three bunks. *One* bath room for the upper servants and *one* for the lower servants …[will be]… quite sufficient. It would be as well if an officers' bath

Table 11: Dimensions of the Queen's Apartments

	Victoria and Albert (II)	*Victoria and Albert* (III)
Breakfast/dining room	22ft x 18ft (mean) = 410 sq ft	23.5ft x 18.5ft = 434 sq ft
Drawing room	26ft x 14.5ft = 380 sq ft	23.5ft x 19ft = 446 sq ft
Bed room	20ft x 14.25ft = 285 sq ft	21ft x 19ft = 399 sq ft
Wardrobe room	12.5ft x 15ft = 190 sq ft	11ft x 16ft = 176 sq ft
Dressing room	13.5ft x 13.5ft = 180 sq ft	14.5ft x 18.5ft = 268 sq ft
Bathroom	–	10ft x 10ft = 100 sq ft

These photos of the drawing room of HMY *Victoria and Albert* (II), taken in 1876, give an idea of the luxurious internal fittings of a Royal yacht of the period. The Queen required that familiar furnishings be transferred to the new ship. (NHHC, NH 110445)

Victoria & Albert As Fitted

Profile

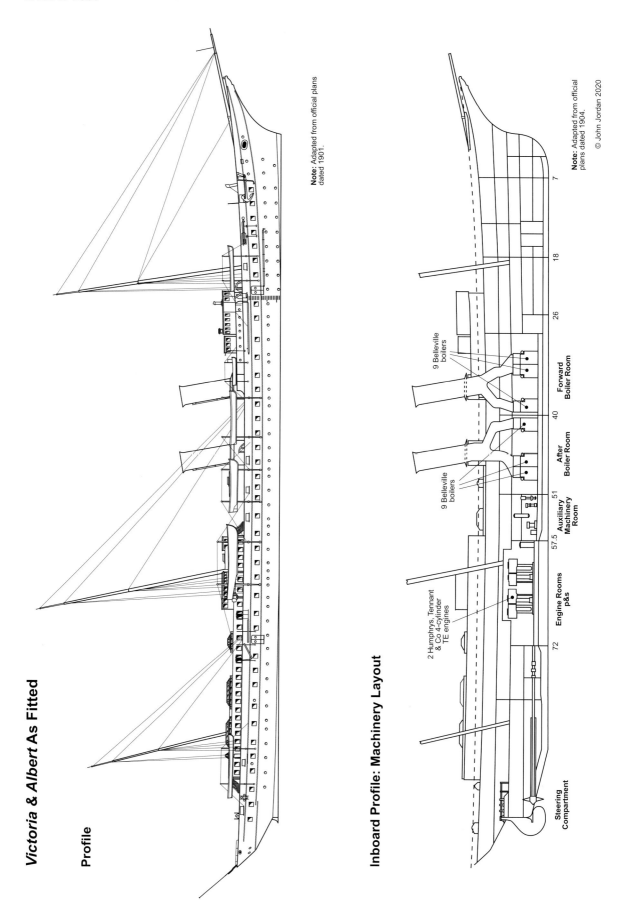

Note: Adapted from official plans dated 1901.

Inboard Profile: Machinery Layout

© John Jordan 2020

Note: Adapted from official plans dated 1904.

2 Humphrys, Tennant & Co 4-cylinder TE engines

9 Belleville boilers

9 Belleville boilers

Steering Compartment

Engine Rooms p&s

Auxiliary Machinery Room

After Boiler Room

Forward Boiler Room

72 57.5 51 40 26 18 7

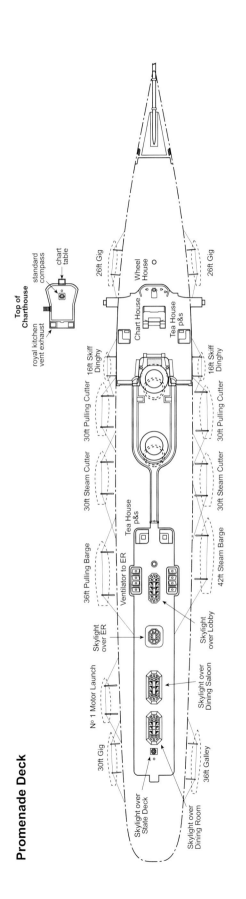

Promenade Deck

Top of Charthouse

royal kitchen vent exhaust

standard compass

chart table

26ft Gig

Wheel House

Chart House

Tea House p&s

26ft Gig

16ft Skiff Dinghy

30ft Pulling Cutter

30ft Steam Cutter

16ft Skiff Dinghy

30ft Pulling Cutter

30ft Steam Cutter

Tea House p&s

30ft Pulling Barge

Ventilator to ER

42ft Steam Barge

Skylight over ER

Skylight over Lobby

Nº 1 Motor Launch

Skylight over Dining Saloon

30ft Gig

36ft Galley

Skylight over State Deck

Skylight over Dining Room

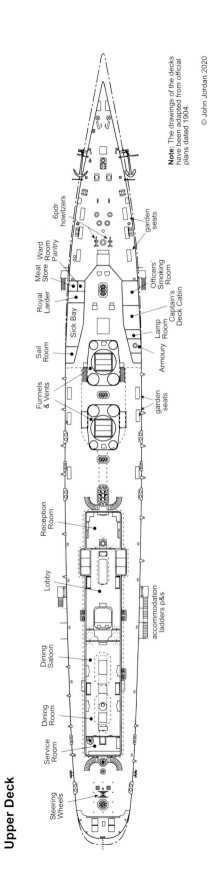

Upper Deck

Note: The drawings of the decks have been adapted from official plans dated 1904.

© John Jordan 2020

6pdr howitzers

garden seats

Meat Store Room

Ward Pantry

Royal Larder

Officers' Smoking Room

Sick Bay

Captain's Deck Cabin

Sail Room

Lamp Room

Armoury

Funnels & Vents

garden seats

Reception Room

Lobby

accommodation ladders p&s

Dining Saloon

Dining Room

Service Room

Steering Wheels

Victoria & Albert: Royal Pavilion

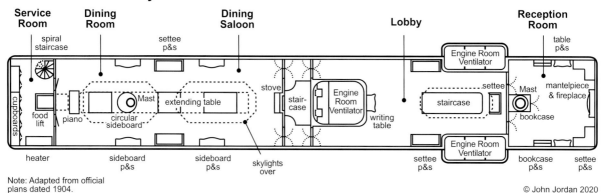

Note: Adapted from official plans dated 1904.

© John Jordan 2020

room can be supplied but it is not absolutely necessary.'

Interior design was entrusted to the firm Waring & Sons, furniture makers, the plumbing to Messrs Broadfoot, with other companies responsible for their specialities.

Construction

The keel was laid 15 December 1897 and the ship launched by the Duchess of York (later Queen Mary) on 9 May 1899, taking the name *Victoria and Albert* (the Queen had first requested the name *Balmoral*).

The launch weight was 2,930 tons, including 2,523 tons for the hull and the cradle; the permanent ways were 2ft 7in wide and long enough to limit the weight to two tons per foot of bearing surface. 160 tons of water ballast were among the weights on board at launch 'to be removed.' The ship was known as 'the new Royal Yacht' or 'the new *Victoria and Albert*' while the *Victoria and Albert* (II) remained in service.[9]

The new yacht was fitted out afloat at Hobbs Point (including installation of 550 tons of machinery in the boiler and engine rooms and the stepping of masts) until 10 July when placed in dry dock for completion, mainly because the finishing berth was required for the cruiser *Spartiate*. No stability problems were reported while the ship was afloat. Weights on board when going into dock included 88 tons of ballast in the double bottom and 50 tons of reserve feed water.

Soon after launch, an assistant constructor prepared a special report for White, noting the ship's condition and comparing actual and estimated weights, as far as could be done. The report stated that the final extra draught on completion would be 3in–4in, corresponding to 100 tons extra displacement, figures satisfactory and acceptable in view of the approved additions and modifications made during construction. White noted later that the report was entirely untrustworthy in regard to the estimate of weights to be added subsequently, before completion. Had fitting out been completed afloat instead of in dock, the problems of excessive topweight and deficient stability would have become evident, and remedial measures taken.

The Accident

The ship was to leave dry dock at 0645 on 3 January 1900, in preparation for steam trials. There was almost no coal in her bunkers, only water in three boilers and she was in many respects still incomplete. At this point, the total weight of the vessel was about 4,550 tons, including 3,264 tons of hull. No inspection or determination of the final draught and trim by an expert official took place shortly before the flooding out began. At 0545, just as the ship was about to float aft, the fore part being already well afloat (the difference in draught fore and aft was about three feet), the keel slid from the after block to starboard, the shores dropped and the ship heeled to about 7 or 8 degrees to port, apparently stopping there.

Lieutenant HB (later Admiral Sir Henry) Pelly, one of a party of officers and men sent to bring the new yacht to Portsmouth, was sleeping on board. 'The first I knew of it was gently slipping out of my bunk while everything in my cabin took charge and went crashing to leeward.'[10] As soon as she began to list the Marine sentry sounded the bugle call and all ports and scuttles were closed; this was just in time as it was soon apparent that the heel was increasing further. The heel continued until the coaling ports and lower deck circular scuttles were submerged, the water reaching the top of the rigols amidships. The ship moved bodily to starboard, her bilge up against the starboard side of the dock.

The time was then about 0615 and the list about 20 degrees. When the starboard docking guys were slackened to relieve strain, the list increased to about 24 degrees and the guys had to be tightened again; it was about high water. It was then considered unsafe to disturb the ship in any way, and to retain her afloat in the dock. However, the caisson could not be closed properly as the ship's stern was resting against it, tilting its lower part away from the water-stop. Every effort was made to supply water from the dockyard main, a fire tender and the tug *Alligator*, but the level fell gradually, although enough was retained to help support the ship's bilge. The ship grounded, supported by stout shores, with the starboard bilge on the side of the dock and the port bilge on the middle line of

HMY *Victoria and Albert* (III) upright in the Pembroke dock in which she almost capsized. (© National Maritime Museum, Greenwich, London, C7391/A)

dock-blocks. As found in the subsequent docking in Portsmouth, the bottom on either side amidships was indented to a depth of about eight inches over a length of about 24 feet. A considerable crushing in of the outer bottom occurred but the inner bottom was only slightly injured. There was no significant free water within the hull. The engines were entirely undamaged and the propellers were uncoupled at the time of the accident.[11]

As the ship continued quite watertight, it was decided to ballast her and refloat her on the evening tide. Two hundred tons of water ballast and 105 tons of pig iron paving blocks from the dockyard were added, distributed as low in the ship and as symmetrically as possible. All double bottom compartments between stations 26 and 51 were filled. The work was not complete in time for the evening tide, but she was refloated successfully on the next morning's tide,[12] still with a list of 10–11 degrees. In that state she was towed out of the dock by tugs and moored to a buoy in the harbour.

Ramifications and Remedies

The yacht lay with a list to port of 10–11 degrees, with the 305 tons of added ballast on board and a mean draught of about 18ft 4in (16ft 11in forward, 19ft 9in aft) corresponding to an approximate displacement of 4,900 tons. From the later enquiry, there was a very small negative metacentric height.[13] Lt Pelly takes up the story. 'Coal lighters were brought alongside and I was left in charge with orders to get the ship on an even keel. That seemed easy enough, but to my horror, when she reached the upright, she started to heel over in the opposite direction. For some seconds it was most alarming, and I was more than thankful when she steadied at about ten degrees.'[14]

White, who had arrived from London during the night to survey the chaos, used the weight of 475 men moving across the ship in batches to determine her stability. Using a 13ft plumb line, and taking 100 men to weigh seven tons, the GM was recorded as 11in to 14in, according to different observations. She was then brought alongside, plant on deck removed and the list corrected by placing about 33 tons of coal in a bunker. Lt Pelly once more: 'Coal lighters had to be shifted round to the other side, and further operations got the yacht on an even keel at last.'[15] Approximate weights recorded out of the ship to 16 January totalled 75 tons 14.5cwt.

The Controller, Rear Admiral AK Wilson, decided (9 January 1900) on a 'complete investigation of the stability and weights on board the new Royal Yacht before the vessel leaves Pembroke.' Chief Constructor WH Whiting was appointed to undertake this work.

Before the vessel left Pembroke, a second inclining experiment took place; draught was 18ft 8in fwd, 20ft 3in aft, mean 19ft 5.5in, displacement 5,314 tons.

Twenty tons of ballast on the upper deck were used with a shift of 36ft and two 15ft pendulums. The mean of four readings gave a GM of 1.95ft.

The vessel left Pembroke on 18 January with Wilson and White aboard. She had 270 tons of coal (120 tons for the journey and 150 tons margin) and about 550 tons of water and iron ballast on board with mean draught 19ft 6in instead of the designed 18ft.

She reached Portsmouth on the 19th. The masts were unshipped to be shortened, and the ship was docked on 23 January, when the hull was examined and it was decided to replace rather than repair the dented plates. Steps were taken to further lighten the ship. Forty-nine tons of silicate cotton were removed; experiments showed that some employed over the engine rooms and on funnel casings to keep the vessel cool could be dispensed with, without interfering with the comfort of the ship and in fact enabling the ventilation of the lower deck to be improved. Substituting wooden shutters for the brass shutters fitted on the pavilion windows saved about five tons, but the brass cabin side-lights were retained. The funnels were shortened. Some wood removed was used to make dummy crew members for the *Belleisle* experiments. Warings, Broadfoot and other contractors were quizzed about the weights added during the Pembroke docking and changes to be made at Portsmouth. About 150 tons of permanent iron ballast would be required.

Interventions continued during the modifications. Fullerton to Wilson (18 February 1900): '[I am told] that you are going to remove the Steel deck under the Bridge as far as the Bulkheads of the Sick Bay and Smoking Room. It *does* seem a pity I think not to make a *job* (underlined twice) of it, i.e. Remove steel bulkheads, cut all the deck away, substitute wooden bulkheads .... It will increase the expense doubtless, but it is *worth it*, and the yacht has to run many years ....'

The yacht was 400 tons overweight in a number of inclining experiments made shortly before sea trials:

a) vessel fully equipped with all stores, 650t coal, 70t fresh water and with 125t permanent ballast in bottom: mean draught 19ft 9in, displacement 5,400t, GM 2.4ft;
b) in legend condition, 350t coal, 70t fresh water and 125t permanent ballast: mean draught 18ft 11in, displacement 5,100t, GM 2.2ft;
c) in light condition, *ie* all coal, provisions, fresh water, reserve feed water and half WOs' and Engineers' Stores consumed and with 125t permanent ballast: mean draught 17ft 6in, displacement 4,585t, GM 1.3ft. and the range of stability over 90 degrees.[16]

She was undocked on 16 July and four days later received new wooden masts, leaving Portsmouth for speed trials on 9 August. She left Plymouth on the 28th for full-power trials, then returned to Portsmouth, carrying out circle, anchor, stopping and starting trials on the way. White reported that after sea trials the excess of draught was 11in, corresponding to an increase of 340 tons displacement. In the load (legend) condition the metacentric height was 2.4ft, when fully laden about 2ft 9in.

Queen Victoria died in January 1901, and the yacht had to be further altered to meet the wishes of King Edward VII. Between 6 May and 15 June 1901 she was again docked at Portsmouth.

Victoria and Albert (III) was commissioned at Portsmouth on 23 July 1901. The Statement of Stability stated that the mean draught in the load condition would be 9.5in greater than originally intended, the increase including the effect of the 125 tons of permanent ballast, and that the ship would have almost exactly the trim by the stern as designed; the corresponding metacentric height would be 2ft 9in, angle of maximum stability about 64 degrees, angle at which stability vanished entirely about 128 degrees. Up to 250 tons of water ballast could be admitted if the Commanding Officer

HMY *Victoria and Albert* (III) at Pembroke. The ventilators immediately abaft the mainmast were not seen on the draughts. (Author's collection)

Table 12: **Excess of Weights**

Errors and deficiencies in directly calculable items	72t
Errors in 'judgment' items (not directly calculable)	436t
Additions authorised during construction including lengthened masts	103t
Additional weights worked in at DY without authority	50t
Additions undefinable under any of above headings	110t
Total	771 tons

Table 13: **Weights worked in/taken out July 1899 – December 1900**

Dates	Weights worked in	Weights taken out
01.07.99 – 30.09.99	234.3t	[none recorded]
01.10.99 – 31.12.99	450t	[none recorded]
01.01.00 – 31.03.00	42.95t	390.089t
01.04.00 – 30.06.00	63.161t	89.528t
01.07.00 – 30.09.00	193.46t	1.9t
01.10.00 – 31.12.00	22.356t	35.818t

judged stiffness needed increasing. Even in extreme light condition the metacentric height would be 1.7ft, with corresponding angles of about 64 and 115 degrees.

The new yacht was used by Royalty for the first time on 9 August, when the King and Queen crossed to Flushing (Vlissingen) for the funeral of the Empress Frederick in Berlin. The Commanding Officer's report on the subsequent trials voyage to Gibraltar without Royalty noted that there was absolutely no hesitation at the end of each roll; the ship was 'extremely lively, perhaps a little too much so from the point of view of people with delicate interiors … There was an almost complete absence of vibrations from the engines.'

Stability and the Blame Game

As calculated by Whiting after the accident, had the ship been completed to the original proposals and with all additional fittings subsequently contemplated, the excess of weight would have been 771 tons (Table 12). The directly calculable items in the Table were those comprising the general structure of the hull, the 'judgment' items those estimated by judgment from similar ships and consisting of fittings, furniture, decoration and ornamentation, panelling etc. The 'judgment' items were the principal source of error and were concentrated in the upper part of the ship. The weights added and removed from the ship over quarterly intervals are listed in Table 13. Between 1 July 1899 and 1 January 1900, 121.55 tons of paint, varnish etc were worked in.

The total loss of metacentric height from accepted conditions would have been six inches, but the effect of all the additions would be to reduce the metacentric height to three inches at 5,696 tons (with 650 tons of coal) and 20ft 5.5in draught (a metacentric height of two inches at 5,686 tons is also listed). With 350 tons of coal, there was no initial stability – the metacentric height was minus two inches, and in the light condition she would have been 'largely unstable.' The estimated weights in this condition are listed in Table 3(b).

To the layman, the ship as floated out on 3 January 1900 was top-heavy. For naval architects, the cause of the accident became all too apparent. The unexpectedly high centre of gravity caused a negative metacentric height and the absence of a righting moment up to the 11 degrees noted when removed from the dock; the ship was unstable when upright and 'lolled' (heeled over) to this angle on either side. Above 11 degrees the metacentric height became positive, and the ship behaved more

normally, although with a low righting moment and slow roll, making her 'tender.' The phenomena of lolling and tenderness were familiar for lightly loaded or unloaded merchant ships, which compensate by taking on water ballast to lower the centre of gravity,[17] but highly undesirable in a Royal yacht. The detailed reasons and responsibilities for the high centre of gravity were the subject of intense investigation and debate.

Oscar Parkes[18] blamed WE Smith,[19] DK Brown[20] blamed WJ Luke[21] – seemingly unjustly as he had left the Admiralty well before the accident. Manning discussed White's responsibilities as technical line manager in his department, and considered that Fisher, who as Controller had discussed with White and Fullerton 'the plans and proposals … to be embodied in the design,' must be assigned 'a certain proportion of responsibility.' 'But White was left to bear the whole responsibility … His failure was really the failure of the existing system.' If a Royal personage suggested improvements or additions, obsequious officials would press for their adoption, and such suggestions were difficult to oppose. In those days 'influence' counted for much in the lives and careers of naval men.[22]

Strikes and lockouts in the Engineering industry (July 1897 – February 1898) and consequent delays in shipbuilding had greatly increased the pressure of work on White and his subordinates. Inquests were inevitable; from 25 October 1900 to June 1901 a Committee headed by the First Lord inquired into the causes of the accident, while from early in 1901 a second Committee chaired by Parliamentary Secretary HO Arnold-Forster considered the arrears in shipbuilding; White was a member of this Committee but in a sense also on trial before it. The Fane[23] Committee (September 1901) considered the requirements of White's Department and concluded that the whole Department was overworked, understaffed and in part underpaid.

The Earl of Selborne, First Lord from November 1900, summarised the situation (House of Lords, February 1901). 'There is nothing new to be said about [the accident]; the whole of the facts are well known. It is a matter of great anxiety and concern, to nobody more than Sir William White. No man less deserved that such an accident should have occurred in the middle of his great career of public service …. The services he has rendered to this country can only be described in the words "very great" (Hear, Hear), and it is indeed unfortunate that, out of the 220 ships which up to the beginning of last

HMY *Victoria and Albert* (III) at Portsmouth in 1937. (World Ship Society)

year had been designed, the only one in which any error in calculation existed was the one most calculated to draw public attention to it.' (Hear, Hear.)

Cost

The 1897–98 Estimates, introduced in March 1897, earmarked £75,000 for the new Royal yacht in the forthcoming Financial Year. In July 1897 First Lord Goschen gave the figure of £250,000 as a non-binding general estimate. For a breakdown of costs at the 'As Designed' stage, see Table 14. The total estimated cost reported in the 1898–99 Estimates was £237,000, the figure rising to £354,000 in 1899–1900 (final estimate) and after the accident to £434,582 in 1900–01.[24] According to the Particulars of Cost dated 12 November 1898, the total estimate, exclusive of the main machinery and the special contracts for Royal and State apartments etc, was £193,448. The cost of the repairs and alterations up to 1 December 1900 was £22,381. The total net cost to date was announced to the Commons in March 1901 as £433,637, the estimated amount to complete the yacht £32,146; to these amounts had to be added £46,251, the estimated proportion of incidental changes upon the entire work, for a total of £512,034. Comparable costs were reported as £220,000 for *Hohenzollern* (II) and £350,000 for *Shtandart*.

Concluding Notes

Subsequent to the accident, very great care was taken in recording weights added. In January 1904, the Secretary of the Admiralty, in approving the supply of extra chairs

for the Royal dining room, stipulated that reports of the additional weights involved and of their centre of gravity, in this and in future additions, be forwarded. As a result, a later list of items added included entries such as 'coffee spoons, weight 1.25lb.'

Immediately before the outbreak of the First World War, the captain, executive officer and most of the crew of the *Victoria and Albert* were transferred to the battleship *Agincourt* for the duration. They returned to the Royal yacht after the war, during which she was laid up with a small care and maintenance party. Postwar proposals for reboilering and/or conversion to oil fuel were not accepted on grounds of expense. An accommodation ship in the Second World War, she was broken up in 1954–55 at Faslane.

Table 14: Estimated cost of Royal Yacht
Information as of 15 October 1897.

	Labour	Material	Sub-total
Hull incl masts, rigging, capstan	£95,000	£65,000	£160,000
Propelling and aux machinery[1]			£85,000
Steam boats			£2,500
First stores fitting			£2,500
Special for decoration			£15,000
Total			£265,000

Notes:

[1] A pencilled note against this entry reads '+20,000?'.

The actual cost of Waring's contract was £28,080 15s. and the total weight of their work 42 tons.
The actual cost of Broadfoot's plumbing was £4,897 and the total weight of their work 26t 18cwt 0qr 4lb.

The next Royal yacht, the *Alexandra* (1907–22, sold 1925) was designed by Watts, put out to tender and built by Inglis of Glasgow, a commercial yard. A *Victoria and Albert* (III) replacement that could be used as a hospital ship in a future war was included in the 1939 Naval Estimates but did not progress because of events. The construction of HMY *Britannia*, the eventual replacement, would be under the close supervision of the DNC.

White had a nervous breakdown from the shock of the accident and years of overwork, and was on sick leave from early March to July 1900. After the First Lord's Committee concluded its deliberations, he was formally reprimanded, and in December 1901 he sent in his letter asking for retirement on grounds of ill-health. White's last day as DNC was 31 January 1902. Sir William died in February 1913; neither *The Times* obituary, the *Trans RINA* obituary nor his entry in the *Dictionary of National Biography* mentions the accident to the Royal yacht.

Acknowledgements:

Grateful thanks to Jeremy Michell, Andrew Choong, Bob Todd, Scarlet Faro and colleagues in the Brass Foundry outstation of the National Maritime Museum, and to Brian Hargreaves of the World Ship Society for photographs.

Sources:

National Maritime Museum
Three volumes of Covers (Nos 150-150B) and three sets of plans, 'As Proposed' (May 1897), 'As Designed' (Oct-Nov 1897) and 'As Fitted' (1901/1904).

The National Archives
– Ships' Books (Series II) ADM 136/20 for *Victoria and Albert* (III) to 1939, and ADM 116/6009 for 1945–55; ADM 136/1 for *Victoria and Albert* (II) to 1875.
– Four photos of Victoria and Albert (III) in ADM 176/753 (two) and 176/1123 (two).

Books
Agar, Augustus, *Footprints in the Sea*, Evans (London 1959).
Gavin, CM, *Royal Yachts*, Rich & Cowan (London 1932).
Manning, Frederic, *Life of Sir William White*, John Murray (London 1923).
Pelly, Henry, *300,000 Sea Miles: An Autobiography*, Chatto & Windus (London 1938).
Phillips, Lawrie, *Pembroke Dockyard and the Old Navy*, History Press (Stroud 2014).

Periodicals
Engineering, Vol LXVII (Jan–June 1899), Vol LXX (July–Dec 1900).
Navy and Army Illustrated, Vol VII, 1898–99 and Vol VIII, 1899.
Files of *The Times*.

Endnotes:

1 Fullerton commanded the Royal yachts from 1884 to 1901 as Captain and from 1893 as Rear Admiral.
2 Fisher, Third Naval Lord and Controller 1892–97, was related to Fullerton by marriage. He was succeeded by Rear Admiral AK Wilson (appointed 16 July 1897).
3 King Edward commissioned HW Cox to install an X-Ray machine in the yacht. As Prince of Wales the King had studied Cox's earlier X-Ray machine fitted in the Second South African War hospital ship *Victoria*.
4 Abdul Karim, the Queen's Indian Secretary, known as the 'Munshi' (teacher).
5 Other European navies did not make serious attempts to construct Royal or Imperial yachts that could double as auxiliary cruisers, although nominal wartime armaments are found in official lists. A few warships were fitted or refitted as occasional yachts (Russian *Svetlana*, White's *Renown*).
6 The approved summary of weight of the allowed armament in *Victoria and Albert* (II) was 4t 17cwt 0qrs 22lb, and weaponry included 21 5.5in SB mortars for firing fireworks, 100 magazine rifles with bayonets, 19 Webley pistols and 60 naval swords.
7 Dockyard practice was to calculate bunkerage by volume for North country coal at 43 cubic feet per ton.
8 When shown the model of the new yacht, the Queen's only request was for plenty of headroom so that she would not bump her head when being carried between decks. White was able to reassure her that the lift would remove the need for carrying her between decks. The lift cost £225. The same chair was used for both carrying and wheeling, but the Queen was only carried when passing over uneven surfaces. The chair was 3ft 1in front to back, 2ft 8in across outside wheels, 3ft 2.5in height of back. When being wheeled an additional 18 inches would be required, if being carried an additional 18 inches at each end.
9 The *Victoria and Albert* (II) was decommissioned on 23 July 1901, paid off into Dockyard Reserve on 3 December 1901 and broken up and burned in 1904.
10 Pelly, *300,000 Sea Miles: An Autobiography*, 64.
11 *Marine Engineer and Naval Architect* 1900, Vol 21, 506, 509.
12 Gavin, *Royal Yachts*, 298–9.
13 Manning, *Life of Sir William White*, 421.
14 Pelly, 65.
15 *Ibid*, 65.
16 Manning, 430.
17 PW (Sir Philip Watts), *Encyclopaedia Britannica* 1911, Vol XXIV, article *Shipbuilding*, 927–8.
18 Parkes, *British Battleships 1860–1950*, Seeley Service (London 1957), 347.
19 In October 1911 the Admiralty announced that WE (by then Sir William) Smith, Director of Contracts since 1902, would succeed Sir Philip Watts as DNC on the latter's impending retirement. Smith's appointment was not confirmed by the next First Lord, WS Churchill, who stated in the Commons (18 June 1912) that his services were more usefully employed as Director of Contracts than as DNC. The Admiralty announced on 29 July 1912 that Watts and Smith would retire on 1 August, and be replaced by EH Tennyson d'Eyncourt and WH Whiting respectively.
20 Brown, David K, *A Century of Naval Construction*, Conway (London 1983), 76.
21 Luke left Admiralty service in mid-1898 and had a distinguished career with John Brown.
22 Roskill, *Admiral of the Fleet Earl Beatty*, Collins (London 1980), 24.
23 Vice Admiral CG Fane, Admiral Superintendent of Portsmouth Dockyard 1892–96.
24 House of Commons 14 June 1901: Debate on Navy Estimates 1901–2.

POSTWAR SONAR SYSTEMS IN THE ROYAL NAVY

Peter Marland follows his previous articles on RN weapons and command systems with a study of the postwar RN development of sonar.

This article, the first in a series on postwar RN sensor technology development, focuses on sonar.[1]

Although notionally the oldest of the sensor technologies (compared to radar or electronic warfare), sonar is arguably the most difficult. This is due to the speed of sound in water being approximately one two hundred thousandth of the speed of electro-magnetic waves in air. This is compounded by an underwater environment that bends sound in different directions, driven by the thermal gradients, and includes a significant amount of background noise, plus reverberation (the 'backscatter' from the transmitted sonar pulse).

There are two fundamental mechanisms: active sonar, whereby 'pings' are transmitted by the searching unit, and received after being reflected from a submarine target (akin to radar), and passive sonar, which relies on machinery noise from the target reaching a receiver in the searching ship or towed behind it (akin to EW). Additional detail is provided in Annex A.

Note that in the immediate postwar period the RN still used the prewar ASDIC designation, but later adopted 'sonar'; both forms exist in parallel in documentation. For consistency, sonar type numbers have been used throughout.

The Situation After 1945

At the end of the Second World War, the RN had good searchlight sonars[2] and ahead- throwing weapons such as Squid, sufficient to counter traditional diesel-powered U-boats. In addition to these equipments, overall ASW capability included critical contributions from intelligence (ULTRA), and from high-frequency direction finding (HF/DF), plus air support from escort carriers and land-based long range maritime patrol (LRMP) aircraft. It is much less clear that this ensemble could cope with the large Type XXI fast battery U-boat or the smaller Type XXVI running on hydrogen peroxide, both using a schnorkel to extend their time at periscope depth, that were entering service with the *Kriegsmarine* at the end of the war. The real fear was that this technology would swiftly transfer across to the Soviet Union.

This was the impetus to the first technology push to move beyond Sonar 170/Mortar Mk 10 (Limbo) through to Sonar 177, which was a fifty-fold improvement on the wartime searchlight sonar (x8 bearing and x6 times range

coverage). This provided the core RN ASW capability through the 1950s, in destroyer conversions such as the Type 15 and the new-construction frigates of the Type 12, 14 and 81 classes. Search tactics were based on groups of ships, formed as a Search Attack Unit (SAU), able to cover the Limiting Lines of Submerged Approach (LLSA) – a limited sector when facing diesel submarines equivalent to the UK *Porpoise* or the Soviet 'Whiskey' classes.

This whole strategy was outflanked by the early nuclear attack submarines (SSN), which were able to use their dived speed to approach a defended main body or convoy from any direction, including directly astern. The RN's ASW capability then stalled: Sonar 184 was in theory able to scan all round, for greater volumetric coverage, but had a much-reduced effective range. This was followed by a very slow recovery via Sonars 2016 and 2050.

The Royal Navy tried Variable Depth Sonar (VDS) but did not persist after Sonar 199 (unlike Canada and France). However, it made a real success of Towed Array (TA). Through to the late 1980s, the RN was the pre-eminent anti-submarine force, tackling relatively noisy Soviet SSNs in the North Atlantic and acting as the ASW screen commander for the NATO and US Navy striking fleets.

During the postwar period, all aspects of ASW were the responsibility of the Torpedo and Anti-Submarine (TAS) branch, with officers, senior rate instructors, and more junior underwater control (sonar operator) or 'quarters' (weapon crew) ratings. However, from the 1970s, the weapons 'quarters' progressively became the responsibility of the Weapon Engineer Officer (WEO), while the sonar operators (and WE personnel in outstations) were directed by the Principal Warfare Officer (Underwater) [PWO(U)] from the operations room.

Surface Ship Sonar

At the end of the Second World War, the RN was left with a considerable legacy of sonar and underwater weapons equipment for anti-submarine (AS) purposes; depth charges, Hedgehog and Squid (AS Mortar Mk 4). Land-based aircraft had also adopted the early American anti-submarine homing torpedo (Mk 24 Fido), while the Germans had developed several sophisticated submarine weapons such as the Gnat and T5.

Fig 0: Hull Outfit 15 in HMS *Blake*. This is a classical 1950/60s 'staybrite' single curvature dome, prior to the introduction of two-curvature GRP sonar domes (Courtesy of Conrad Waters)

The wartime sonars were the searchlight Type 147 with the Q2 deep target attachment, and Type 164, both working at relatively high frequencies (14–50kHz) and with 5-degree step search patterns. These used a combination of manual training of the transducer, aural signal detection, and paper range and bearing recorders which marked on moist paper, like contemporary echo sounders. One area in which the UK led the US was the streamlined

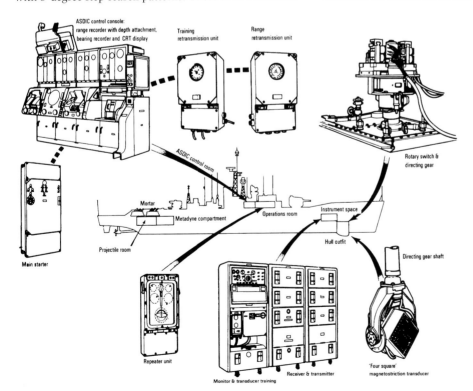

Fig 1: Sonar 170 installation. (Hackman, *Seek & Strike*)

'staybrite' steel domes and retractable hull outfits (HO) that allowed the sonar to be used effectively at relatively higher search speeds than the equivalent US spherical rubber domes. RN ships also used towed pipe noise-makers such as Unifoxer to counter homing torpedoes.

Hull Mounted Sonars

Sonar 170

This was the first significant postwar development. The associated Mortar Control System 10 (MCS10) carried out fire control calculations, calculating trajectory bearing, range and depth to be set on the projectiles. Sonar 170 used a 'four square' transducer, with both tilt and train motions which enabled it to track targets in three dimensions, had CRT displays in addition to the paper recorders, and a mechanised search pattern. A separate monitor transducer (in the Hull Outfit 5 'Sword') allowed the performance and beam patterns to be monitored.

Sonar 177M

More significant was the development of the first true medium-range sonar, Type 177M. Compared to the wartime sonars, this moved the operating frequency down to 6–9kHz, achieved sector search up to 40 degrees wide, and with high power increased range out to a maximum of 20,000 yards. The system had a large (1.75/2.00-tonne) trainable transducer (stabilised in bearing) that formed five transmit beams each 10 degrees wide, and four 10-degree receive beams. It used combinations of pulse lengths within the ripple transmission, and also hosted doppler processing. Displays were sector CRT, doppler CRT and a paper range recorder. Standard procedure with 177M was four pings per bearing (4 x 24 seconds at maximum range). Each 40-degree sector therefore took 90+ seconds to search, so it took almost twelve minutes to cover a full 360 degrees.

Sonar 177M was best suited for team searches against a single diesel attack submarine (SSK) where the units deployed had a reasonable idea of the target direction. It became progressively outflanked when the number of ASW units was reduced (as GP frigates replaced both ASW frigates and AA destroyers) and the threat became a nuclear-powered SSN able to approach from almost any direction, rather than being constrained to the LLSA

Fig 2: Sonar 184 console in HMS *Bristol*.
The Sonar Control Room was forward of the Ops Room, and had Doppler, PPI and HE displays. Each display had a desk with rollerballs for trackers. (Peter Marland)

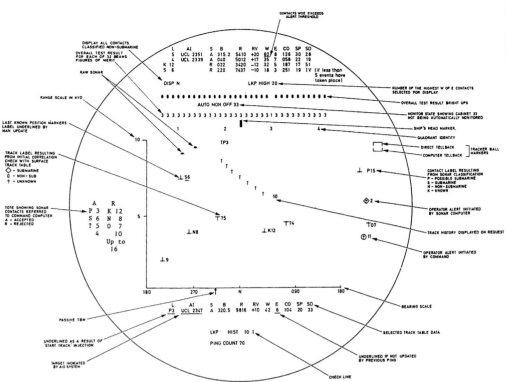

Fig 3: Typical 2016 surveillance display format. (NER Vol 37 No 4)

that confined the diesel SSK to a threat sector. 177M did give better detection ranges that the later 184M, and the author recalls being in a 'Tribal'-class frigate in the early 1970s, tracking an American SSK up and down the entire AUTEC range without losing contact for a whole day's exercise.

Sonar 184

This was initially developed as a torpedo warning system for capital ships, and used a large cylindrical transducer array; it suffered badly from reverberation, and 184M added the frequency-modulated (FM) pulse in ripple or omni transmission modes, displaying the results on CRTs as a circular PPI display, with both doppler and Hydrophone Effect (HE) displays. HE was a passive receiver and display intended to capture machinery noise close in, especially of torpedoes, as a torpedo warning.

Despite similar operational frequencies to 177M, 184 was never a convincing performer, and overall ASW effectiveness languished throughout the 1970s. In later life, 184M had multiple problems with valve electronics, high-speed mechanical commutator wear and analogue doppler filters working at the bounds of their component tolerances, plus self-noise due to extra high tension (EHT) wiring. This relative period of stagnation led to a rolling programme to modernise the system by small steps as 184P→R, modelled on the solid-state Graseby GI750.

Sonar 184 used a cylindrical array 3.5 feet in diameter and 4 feet high, with 32 staves. There were two pulses: a PPI pulse of 45ms which was frequency modulated to limit reverberation, and a longer continuous wave doppler pulse (DD) of 150ms. These were both radiated in an omni mode, or the PPI transmission could be rippled using sequential 11.25-degree beams, with the DD transmission over a five-beam sector of 56 degrees. The HE beam was a passive receiver, covering separate 11.25-degree beams for aural and CRT display. The PPI system formed eight receive beams (184M sets) upgraded to 16 beams (as 184P).

Sonar 2016[3]

This was the result of a 1971 staff requirement[4] that led to early trials in HMS *Matapan* in 1973, and acceptance for the system in Type 22 Batch 1 frigates and Seawolf *Leander*s from 1979–82. The result was a 5.5–6.5–7.5 kHz active sonar with a 2.1m diameter array 1.66m high, with 64 staves each of twelve transducer elements. The initial arrays were hydraulically stabilised in roll within a large HO29 dome, though later fits (and the successor 2050) had a fixed transducer with full electronic stabilisation. The array also contained a near-field monitor rotating around the main transducer, and replacing the HO5 in earlier ships.

The system used a three-bay display console (usually in the ops room, except for Type 42 retrofits), and a total of thirteen cabinets in the sonar instrument compartment. There were three main displays: classification (with an adjunct history display), surveillance, and passive and external data (also used by the sonar controller) – see Figure 3.

The set used a linear-period FM pulse approximately one second in length, in either normal mode (horizontal), or a 7-degree dipped mode for convergence zone (CZ) operation. The FM1600B computer permitted replica correlation processing of received pulses in the active receive function, plus separate passive and audio receivers. The computer processing also supported clustering, echo shape recognition, and ping-to-ping association via weight of evidence. Sonar 2016's large-bandwidth long FM pulse gave good detection performance in both deep and shallow (high reverberation) environments; however, some combinations gave problems of classification and high false alarm rates. In the future, submarine target strengths were projected to reduce; these shortcomings led to the proposal for 2050, against NSR7734.

Sonar 2050[5]

This used the same array as 2016 with additional pulse types, in order to optimise the performance in both reverberation- and noise-limited conditions. The computer detection and tracking processes were essentially derived from 2016, but manual operator classification aids were replaced by a new automatic active classifier developed by DRA Portland. Sonar 2050 was fitted in a new bow dome in Type 23 frigates, and was also retrofitted in place of 2016. It used the same fixed array (electronically stabilised) and operated between 5.4 and 7.75kHz with 2-degree normal and 7-degree depression angles for CZ conditions, giving range scales of 4,000, 8,000, 16,000, 24,000, 48,000 and 64,000 yards.

The system retained the same long FM pulse as 2016, but added a continuous wave (CW) pulse to exploit doppler processing and to improve performance in very high reverberation areas. Long and short FM pulses with the same bandwidth, and long and short CW pulses could be combined in five active operating modes. The 2016 replica correlation only handled a finite dynamic range and required Automatic Gain Control (AGC). In contrast, 2050 is a fully linear system with a wide dynamic range and no AGC; this doubles the reverbera-

Fig 4: Sonar 2050 display console. (JNE Vol 33 No 3)

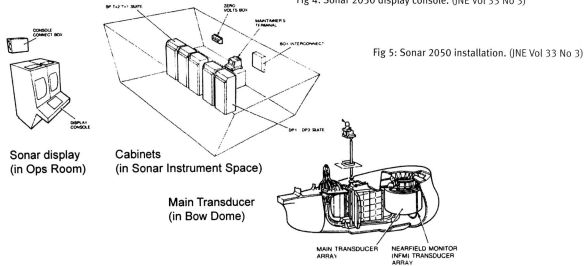

Fig 5: Sonar 2050 installation. (JNE Vol 33 No 3)

Sonar display (in Ops Room)

Cabinets (in Sonar Instrument Space)

Main Transducer (in Bow Dome)

Table 1: Sonar Set Comparisons

Parameter	170M	177M	184M	2016	2050
Frequency	15–30kHz	6–9kHz	6–9kHz	5.5–7.5kHz	5.4–7.75kHz
TX power	2kW	24kW	32kW	44kW	–
RX process	analogue receivers/video + doppler filters			one FM1600B	25 M700
Displays	2 paper 1 CRT	1 paper 2 CRT	3 CRT	3 CRT	2 Raster
Operators		3 each + SC*	3 + SC	2 + SC	1
SIS Cabinets	3	14	15	13	5

* SC = Sonar Controller

tion-limited active ranges. The system has a single operator console position, with dual TV raster displays as shown in Figure 4.

The system uses twenty-five M700 processors (vice the single FM1600B), and the time-domain beam-forming and processing algorithms draw on the same Curtis architecture work used in towed arrays. This reduces the cabinet count from thirteen to five but gives much-enhanced facilities that include oceanography, target motion analysis and wreck charts. Sonar 2050 was granted provisional acceptance on 1 April 1999. Figure 5 shows the installation, whilst Table 1 summarises the development trajectory.

AD Waite was a key contributor to this sequence,[6] but his conclusion as sonar 2050 was entering service was that:

> The detection ranges surface ship active sonars achieve in *Controlled Trials* have not significantly changed over 40 years … The *Operational ranges* achieved by the early sonars were disappointing – typically only one third of the expected detection ranges. The introduction of computer assisted detection and tracking and improved displays has much improved operational performance … together with the good performance of the wideband FM pulse ….. are responsible for the consistently good performance of modern duct sonars.

Other Approaches

The UK has never attempted the brute-force approach of the American AN/SQS-26 or -53 series of hull-mounted sonars. SQS-26 operated on 3.5kHz (active) using an array 16 feet (4.88m) in diameter and 5ft 9in (1.75m) high, weighing 30 tonnes. This had 72 staves and 8 elements, with beam steering/depression for bottom bounce (BB) and CZ modes and source levels of 233dB or 226dB.

Australia developed the Mulloka sonar as a special case, using a higher frequency (18kHz) and a 96-stave array to give narrow vertical beams that were less susceptible to reverberation in tropical waters. The system was trialled in 1975 and fitted from 1979, but pre-dated large-scale replica correlation by computer. Mulloka gave better coverage than traditional LF sonars, out to a guaranteed 10,000-yard range to dissuade torpedo firers. The other case is the French Spherion sonar with a spherical array that eased the computational load of beam steering and mimicked the much larger US SSN sonars. Spherion is not especially high-powered, but is fitted in corvettes and light frigates as a combined search and attack set.

Given the so called 'afternoon effect', when a warmer sea surface deflected sonar beams downwards, or surface ducting captured the sonar energy into a shallow surface layer, there was much interest in Variable Depth Sonar (VDS). This would have the ability to position the transducer deeper in the water column to look below thermal layers. The UK trialled a large VDS similar in size to 177M (2 tonnes) as Beta or 192. Ship studies show the

body retracting into a keel recess, not being brought aboard via an overside rig. Despite solving many of the technical problems, the UK eventually adopted the much smaller Canadian-developed CAST-1 VDS as type 199, which was fitted in *Leander*s and some 'Tribals'.

By way of international comparison, both Canada and France persisted with large VDS, with similar size arrays in the towed body to those fitted in the hull-mounted sonar fed by a single integrated transmitter, and a suite of displays to achieve good shadow zone coverage (French DUBV-23 and -43, and Canadian SQS-505). The rationale for the French development was to cope with the thermal layers in the shallow waters of the Mediterranean. The US Navy went further with its AN/SQS-89 integrated sonar and display suite, joining hull-mounted sonar, towed array and the SH-60 LAMPS helicopter (equipped with sonobuoys) via downlink, with a common display family.

Sonar Trials

RN sonar development has made use of a series of trials ships to research new technology, or try out early model (ie laboratory standard) sonars. Examples include:

– HMS *Brocklesby*: trials of an abortive large VDS Beta, or 192.
– HMS *Verulam*: trials of 2001X.
– RDV *Crystal* (afloat in Portland harbour): near-field trials, especially of full-size transducer arrays.
– HMS *Scylla*: trials of 2050, fitted in a portacabin on the forecastle, linked to the existing 2016 array.
– The trawler *Northern Horizon* (taken up from trade): trials of Single Role Minehunter (SRMH) variable depth sonar (2093).

The most complex trials ship was the former 'Battle'-class destroyer HMS *Matapan*, which underwent conversion to support the Phase 1 trials of 2016 and the Phase 2 trials of the Joint US/UK Sonar Programme and planar array.

HMS *Matapan*

The Memorandum of Understanding drafted in 1969 defined joint development of a large, low frequency, very-high-power planar sonar array. This would primarily exploit the bottom-bounce mode, but it would also take advantage of ducts and convergence zones and of passive capabilities exceeding those of existing surface ship systems.

The US provided the array and transmit subsystem and was responsible for the computer-controlled instrumentation centre. The UK provided the receive and stabilisation subsystems and was responsible for the hard-copy and analogue recording facilities. The starboard-facing array was mounted on a skeg under the keel, and three smaller noise measurement arrays facing to port were fitted for experimental purposes.

The conversion involved extensive restructuring; when finally completed, the full-load displacement had

HMS *Matapan* photographed out of Portsmouth following her reconstruction as a sonar trials vessel. (C & S Taylor)

The large bow dome for the prototype Type 2016 sonar is on prominent display in this photo of *Matapan* in dry dock at Portsmouth. (C & S Taylor)

increased from 3,430 tons to approximately 5,000 tons, and the draught from 5.3m (17.5ft) to 7.9m (26ft). The major part of the work was completed in three years from 1970 to 1972. The ship was docked on 4.3m (14ft) blocks, all superstructure was removed and interior equipment stripped out except for the main engines. The reconstruction incorporated an extra deck, compartments for trials equipment, and accommodation for 22 scientific personnel.

The conversion was followed by the Phase 1 duct sonar trials (with 2016). The joint US/UK system was then installed; these trials were originally scheduled to end mid-1976, but an extension to mid-1977 was granted. HMS *Matapan* was de-equipped in 1978 and sent to the breakers in August 1979.

The array, weighing about 100 tons, was mounted on the skeg, tilted downwards 26 degrees from the vertical. The transducer consisted of 2,128 elements arranged in a configuration of fourteen rows by 152 columns, forming a planar surface 26m (85.5ft) long by 2.4m (8ft) high. It was covered by a 32m x 3m (105ft x 10ft) one-piece fibreglass dome. The 100 electronic cabinets were installed in the Upper and Lower Cabinet Spaces and the Instrumentation Centre, and other trials support equipment in the Sonar Operations Room.

The principle sonar parameters were:

- Surface duct, bottom bounce, convergence zone, active or passive.
- Frequency: active band 2.5–3.5kHz, selectable; passive band 0.4–2.0kHz or 2.0–5.0 kHz.
- Range scales: 5,000/10,000/20,000/40,000/80,000 yards.
- Ripple transmission: 1, 2, 4, 8, 16 or 32 pings per

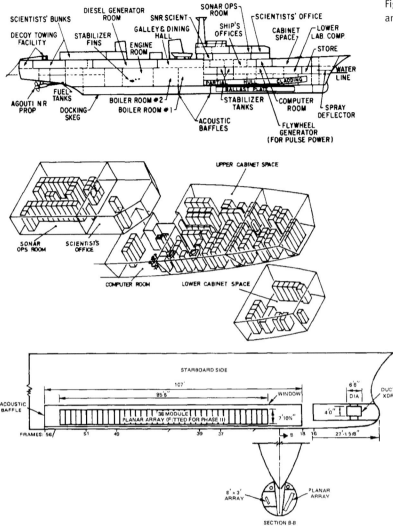

Fig 7: HMS *Matapan*: profile, sonar installation, and skeg array. (JUWAC Vol 31 No 1)

cycle; series of primary pulse codes of Linear Period Modulation (LPM), and CW pulses.

- Transmit: single steerable beam; receive: 2 fans of 16 contiguous beams; each fan independently steerable.
- Coverage: 000–180°G, at 0–45° depression.
- Source levels: beam was either shaded or broadened, using full, half or one quarter of the array. The maximum source level on the narrowest beam (1.2° wide and 13° high) was 257dB re 1μPa at 1m.

This astronomic source level required huge pulses of energy, and the transmitters were powered by two flywheel motor-generator sets with 4.5-ton flywheels. These stored 15MJ of energy at 1800rpm; 6.5MJ could be extracted as pulse power while the machine slowed.

The sonar was controlled by sixteen DEC PDP/11 computers, and the receiver included a FM1600 computer; the sonar beams were electronically stabilised in both azimuth and elevation by a Mk 19 compass.

Figure 7 shows the profile of the ship after its conversion, the scale of the sonar equipment, and the skeg array.

The trials used a variety of consorts: initially repeater targets lowered from a RMAS support vessel, then UK diesel SSKs, and lastly USS *Tullibee* (SSN-597) in 1976, after which *Matapan* visited the submarine's homeport at New London.

The single stabilised transmit beam used coded pulses, with fan arrays of receiver beams, and allowed a wide range of search strategies. Overall, the US/UK planar sonar in *Matapan* did not hugely extend BB ranges compared to the previous benchmark set by SQS-26. However, it lifted the reliable performance from under 50 per cent to well over 80 per cent in much more challenging sea states and environments. It therefore reset expectations of what could be achieved, but at a cost of a seven-fold increase in the equipment required (30-tonne SQS-26 array, up to 200 tonnes for a fully populated planar array, plus 15 electronic cabinets, rising to 100). This was not repeated, and after *Matapan*, the Allies' emphasis shifted towards passive sonar operations.[7]

Towed Array Development

The UK gained initial experience with towed arrays (TA)

Fig 8: Sonar 2031Z: signal processor rack (left) and (right) SDR displays. (IDR/ commercial brochure)

via bi-lateral research exchanges, and the incorporation of US BQR-15 components with a nationally-developed 'wet end' into sonars for SSBNs. The US had long surveillance (SURTASS) and much shorter tactical (TACTAS) arrays,[8] but the UK successfully combined both functions, developing both a 'clip on' array for UK submarines, and 2031 for surface ships.

Sonar 2031 was developed against NSR7638. Prototype 2031X was trialled in HMS *Lowestoft* between 1978 and 1985, using a 1,830-metre tow cable plus 663m five-octave array, with a GEC Marconi special-purpose processor. This was followed by four 'production' 2031I equipments in the *Leander* Batch 2A towed array conversions, and the prototype was upgraded and moved across to *Arethusa* as 2031Y. UK systems were Critical Angle Towed Array Systems, with depth set by speed and length of the tow cable.[9] In contrast, most US towed sonars were Depressed Towed Array System using the SQS-35 VDS body as a depressor, at shorter scope.

Sonar 2031X-Y-I had hard-wired special-purpose signal processors from GEC that required four racks. This was superseded by 2031Z in Type 22 Batch 2/3 and Type 23 frigates. Sonar 2031Z used the novel 'Curtis Architecture' signal processor to shrink the electronics to

a single rack, and a new wider, lower winch, fitted further aft on the quarterdeck. The architecture was the product of the AUWE research programme, and used general-purpose computing with special-purpose digital signal processor chip sets and pipelining under software control[10] to achieve the Fast Fourier Transform (FFT) signal processing to generate the 'waterfall' formats on raster scan displays. Sonar 2031Z reached Fleet Weapon Acceptance in late May 2002. The new architecture radically shrank the volume and cost of the processing hardware. The system had four acoustic data displays in two dual consoles, with three operators, and broadband information was on hard copy recorders at right (see Figure 8).

Sonar 2031E was an enhancement designed to improve the performance prior to the introduction of the low-frequency active 2087 (see below), and provided enhanced Vernier processing with higher update rates, higher resolution and SWATH processing; it is a stand-alone flat panel display, fitted alongside the existing equipment, and is essentially a Curtis private development.[11] The new system superseded most of the legacy signal processing, and required one sixth of the electronics volume compared to its predecessor. Sonar 2031E was introduced from 2001 and was installed as an 'appliqué' to 2031Z.

Fig 9: UK towed array winches: 2031I (left) and (right)2031Z. (JNE Vol 29 No 2)

Signal Processing Algorithms

Passive sonar systems provide broadband, narrowband processing, DEMON analysis, transient detection, plus intercept and recording. Processing is in both time (broadband) and frequency domain (narrowband), the latter usually by Fast Fourier Transform. All the above processes are required simultaneously for target detection, classification and tracking; systems require very high throughputs – typically up to 10^{12} arithmetic operations per second.

TA Operations

The system was capable of multiple CZ ranges in the Norwegian Sea against suitably noisy Soviet submarines. Operations through to the late 1980s frequently used Sound Surveillance System (SOSUS) information to cue contacts for TA ships that were then handed over to SSNs or LRMP for further trailing. Passive TA contacts required a change of heading to resolve bearing ambiguities. In order to resolve potential ranges, a contact evaluation plot or bearings-only plot was used for Target Motion Analysis, based on comparing bearing shift and potential target course and speeds (derived from the blade rate) using '1939' dividers. Despite topweight

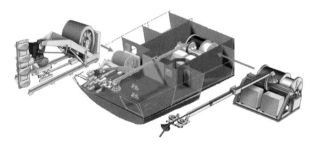

reductions, the drift part of 'sprint & drift' TA operations at slow speed led to severe strain imposed on the *Leander* hulls, with the array/cable loading 'tail wagging the dog', leading to stress cracking, splits and structural failure that necessitated extensive strengthening.

Array and Winch

The 2031 array used 89mm-diameter oil-filled acoustic module sections with a 40mm tow cable, Vibration Isolation Modules plus a rope tail (overall length >2km). The 2031I handling machinery used a special fairlead and a tall drum, but with 2031Z became a wider, flatter drum, where it was fitted below the flight deck in the Type 22 Batch 2/3 and Type 23 frigates (see Figure 8).[12]

Sonar 2038 (and 2057) were intended to use much larger arrays. Later systems proposed a 2.3km acoustic array, eventually reduced to 1,280m. The 2038 array was therefore potentially four times larger than 2031.

In 1988, a senior Ship department spokesman proudly extolled the virtues of the new Type 23 frigate design to the whole RN Staff Course at Greenwich, but was blissfully unaware that the next towed array (2038) was to be three time longer than 2031 and therefore risked inflicting similar damage on the Type 23 frigate as 2031 had caused in the Type 12 and *Leanders*.

Low Frequency Active Sonar (LFAS)

Plans for the 2057 TA and the 2080 LF TX active set were merged as 2087, to be fitted in Type 23 in place of 2031Z. The 2087 active frequencies are 900–2100Hz, using pulse lengths of 1–16 seconds to achieve range scales of up to 144,000 yards.[13] The transmit element (the yellow body) and the passive 'triplet' array are towed from separate winches and cables (see Figure 10), and the

Fig 10: Sonar 2087: quarterdeck winches (above); SD and 4 Operator MFC displays (below left/right). (Thales Captas 4 brochure)

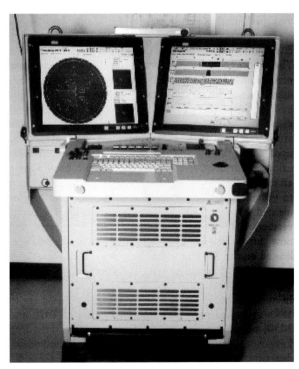

displays include a Sonar Director's position and four multi-functional consoles (MFC) for the Sonar Operators. Early LFAS trials by the NATO SACLANT ASW Research Centre and the French (with SLASM) pointed towards bistatic use, with distributed receivers away from the transmitting ship. 2087 was studied in a bistatic context that required significant inter-unit exchange of ping, part-processed data and contact information, but in the event MoD descoped the SR(S)7590 requirement, and 2087 has been accepted into service as a monostatic capability for force defence, rather than as an area sensor.

Ancillary Sonars

In addition to the main suite covering search and attack, postwar ships also had ancillary sonars such as 176 for torpedo warning/hydrophone effect (HE) when not part of the main search set, bottom contact classification (162), underwater telephone (185 then 2008/9/10), and the bathymetry/environmental package (2013/4/5). Given that ships faced attack by homing weapons, they required decoys; sonar 182 was a towed body noise source against passive homers, with swept tones designed to disrupt active homers.

Minehunting

There were several unsuccessful postwar minehunting sonars intended to find ground mines on the seabed using the small 'Ley' class inshore hunters. The first really successful sonar was the Effie project (193), a very high frequency/high definition sonar that went into 'Ton' class minehunters. This was updated as 193M in the 'Hunt' class, followed by the sophisticated 2093 VDS in *Sandown* and the ultra wideband 2193 sonar retrofitted into the 'Hunt' class.

Environmental Prediction

The profound impact of environmental factors on sonar performance forces ships to measure their local environment and to forecast sonar performance in order to optimise tactics and force dispositions. The initial postwar manually-streamed bathythermograph measured temperature and depth on a smoked glass slide. This was updated as the 2013/14/15 environmental suite, measuring sea surface temperature and the depth profile using an expendable probe, and the acoustic ray trace indicator to visualise the impact on sonar rays. This was supplemented by desk-top calculator programmes for range prediction. This set of applications has been updated, and now forms part of the integrated SEPADS package, hosted on the RN Command Support System.

Calibration

In addition to the HO5 monitor transducer used up to 184M, and the near field monitor as part of 2016 and 2050, other test and calibration facilities included the Ship Launched Underwater Transponder Target, hung under a danbuoy. Ship sonar systems were calibrated at either US or NATO FORACS facilities, and could be evaluated during full-scale trials on the AUTEC range in the Bahamas, or the similar UK BUTEC range at Rona (Scottish Inner Hebrides), both of which allowed for firing of practice weapons against 'padded' target submarines.

Trainers

Operator trainers have AS numbers, ranging from: the AS1070 echo injector for 170, to whole command team trainers such as AS1077 in 177/184 ships with MATCH, and AS1092 for 2016.

Hull Outfit

An important aspect of a sonar set is the hull outfit (HO) used to mount the transducer to the ship or submarine. These are retractable or fixed domes, and there has been a progressive move from 'staybrite' steel to double-curvature fibreglass. The dome surrounds the transducer with still water, and smooths the flow due to the hull's forward motion, thereby avoiding turbulence, cavitation and noise.

Recent Trends

With the end of the Cold War, the First and Second Gulf Wars, and the UK's 'main effort' in Iraq and Afghanistan, the RN's emphasis shifted to anti-air warfare (AAW) and anti-ship missile defence (ASMD) in the Littoral. During the same period, the RN's ASW capability atrophied through a host of savings measures taken against training and expendables, and was also hamstrung by there being virtually no running SSNs to act as targets. Capability is now limited to a few ships with Sonar 2087, and a much-reduced number of Merlin helicopters, plus the promise of P-8 Poseidon maritime patrol aircraft (MPA) from 2019, after an eight-year capability holiday when Nimrod was axed. At one stage Type 45 was merely going to be *fitted for but not with* sonar, but eventually entered service with a 'toy' sonar from EDO.

This article has shown that while growth in the size of hardware (such as transducer arrays) is often the limiting factor when fitting sonar into ships, there has been a parallel reduction in the size of the data processing equipment as electronics have shrunk. Navies have not yet managed to enjoy the 'internet of things' by linking together all the remote sensors such as sonobuoys or bottom arrays that are scattered across the maritime domain, due to limitations on battery life, the cost of expendables, and the real lack of connectivity/bandwidth at sea compared to a similar areas ashore.

Conclusions

In the postwar period, 177M was the first successful medium range sonar, where fast battery-driven SSKs were the main threat. This was outflanked by the first generation of SSNs (USS *Nautilus* onwards), which were able to approach and overtake a task group from astern, completely bypassing the 'forward looking' Limited Lines of Submerged Approach.

Variable depth sonar in the RN (the small 199) was essentially a failure (unlike Canada and France, which continued to develop well-integrated suites of large VDS and hull mounted sonars). This was overtaken by 184M, which had been sold on better volumetric search cover, but at the expense of a pitiful operational range. The RN slowly regained active sonar capability with 2016, then 2050.

The real RN renaissance was with Towed Array sonars such as 2031, able to track and counter Soviet SSNs in the North Atlantic via integrated operations with Allied SSNs and MPA. Therefore, by the end of the Cold War, the RN had a relatively robust and widespread surface

Table 2: Postwar RN Surface Ship Sonars
Missing numbers include systems in helicopters and submarines.

Sonar	Description
147	Depth predictor for Squid, 50kHz, to 1,500 feet; tiltable 'sword' transducer.
162	Bottom-search sonar ('Cockchafer'), 50kHz, 1,200yd slant range. Updated as 162M.
164	Updated 144 searchlight sonar for Squid, 14–22kHz to 2,800 yards; Q2 attachment 38.5kHz 'fan beam' to maintain close contact with deep targets.
170	Short-range attack set for Mortar Mk 10: 15–30kHz, to 3,200 yards.
174	Searchlight sonar, 14kHz to 2,800 yards, 5-degree step angle; in ships with 164 and Q2.
176	Torpedo warning, 12.5 or 15kHz, up to 5,000 yards depending on sea state.
177	Hull-mounted medium range sonar: 6, 7.5 or 9kHz; trainable flat transducer array; five TX beams (50° wide), four RX beams (40° wide); range scales 5–10–20kyd.
181	Towed Acoustic Repeater Target (TART): 10–15kHz repeater for attack sonars.
182	Towed torpedo decoy: 19.5 to 80kHz; body runs at 100ft (5 knots) to 30ft (25 knots).
183	Emergency underwater telephone; 183P with fleet diving teams.
184	Hull-mounted medium-range scanning sonar: PPI 6 or 9kHz, doppler 7.5kHz; cylindrical array, ripple or omni transmissions, doppler and HE. Enhanced to 184P.
185	Underwater telephone, 8.0875kHz CW Morse/9.5kHz TX, 120° sectors or omni.
189	Cavitation indicator.
192	Proposed large VDS (Beta) similar to 177M; trialled but abandoned in 1958.
193	Minehunting sonar in 'Ton' ('Effie'); 100 or 300kHz, to 450 yard range. Updated as 193M in 'Hunt'.
199	CA-developed VDS in Leanders and some 'Tribals': 9.5 or 11kHz (CA uses 12.5 or 14kHz), 450ft cable for 250ft max body dept; range scales 5–10–20kyd.
2008	New underwater telephone: 2008(1) high power with trainable and auto-tracking transducer; (2) is low-power replacement for 185.
2009	Underwater acoustic recognition system (UARS) coded challenge + response.
2010	Acoustic RATT communication teletype (ACUTE) link.
2013	Sea surface temperature recorder (SSTR).
2014	Acoustic Ray Trace Indicator (ARTI).
2015	Expendable bathythermograph (XBT).
2016	New frigate duct sonar: roll stabilised transducer, FM1600B processor.
2028	Proposed half-size version of 2016, not taken forward.
2031	Passive towed array in Lowestoft 2031X, followed by 2031I in TA Leanders and 2031Z in Type 22 B2/B3 and Type 23.
2033	Hydrographer: Simrad Hydrosearch.
2034	Hydrographer: Waverly towed dual sidescan sonar.
2038	Intended replacement for 2031 with larger/longer array, but not taken forward due to relative success of 2031Z with Curtis architecture.
2050	New duct sonar, bow-mounted in Type 23, and back-fitted into some Type 22 and Type 42; same fixed transducer as 2016, full electronic stabilisation and modern digital processing.
2053	Lightweight sidescan sonar.
2056	Portable underwater telephone for Subsunk teams.
2057	New towed array: combined with 2080 as 2087.
2058	Towed source for TA calibration: broadly similar to 182, but software-generated tonals.
2059	Underwater tracker in MCMV: for control of PAP RCMDS vehicles.
2060	Replacement ship XBT/XSV.
2068	SEPADS environmental range prediction suite.
2070	Surface Ship Torpedo Defence (SSTD), replacing 182.
2079	Wide swathe ocean sounding system (SASS) in HMS Scott.
2080	LF Active Sonar (LFAS): merged with 2057 as 2087.
2087	New combined LFAS, with active transmitter in VDS body plus towed RX array.
2090	New XBT recorder.

ASW capability, based on all escorts having 2016 or 2050, plus the airborne Sea King Mk 6 and Merlin Mk 1 helicopters combining dipping sonar and sonobuoy 'sonics' fit.

Post-Cold War, and by the time of the First and Second Gulf Wars the RN's focus had moved towards the Middle East, where there was essentially no (or only a minor) diesel SSK threat. ASW was de-emphasised, with sonar operator billets left unfilled in AAW ships such as Type 42, with the sonar equipment essentially in care & maintenance.

The UK's ASW capability was stripped out because the RN had virtually no running nuclear submarines to act as targets; training was limited to an occasional presence of a NATO SSK in the FOST work-up area, to offset costs via 'burden sharing'; some area capability training was carried out in deeper waters via Joint Maritime Courses. This was compounded by the MPA 'capability holiday', and Merlin ASW effectiveness has also been reduced as numbers to be upgraded were pruned and large ship deck 'spots' were lost.

Capability is not just the equipment, but is crucially dependent on Defence Lines of Development (DLOD). The other aspects that influence operational capability are covered by the TEPIDOIL acronym: Training, Equipment, Personnel, Information, Doctrine & Concepts, Organisation, Infrastructure, and Logistics. With ASW there is simply no substitute for: 'practice – practice – practice'. ASW capability cannot be re-grown by a few sessions on a generic computer-based trainer in a classroom ashore; it requires time at sea, with fuel, consorts, aircraft/helicopters, expendables, and a target; it could take a <u>very</u> long time to restore the RN's former pre-eminence at ASW. There has been a slight re-awakening of interest in countering SSKs with Sonar 2087, but the UK remains very vulnerable to a resurgent Russian nuclear submarine capability, or to long-range submarine-launched weapons wielded by third-world SSKs.

This has been a typical roller-coaster ride, noted in previous articles. The key themes have been some outstanding research and development, led by key individuals, and a free interchange between the RN, scientists and the Procurement Executive (PE), but this has now ossified, with narrow silo procurement, rule-based acquisition, not carrying 'Lessons Learned' forward, stop-go, and outsourcing. UK plc is collectively on the verge of abandoning its former position of being an intelligent customer.

Annex: Sonar Parameters

The classic text on acoustics is RJ Urick's *Principles of Underwater Sound for Engineers* (1967), or the more recent *Sonar for Practising Engineers* (2002) by AD Waite.

Like radar, the performance of sonar is predicted by a range equation that takes account of source level, target signature, transmission losses and detection (recognition) differentials. Unlike radar (where anomalous propagation is not the norm), environmental effects and back-

Table A1: Source Levels

Noise source / Pressure level dB re 1µPa	dB
Faintest audible sound	26
Rustling leaves	36
Quiet suburban street	56
Conversational voice at 12 feet	76
Loud peal of thunder	96
Very heavy traffic	106
Loud car horn at 23 feet	126
Rock music with speakers at 4–6 feet	146
Pneumatic riveter	156
Jet plane at 100 feet	166

* Most noise in air is referenced to 20µPa. The figures here have been re-referenced to 1µPa used for the equivalent sound level in water (by adding +26dB). The faintest audible sound is therefore 26dB higher than the usually quoted figure of 0dB re 1µPa.

Sonar	Transmission	Transmission Source Level dB re 1µPa
177	6kHz	234 sector & ripple
	7.5kHz	234
	9.0kHz	230
184	6.0, 9.0kHz	234 ripple, 224 omni
	7.5kHz DD	all −10dB in low power
195	9.5/10/10.5kHz	227
199	9.5 or 11.0kHz	229 ripple, broadcast
2001	3.25kHz	228 ripple, broadcast
SQS-26CX	3.5kHz	226 omni
		233 sector

ground noise levels have a profound effect on performance. Depending on the temperature & salinity profiles, sound waves can be bent upwards, downwards, or 'ducted', giving a range of 'shadow zones' which are difficult to access.

In the case of active sonar, apart from transmitted pulses reflected from a target, the returning echoes are competing against background noise (due to weather, marine wildlife, or distant machinery), plus reverberation (backscatter from the transmitted pulse). The acoustic environment can also allow Convergence Zone (CZ) or Bottom Bounce (BB)[14] modes for long ranges. Examples of these, with the sonar array positions, are shown in Figure A1.

There are physical limits to the ability to generate sound in the sea, and excess power causes cavitation or boiling at the transducer face. Target Strength (the reflectivity of the target) can be reduced by anechoic tiles. Active sonar may be either noise limited or reverberation limited. Reverberation is somewhat like night driving in fog; more light equals more backscatter, with better visibility from a dipped beam at lower power. Reverberation is also the rationale for Frequency Modulated (FM) 'chirp' pulses, which spreads power density that can be regained in post-processing using correlation. The USN has traditionally had more of a focus on open ocean

Abbreviations

A/S or AS	Anti-Submarine, or (AS) trainer outfit
ASW	Anti-Submarine Warfare
AUWE	Admiralty Underwater Weapons Establishment, at Portland (succeeded by QinetiQ and Dstl at Winfrith)
BB	Bottom Bounce
COSH	Control Ordered Sonar Hardware (technique pioneered by Dr T Curtis)
CRT	Cathode Ray Tube (display)
CW	Continuous Wave (single tone pulse)
CZ	Convergence Zone
DD	Doppler Display pulse
Dstl	Defence Science and Technology Laboratory
EHT	Extra High Tension
FFBNW	Fitted For But Not With (ie fitted to receive equipment at a later date)
FFT	Fast Fourier Transform (software signal processing technique)
FM	Frequency Modulated (pulse)
HF/DF	High Frequency Direction Finding
HO	Hull Outfit (housing sonar transducers)
LFAS	Low Frequency Active Sonar
LLSA	Limiting Lines of Submerged Approach (ASW tactic)
LPFM	Linear Period Frequency Modulated pulse
LPM	Linear Period Modulation
LRMP	Long Range Maritime Patrol Aircraft
MoD	(UK) Ministry of Defence
MPA	Maritime Patrol Aircraft
PE	(MoD) Procurement Executive, succeeded by Defence Equipment & Support
PPI	Plan Position Indicator (display)
PWO(U)	Principal Warfare Officer (Underwater) – successor to TASO
SAU	Search Attack Unit (detached group of escorts to prosecute contact)
SCR/SDR	Sonar Control Room (active sonar)/Sonar Display Room (passive sonar)
SOSUS	(US) Sound Surveillance System
SRMH	Single-Role Minehunter (Sandown class)
SSK	Submarine, Hunter Killer (diesel-powered)
SSN	Submarine, Nuclear (attack submarine)
SURTASS	(US) Surveillance Towed Array Sonar System
TA	Towed Array
TACTAS	(US) Tactical Towed Array Sonar
TAS	Torpedo and Anti-Submarine branch or officer (TASO)
ULTRA	UK intelligence product in WW2 by Bletchley Park (succeeded by GCHQ).
VDS	Variable Depth Sonar
WEO	Weapon Engineer Officer

(rather than littoral) ASW, and has tended to emphasise BB and CZ modes in deeper open ocean waters.

Lower frequencies such as machinery noise can propagate for long distances, making passive operations using a towed array worthwhile. Modern processing has also made bi-static configurations possible, again using lower frequency for maximum range.

The speed of sound in sea water is approximately 5,000 feet per second (1,524m/s), or about 1/196,850 of the speed of EM waves used by radar. Target motion leads to a doppler frequency shift in a received Continuous Wave (CW) sonar pulse of 0.7Hz per kHz per knot. As an example, a 7.5kHz signal is shifted by approximately 5Hz/knot.

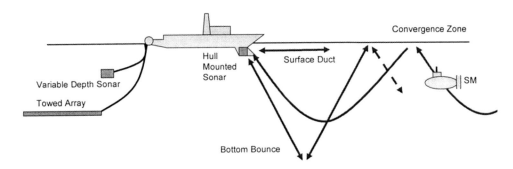

Fig A1: Sonar transmission paths and array positions. (P Marland}

Sound in water is measured as pressure fluctuations, usually converted to sound pressure level (SPL), a logarithmic measure of the mean square acoustic pressure. Active sonar source levels are specified in decibels (dB) with respect to 1 micropascal (µPa) at 1 metre (dB re 1 µPa). Table A1 has comparisons between noisy activities in air, and sonar source levels in water.

There are three differences in interpretation. The scale for acoustic pressure in water differs from that used for sound in air, where the reference pressure level is 20µPa rather than 1µPa; the intensity of an equivalent plane wave makes SPL +61.6dB higher in water;[15] and there are also significant variations in the hearing sensitivity of divers *vs* marine mammals or fish. This would tend to reduce an apparent 230dB sonar transmission to just under 170dB in air, or approximately the same as a jet engine.

Acknowledgements:

Lt-Cdr Clive Kidd RN Rtd in HMS *Collingwood* museum for access to BRs, CBs, Naval Electrical Review (NER) and Journal of the RN Scientific Service/Journal of Naval Science (JRNSS/JNS) articles.

RNE website for Review of Naval Engineering articles.

Major sources:

BR2177. Junior TAS Manual (1965), Admiralty, London.

Hackmann, Willem, *Seek & Strike*, HMSO London (1984).

Kirby, Geof, 'History of the Torpedo', *RNSS Journal*, repeated in *Naval Electrical Review* Vol 26 No 3 to Vol 27 No 2; available online at www.geoffkirby.co.uk/TorpedoHistory 1972.doc. The site also hosts 'The Development of Rocket Propelled Torpedoes' (2000), as rocket-torpedoes.pdf.

Parry, Cdr David, 'The History of British Submarine Sonars', rnsubs.co.uk/articles/development/sonar.html.

Robert Urick, *Principles of Underwater Sound for Engineers*, McGraw-Hill (New York, 1967).

Waite, Ashley, 'Sonars, Reverberation or Noise Limited: An Historical Perspective', *Journal of Naval Science*, Vol 17 No 2, May 1991, 118–126.

Waite, Ashley, 'Sonar 2050', *Journal of Naval Engineering* Vol 33 No 3, May 1992, 653–667; available online via JNE website.

Waite, Ashley, *Sonar for Practising Engineers*, Wiley (Hoboken NJ, 2002).

Endnotes:

1 The best reference to the wartime and early postwar work is Hackmann, *Seek & Strike* (*op cit*), which covers the period to 1954. The other useful source is a series of articles by GJ Kirby, covering the History of the Torpedo – see RNSS Journal, repeated in NER Vol 26 No 3 to Vol 27 No 2, available online at www.geoffkirby.co.uk/TorpedoHistory 1972.doc. The site also hosts 'The Development of Rocket Propelled Torpedoes' (2000), as rocket-torpedoes.pdf.

2 'Searchlight' sonars typically had beams only 5 degrees wide, making it very difficult to search a volume of water at significant range.

3 NER Vol 38 No 4, 26–37.

4 NSR7683 for sonar 2016. See DEFE 10/953 for OR20/73 of 17 April 1973. Original requirement endorsed in 1970; cut from the programme in July 1971, reinstated in July 1972.

5 JNE Vol 33 No 3, 653–667.

6 AD Waite, Sonars, 'Reverberation or Noise Limited – An Historical Perspective', JNS Vol 17 No 2, May 1991, 118–126. The article has detailed background on the 184>184M saga, and spans sonars 177–184–184M–2016–2050, and flags search rate/volume against fast submarines (*ie* SSNs) as the justification for the move away from 177 and towards 184.

7 See US DoD Budget Books, FY73 onwards. TNA files ADM 53/180161 onwards cover the monthly ships' logs from 1972 to 1977, plus ADM 53/193134 for the navigational data book (of ships' handling information). Motion, wetness in head seas, and rolling are reported at ADM 302/5911 and /601 with pictures. The guidance booklet for visiting trials personnel are at ADM 226/1053 and updates at /1073, with drawings. The UK Phase 1 trials results with sonar 2016 are at ADM 302/666, range & bearing accuracy at /627, Phase 1 'clustering' at /780 and submarine echo measurements at /241. The Phase 2 trials of the Joint US/UK sonar programme are covered by JUWAC Vol 31 No 1 dated January 1981.

8 The US surveillance systems were SQR-14 and -15 TASS, which entered service in 1973. The shorter tactical systems were SQR-18 TACTAS from 1980 (with a 223m array on a 1,525m cable), and the -19 from 1983, with a 1,700m cable.

9 For example, 825m deep at full scope and 3 knots, reducing to 20m depth at short scope and 24 knots.

10 TE Curtis, JT Wickenden and AG Constantinides, 'Control Ordered Sonar Software', IEE Proc Vol 131 Part F No 6, Oct 1984, 584–592.

11 See Curtis Technology (UK) Ltd, website http://www.curtis-tech.co.uk/ and Compact Sonar Surveillance Processing Systems paper (csonar.pdf).

12 See article on towed array handling equipment and the creeper hauling machine (JNE, Vol 29 No 2, 329–337).

13 See Thales Captas-4.pdf brochure for 2087, dated 23 July 2012.

14 A convergence zone (CZ) is an annular 'doughnut' of coverage, formed when a sound channel is bent back up towards the surface, typically at 20–40nm range. Bottom Bounce (BB) deliberately depresses the ships transmission, using both the sea bed and then the sea surface as 'mirrors', to give coverage closer-in than a CZ.

15 https://en.wikipedia.org/wiki/Underwater_acoustics] – measurements section.

CHITOSE AND CHIYODA

Authorised as seaplane tenders, the two ships of the *Chitose* class had features incorporated into their design that would enable them to serve as mother ships for midget submarines, as auxiliary fleet tankers, and even as hybrid or through-deck carriers, all with a minimum of dockyard modification. **Hans Lengerer** gives an account of the complex design process and the technical problems encountered.

The extension of tonnage limitations to the cruiser, destroyer and submarine categories at the conclusion of the London Treaty of April 1930 resulted in a major crisis in Japan between the 'fleet faction' and the Government. The reduction in the strength of the Imperial Japanese Navy (IJN) below the 70% ratio when measured against the US Navy was widely viewed as a restriction on Japan's legal rights that aimed to reduce her to the status of a second class power. It was argued that the forces permitted by the Washington and London Treaties would enable the IJN neither to prevent the advance of the US Pacific Fleet into Japanese waters nor to avoid defeat in the 'decisive battle'. The Government was compelled to enforce budgetary restrictions, and the IJN responded by resorting to unconventional measures. One of these measures was the construction of so-called 'concealed' ships; this involved preparations for the conversion, in the event of a national emergency, of a ship type excluded from the treaty restrictions into a category currently forbidden. Among these ships the *Chitose* class were prominent.

Requirements and Design

The *Chitose* class was designed in 1933 as a seaplane tender because, according to Article 8 of the London Treaty, this type did not fall into the 'fighting ship' category; however, this classification was disingenuous, and was chosen to conceal the true purpose of the ships. The design was developed in parallel with a midget submarine intended for a surprise attack on the US fleet immediately prior to the gunnery engagement between the capital ships. It was argued that the midgets would be difficult to locate, and that many of their torpedoes would hit the mark, damaging or even sinking several ships and promoting confusion in the enemy formation.

The design of the midget submarines, designated *Kōhyōteki* ('A' Target), was begun in 1931. They had a high submerged speed, but endurance in both the surfaced and submerged conditions was limited, so they had to be transported to the operational area for launch. Surprise was considered essential for success, so the existence of the midgets and their mother ships had to be

Chitose during her full-power trial off Sadamizaki on 18 July 1938. Note the prominent diesel exhaust emanating from the pipes on the after posts of the flying deck (see also the USNTMtJ section drawing). The combined turbine and diesel propulsion plant was adopted in order to minimise fuel consumption. The two twin 12.7cm HA gun mountings are trained to starboard; two Type 95 floatplanes can be seen above the stern. (Author's collection)

kept secret. For this reason *Chitose* and *Chiyoda* were laid down as seaplane tenders but were designed for rapid conversion to midget submarine carriers, and were also to be usable for other purposes.

The designers had to take into account no fewer than five possible configurations, and this fact alone makes them remarkable ships. They were to be employed as seaplane tenders until immediately before the outbreak of war when, following modification, they would have a dual function as midget and seaplane carriers; they would also be capable of operating as fast fleet tankers, of conversion into hybrid carriers with a landing deck for wheeled aircraft, and of a more radical reconstruction as light fleet carriers. Each of these envisaged conversions was to take no more than a month, including the full conversion to a through-deck carrier – a requirement that proved to be overly optimistic.

The biggest problem faced by the Japanese constructors was the design of equipment and fittings to fulfil multiple functions.[1] In particular, the design of the machinery had to take into account operation not only as an auxiliary warship (AV or AO), with a maximum permitted speed of 20 knots, but as midget submarine carriers and air-capable ships capable of much higher 'fleet' speeds. Even though the failure of such a project had been anticipated by some in the constructors department, a satisfactory solution was eventually evolved, although only after the Naval General Staff (NGS) had agreed to modifications in their requirements.

The construction of *Chitose* and *Chiyoda* was duly authorised as part of the Second Naval Replenishment Programme of 1934. They were classified as No 1 and No 2 'A' transports (*Dai ichi*, *Dai ni kō unsōsen*) at the launch ceremony. Their treatment in the course of drafting this programme and thereafter has been detailed by Fukui Shizuo.[2] A brief outline follows:

- On 14 Jun 1933 the draft of a new construction programme was submitted to the Navy Minister by the Chief of the Naval General Staff (NGS). In the section concerning ships outside the treaty restrictions three seaplane tenders (*Suijōkibokan*) were listed, with a standard displacement of 9,000 tons, a speed of 20 knots and a range of 8,000nm at 16 knots; no armament was specified, but a note stated that this would be determined at a 'special conference'.
- In the explanatory text for the 1934 programme it was

stated: one ship of 9,000 tonnes, 22 knots; two ships of 10,400 tonnes, more than 28 knots. The latter two ships were each to operate 24 seaplanes, but no catapults were be fitted in peacetime; in wartime they were to be armed with four 12.7cm twin HA guns and five 13mm quadruple MG.
- On 23 Nov 1933 the Vice-Chief of the NGS submitted to the Vice-Minister of the Navy the following change in requirements: 'In wartime: *Hyōteki* [midget submarine] carrier; speed with turbines "more than 28 knots" to be amended to "more than 30 knots"'. Shortly thereafter the plans were again changed with regard to displacement and engine power.
- In 1936, when both ships were already under construction, general characteristics for the configuration of 'transports' A [*Chitose* class] and B [*Mizuho*] were drawn up (see accompanying table).

When the NGS worked out the staff requirements it had to consider several treaty restrictions to ensure that the design fell within the 'auxiliary warship' category. One was the comparatively low speed of 20 knots. In order to operate the ships as midget submarine or aircraft carriers a speed of 30 knots or more was considered necessary; on the other hand 20 knots was sufficient for the fleet tanker role, while 30 knots would be uneconomical. The Preliminary Design Section of the NTD proposed the abandonment of the tanker requirement, but the NGS insisted on its retention because (i) the Combined Fleet urgently required fast tankers and (ii) the budget did not allow for the construction of a specialised fleet tanker, nor would this situation change for several years.

This decision rendered the task of the designers more difficult because they had to provide for the installation of more powerful propulsion machinery to deliver the 30+ knots that would be required in the event of conversion to an aircraft carrier. Moreover, the stability investigation that followed the *Tomozuru* Incident in March 1934[3] led to changes in the principal dimensions of the preliminary design: waterline length was reduced by 10 metres, draught increased, and the area exposed to wind pressure was reduced. These changes resulted in a lower maximum speed of 29 knots and a reduction in fuel bunkerage – the figure of 5,000 tonnes originally proposed was already viewed as excessive.

In order to simplify the design, which was executed in close cooperation with the NATD, it was decided (i) to

Table 1: General Characteristics for 'Transports A & B'

	Unsōkan (Kō)	*Unsōkan* (Otsu)
Displacement (trial)	11,500 tonnes	9,400 tonnes
Speed	20 knots	22 knots
Horsepower	16,000bhp	20,000bhp
Fuel	1,500 tonnes	750t + 3,000t (supply)
Endurance	8,000nm at 16kts	8,000nm at 16kts
Armament	2 x II 12.7cm HA	–
	6 x II 25mm MG	–
	24 (+ 6) floatplanes	24 floatplanes

A similar view to the previous photo, this time of *Chiyoda* during her full-power trials on 10 November 1938. (Author's collection)

complete the ships as seaplane tenders, (ii) to utilise them as fast fleet tankers with much reduced fuel for supply, (iii) to investigate the mounting of a landing-on deck above the weather deck at the preliminary design stage and to have the necessary structures in place to facilitate rapid construction, (iv) to provide for propulsion machinery that would be capable of the higher speeds required for use as mother ships for midget submarines, and (v) to prepare for the ships' conversion to light fleet carriers, the work to take three months or more.

These decisions eased the work of the designers. The combination of the midget carrier and seaplane tender roles, with the primary emphasis on the former function, meant that the smaller machinery plant and the associated reduction in heavy oil and diesel stowage allowed the arrangement of several tanks below the midget hangar. These tanks were to serve for to improve stability and for weight compensation when no midgets were carried, but were also important for the launching of the midgets and for the conversion of the ships into hybrid or through-deck carriers. However, the designers recognised that this measure alone would not secure sufficient stability and made further structural modifications.[4] Conversion into an aircraft carrier had to be considered, but because this was now seen as requiring major dockyard work the designers could prioritise other roles.

Irrespective of this retreat from the initial NGS stipulations, there were no other Japanese ships designed under more complex and differentiated requirements, and even before the end of the preliminary design phase a number of problems emerged, some of which are outlined below.

Problems Appearing During the Design
Configuration as Midget Carrier
Although the official designation of the ships as initially conceived was seaplane tender, the design was determined by their primary function as mother ships for midget submarines. The following characteristics had to be studied:

- a crane system for embarking the midget submarines, each of which weighed in excess of 40 tonnes and had a length of *c*24m with a hull diameter of 1.8m
- a transport system within the midget hangar
- facilities for the launch of the midgets.

The after half of the hull below the upper deck was employed as a hangar for midget submarines. Above the forward part of the hangar, the upper deck was reinforced in order to permit the installation of a broad hatch 26m x 6m for embarking the midgets. The hatch cover comprised several sections in order to facilitate opening and closing. Above this hatch was a platform some 40 metres length and almost as broad as the hull, supported at the four corners by thick pillars. This embryonic flight deck (see Hybrid and Light Fleet Carrier sections) served to conceal the dimensions of the hatch and was used to mount light AA guns and searchlight projectors. Each of the four pillars supported a 20-tonne derrick on its inboard side. A midget submarine was lifted by two derricks, swung beneath the platform and embarked in the hangar via the hatch.

On the hangar bottom there were four parallel sets of

Chitose as Seaplane Carrier

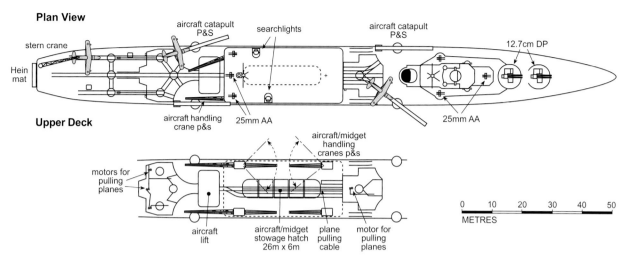

Plan View

Upper Deck

Chiyoda as Midget Carrier

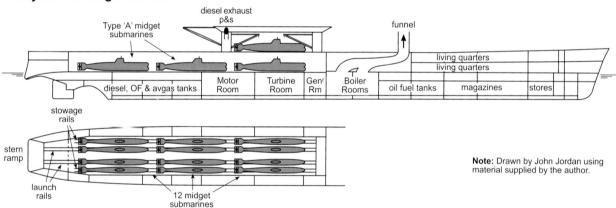

Note: Drawn by John Jordan using material supplied by the author.

The upper drawing has an overhead schematic of *Chitose* as a seaplane carrier, together with a plan of the upper deck showing the embarkation hatch, the aircraft lift, and their associated rails and handling cranes. The lower schematic shows *Chiyoda* as converted to a midget submarine mother ship. The twelve Type 'A' submarines were embarked via the large hatch concealed beneath the partial flight deck and launched via a stern ramp. (Drawn by John Jordan)

rails. Three midgets could be accommodated on each set of rails on articulated four-wheeled carriages.[5] The two inner rails were used for launch, the outer ones for stowage. Between the inner rails there was a deep recess for an endless chain connected to an electric winch located at the forward end of the rails. When the chain was connected to the carriage of a midget it could be moved longitudinally.

The outer rails on both sides ended at the after hangar bulkhead, while the inner rails were continued to the stern. The bulkhead had two openings, each 4 metres high and 2.5 metres wide and closed by watertight doors. From here to the stern the bottom of the hangar, which was about 1m above the waterline, was inclined downwards at an angle of c10 degrees to end below the waterline at the stern. For the launch of the midgets there were two openings in the outer stern plating, each of which was 4 metres high and 3 metres wide and closed by hinged watertight doors.

Inside the hangar batteries could be charged, air flasks filled and torpedoes adjusted and loaded. Each boat had a device for the extraction of battery gases, and overhead cranes were mounted for the loading of torpedoes. Telephone communication between the boats and the bridge of the mother ship was possible up to the time of launch.

The midgets were launched in two groups: the six boats lying on the inner rails first, followed by the six on the outer rails. For this purpose the carriages of the six boats stowed on the outer rails were connected to a steel hawser and were moved using the winch to the inner (launch) rails via a turntable. The connection between carriage and transport chain was automatically detached when the boats passed the inner door and the midget, still on its carriage, ran down the incline by force of gravity. When the midget entered the water the carriage separated and sank to the bottom of the sea.

Because the midget entered the water stern first, the

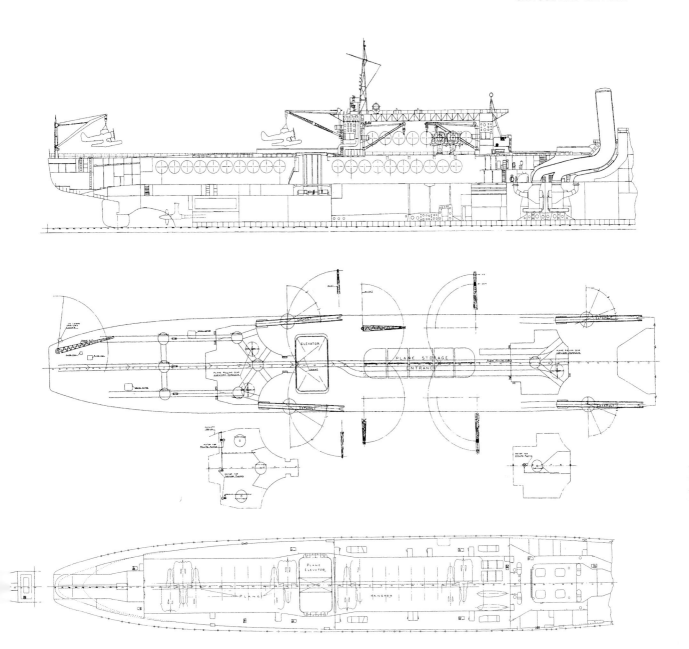

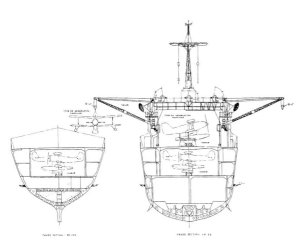

These four sketches were published in the report of the US Naval Technical Mission to Japan. The inboard profile shows stowage for up to 20 floatplanes with wings folded in the lower (auxiliary) hangar and a further eight in the open hangar beneath the flying deck; in reality these ships could operate a maximum of 20 floatplanes in the seaplane carrier configuration. In the plan view note the large hatch for the embarkation of floatplanes (and subsequently midget submarines) and the separate aircraft lift. Note also the system of rails and turntables for the movement of the floatplanes on deck, the four catapults, and the crane at the stern for lifting the aircraft on board after landing. The section through the after superstructure shows the stowage of the planes in the main and auxiliary hangars, the cranes on the support posts for the flying deck, the mainmast, the 25mm AA guns and the exhaust pipes for the diesels. (USNTMtJ)

most favourable angle and speed had to be carefully calculated. The NTRI undertook experiments using small models followed by experiments using a full-scale model; launch trials were then conducted using *Chiyoda*. The trials confirmed the results of the model experiments from which the final configuration had been derived; the launch of the boats met expectations, and the report did not record any serious defect.

It was calculated that a distance of 1,000m was required from the launch of the midget until it surfaced. If the carrier steamed at 20 knots it travelled this distance in 100 seconds, so one midget could be launched every 100 seconds and the full complement of twelve in 20 minutes. This period could be slightly reduced if the carrier ran at a higher speed, had the optimal flooding of the trim tanks (which dictated the angle of launch), and had a well-trained crew. In this case the midgets, whose launch was subject to a control device, did not dive so deep, surfaced more quickly, and the carrier needed less time to cover the planned distance.

When designing the midget submarine carrier the naval architects had placed particular emphasis on the stability of the ship when the stern doors were open and on the launch method of the midgets. The 1-metre height of the hangar deck above the waterline took into account the following factors:

– maintenance and other work on the midgets in heavy seas, causing the carrier to yaw
– the position of the centre of gravity (CG) and its influence upon stability
– the need for a positive metacentric height (GM) during the entire launch process (the movement of the midgets from roughly amidships to the stern was nearly comparable to the effect of free fluids).

Key to the launch method was the automatic compensation system adopted for the fuel tanks. When the midgets were launched, the draught and trim conditions of the carrier changed. Weights in the after part were reduced by about 600 tonnes, and this had to be compensated by the automatic flooding of the tanks beneath the hangar.

In wartime twelve seaplanes were to be transported and handled on the upper deck. The aircraft were to be mounted on rails with crossover points and a turntable. This in effect served to disguise the true purpose of the ships.

Configuration as a Seaplane Carrier

If the ships were employed as seaplane carriers the midget submarine hangar was to become the seaplane hangar, the derricks were to be used for large landing craft (*Daihatsu*) and the ships' boats, and the hatch for the

Taken on 18 July 1938, this photo provides a good view of the high freeboard, the flush deck, the bridge structure and the flying deck. The latter was to be part of the landing deck in the hybrid carrier configuration; it served as a test of structural strength and also to investigate its effect upon stability. When serving as a seaplane carrier the flying deck was used to mount 25mm AA guns and searchlights. (Author's collection)

midgets for the movement of seaplanes between the hangar and the upper deck; the rail system on the upper deck was to be retained in its current configuration. Two cranes with lattice booms were fitted on the after side of the after set of pillars for aircraft handling. A third crane, mounted above the stern to port, was associated with the Hein landing mat that enabled the floatplanes to be recovered with the ship steaming at 10–15 knots; it lifted the floatplane on board and onto the rails set into the upper deck.

As long as the treaty was in effect only two catapults could be mounted, but in wartime two more were to be fitted to permit the catapulting of all 24 seaplanes (twelve carried on the deck plus twelve in the hangar) in 30 minutes.[6] This allowed for a take-off interval of not more than 6 minutes. The designers believed this could be achieved with a well-trained crew and reports indicate that it was obtained, and even exceeded under training conditions.

Because of the previously stated reduction in the length of the hull, the construction of the platform amidships and the need to move the after wall of the hangar comparatively far forward, stowage was insufficient for the 24 floatplanes it was planned to operate; only 20 could be accommodated. The NGS accepted the reduction because the necessary space could have been gained only by an increase in hull dimensions. The reduced area available also prompted the decision to ditch damaged aircraft overboard because there was no space for repair facilities.

The complement differed according to the role of the ship, so the conversion of the living spaces was another aspect to be considered. The air crews and the maintenance personnel for the floatplanes required much more space than the crews of the midgets. The seaplane carrier configuration also required a large air engine maintenance and test room, while this was unnecessary in the case of a midget carrier.

Operation as a Fleet Tanker

Owing to the agreement of the NGS to reduce fuel capacity, the preliminary design aimed not for a specialised fleet tanker but for a seaplane carrier with exceptional fuel stowage capable of supplying other ships. The arrangement of the fuel tanks, piping and pumping had to be compatible with the ships' primary role as midget submarine mother ships and seaplane carriers. The derricks intended to lift the midgets and the floatplanes were to be used for the suspension of the supply hoses, and supply connections, the pump room and stowage for the hoses were carefully designed so that other functions were not impaired.

Employment as a Hybrid Carrier

The installation of a landing deck offered carrier-borne aircraft the opportunity for landing on in the event of damage, wounding of the pilot, lack of fuel or if the mother carrier had opened the range to avoid a counter-attack. The midget submarine carrier role involved oper-ating close to the line of battle, while the fleet carriers were intended to launch their aircraft at extreme range. The availability of a flight deck close to the point of attack offered a good chance of survival for air crew and was also favoured by the NAD, which stipulated a minimum length of 100m and minimum width of 20m for the landing deck. This meant the investigation of two particular problems, namely structural strength and the effect on stability.

The landing deck had to be located in the after two thirds of the ship's length because the fore part had to be employed for the navigation bridge, the HA gun mountings, and the anchors and cables. The construction of the deck and its supports had to be sufficiently strong to take the high stress of the landing of a wheeled aircraft. A network of longitudinal and transverse girders and beams similar to that used in the construction of the flight deck of a fleet carrier therefore had to be adopted, and the deck supported by strong pillars or heavy frames. In order to test the principle the designers planned the experimental construction of a section of the landing-on deck amidships and to use the pillars as derrick supports. In peacetime the deck was to serve as a platform for light AA guns, and the ships' boats were to be stowed beneath. The increase in the displacement and the increased height of the centre of gravity (CG) was to be compensated by embarking fixed ballast if necessary. Because the hatch for the midget submarines would also be protected by this deck, the latter had to be raised well clear of the upper deck. However, this was not the only reason for the height of the deck above the waterline, as this was also necessary to permit aircraft to be landed on in adverse sea conditions. After the effect on the CG and GM had been calculated, the designers found it necessary to fit bulges as compensation, thereby increasing beam.

Conversion into a Light Fleet Carrier

In the event of a decision to convert the ships as light fleet carriers, two alternative proposals were advanced: conversion within a very short period (about three months); and conversion over a much longer period. The first solution would have required the incorporation of features in the initial design that would conflict with the arrangements for other missions, so planning for this eventuality assumed a low priority. Preparations were reduced to the construction of the flight deck with all the fittings and equipment for aircraft operation, the placement of the hangar, the uptake and funnel arrangement, and the shape and location of the bulges. Reconstruction as a light fleet carrier was to be realised in the course of a major conversion.

Propulsion Machinery

Chiyoda and *Chitose* had two-shaft propulsion machinery, with a single diesel engine and a set of turbines connected to each shaft. The diesels were to be used for cruising, and were operated in conjunction with the turbines for higher speeds. This combination increased endurance at cruise speed, while minimising oil fuel consumption in favour of the supply of other ships. The diesels drove the shafts directly via Vulkan gearing and were connected to the turbine sets via a clutch between the reduction gears and turbines for higher speeds. This arrangement was a characteristic feature of *Chiyoda* and *Chitose*; the seaplane carrier *Mizuho* that followed was powered exclusively by diesels.

The diesel engine No 11 Type 10 was a two-cycle, ten-cylinder model designed to deliver 6,800bhp, but achieved a maximum of only 6,400bhp (total: 12,800bhp at 343rpm). Because the output was insufficient for 16 knots, cruise turbines each rated at 3,000shp were fitted.

Each of the main turbine sets comprised one high-pressure (HP) and one low-pressure (LP) turbine. These were the same turbines fitted in the destroyers of the *Hatsuharu* class;[7] maximum output was 22,000shp per set. The designed total combined horsepower of the diesels and the main turbines was 56,800shp at 290rpm.[8]

The four Kampon Type B water-tube boilers were fitted with air preheaters and superheaters and had a working steam pressure of 22kg/cm² steam pressure and a maximum steam temperature of 300°C. Because of the restriction to 20 knots maximum speed in response to treaty requirements, it was planned to fit only two boilers in the first instance and to employ the empty boiler rooms as fuel tanks until it became necessary to install the other two boilers. However, in the event all four boilers were fitted when the ships were completed. There was a single slim inclined funnel forward housing the boiler uptakes, while the diesel exhausts were led up though the after pair of pillars supporting the raised platform.

Between the four boiler rooms and the four engine rooms (two for the diesels, two for the turbines sets) there were three diesel generators in two compartments.

Conclusion

Every warship is a compromise between contradictory requirements, but warships designed for multifunctional use can never be equally well-suited to perform all their

Table 2: **Building Data**

	Builder	Laid down	Launched	Completed
Chitose	Kure NY	26 Nov 1934	29 Nov 1936	25 July 1938
Chiyoda	Kure NY	14 Dec 1936	19 Nov 1937	15 Dec 1938

Chiyoda saw service during the Second Sino-Japanese War (China Incident). This photo was taken at Woosung in April 1940. Note the floatplanes on the after deck. (Author's collection)

Two photographs taken by the US Navy at Woosung, China, in April 1940, with excellent close-up views of the flying deck and the floatplanes. In the first photo, note the heavy cranes attached to the supporting pillars for the flying deck, which were designed to embark midget submarines (*Kōhyōteki*) via the centreline hatch. (NHHC, NH 82445/51)

intended roles. The biggest problem for this class was speed. The speed of the initial design had to be limited to 20 knots in response to the treaty stipulations, but would need to be raised to more than 30 knots if the ships were to be later employed as light fleet carriers.[9] The latter requirement meant a combined propulsion machinery plant and a corresponding hull form. The IJN correctly recognised that a ship could be converted to perform many roles provided the machinery and the form and structure of the hull below the upper deck remained unchanged; otherwise a total reconstruction would be required. The conversion of cruiser hulls into light fleet carriers by the US Navy during the Pacific War bore this out. However, the concept of the flexible multi-role vessel proved difficult to realise; the designers were compelled to execute several changes before the ships were laid down, and the conversions required much more time than anticipated. Stability and structural weaknesses required improvement, and the diesels proved less reliable than the turbines.

Employment before Conversion into Light Fleet Carriers

After completion both ships were used as seaplane carriers. *Chiyoda* was converted into a midget submarine

carrier in the summer of 1940, when the rails, endless chain system and the launch platform at the stern were fitted. The launch trials took place in the western part of the Inland sea (Iyo-nada) in July/August 1940. Admiral Toyoda Soemu, who in 1944 was to become C-in-C Combined Fleet, called the trials later 'a truly heroic spectacle'.

Chitose was not converted, as her operational use testifies. This suggests either that the tacticians had little confidence that the use of midget submarines on the high seas would bring about the expected results, or that they came to doubt the concept of the early 'decisive battle' The same doubts cut short the proposed conversion of the older light cruisers into heavy torpedo ships.[10]

Conversion to Light Fleet Carrier

The loss of four aircraft carriers in the Battle for Midway prompted a review of shipbuilding policy and brought about a major revision of the Fifth Naval Replenishment Programme, with the Urgent Aircraft Carrier Reinforcement Programme as its main focus.[11]

The conversion of *Chitose* at the Sasebo Navy Yard began on 16 January 1943 and was completed on 1 January 1944. The corresponding dates for *Chiyoda*, whose conversion was assigned to Yokosuka NY and

An excellent overhead view of *Chitose* taken in 1941. Her sister *Chiyoda* was converted to a midget submarine carrier and the flying deck served to conceal the large hatch in the upper deck for the embarkation of the midgets. Note the searchlights and 25mm MGs mounted on the flying deck, the crutches for stowing the floatplanes beneath the flying deck, and the transom stern. (Author's collection)

Table 3: **Characteristics**

	As Seaplane Carrier	As Light Fleet Carrier
Displacement	11,023 tons standard	11,190 tons standard
	12,350 tonnes trial	13,431 tonnes trial
Length	174m pp, 192.5m oa	
Beam	18.8m	20.8m
Draught	7.21m	7.51m
Propulsion	two-shaft combined turbine & diesel	
Boilers	four Kampon boilers, 22kg/cm^2	
Engines	two sets Kampon turbines (HP + LP + cruise)	
	two No 11 Type 10 2-stroke, 10-cylinder diesels	
Horsepower	56,800bhp	
Speed	29 knots	
Fuel	3,600 tonnes oil/diesel	
Endurance	8,000nm at 18kts	
Armament	4 x II 12.7cm HA	
	6 x II 25mm MG	10 x III 25mm MG
Aircraft	24 E8N 'Dave' floatplanes	21 A6M 'Zero' fighters
		9 B5N 'Kate' torpedo-bombers

influenced the reconstruction of the 'super-battleship' *Shinano*, were 16 January and 26 December 1943.[12] The orders for the conversion had been placed with the naval dockyards on 30 September 1942. The work was completed roughly four months before the official dates: *Chitose* embarked on her 10/10 full power trial on 31 August 1943 and undertook take-off/landing and other trials from September to December 1943. In his *Carrier Album* (page 189) Fukui Shizuo suggests that her completion could have been September 1943 if she were a new-build ship rather than a conversion. In contrast to all other carrier conversions *Chitose* and *Chiyoda* kept their original names, in response to the request of the COs and the crews.

The accompanying drawings are self-explanatory and provide a complete view of the ships as carriers. The key elements of the conversion were as follows:

– Removal of all superstructures and weapons on the upper deck except the 'landing deck', which had been designed to form an embryonic flight deck and was already present in the seaplane tender configuration, and construction of the upper hangar on the former upper deck, which became the hangar deck. The foam fire fighting system visible in the drawings was introduced following the Battle for Midway, and was based largely on the system installed in the German passenger ship *Scharnhorst* – purchased by the IJN and converted as the escort carrier *Shinyō*.

– Conversion of the former midget submarine hangar into a short lower aircraft hangar and the fitting of two aircraft lifts to connect the hangars with the flight deck. The after lift, which was 13m long and 12m wide, divided the length of the lower hangar into two sections, permitting the stowage of four aircraft forward of the lift and two aft. The forward lift, which was 12m (L) x 13m (W), served only the upper hangar and flight deck.

– Construction of the flight deck by extending the existing raised platform to the bow and stern. Dimensions were now: length 180m and maximum beam 23m, tapering to 13m at the forward end and 16m at the after end. At the bow and stern the flight deck was supported by two pairs of pillars fore and aft of the upper hangar. Take-off and landing equipment comprised seven arrester wires, two crash barriers and the usual landing guide lights, but there was no wind shield. At the starboard after corner of the forward aircraft lift a 110cm searchlight was fitted, and forward of this lift on the centreline was the antenna for a No 21 radar; both were in recesses and were retracted when flight operations were taking place. The positions of the searchlight and the radar antenna differed from those in earlier carriers, in which they were normally located aft.

– The air complement comprised 21 'Zero' fighters and nine 'Kate' torpedo bombers, but seven fighters had to be accommodated on the flight deck in a permanent deck park due to the limited capacity of the hangars. On the other hand, these planes could be quickly readied for take-off.

– Four avgas tanks were fitted. The forward group was located forward of the boiler rooms and the after group abaft the motor room for the diesels. The former was located on the double bottom, the latter in the hold. It has been widely recognised that the locations of the tanks were disadvantageous and their protection inadequate.

– The bridge was fitted below the flight deck forward of the upper hangar but projected to both sides in order to accommodate the necessary equipment.

– Because these ships lacked any direct protection, considerable attention was paid to the subdivision of the lower hull into watertight compartments and the arrangement of escape routes for the crew. The former midget submarine hangar in particular was tightly

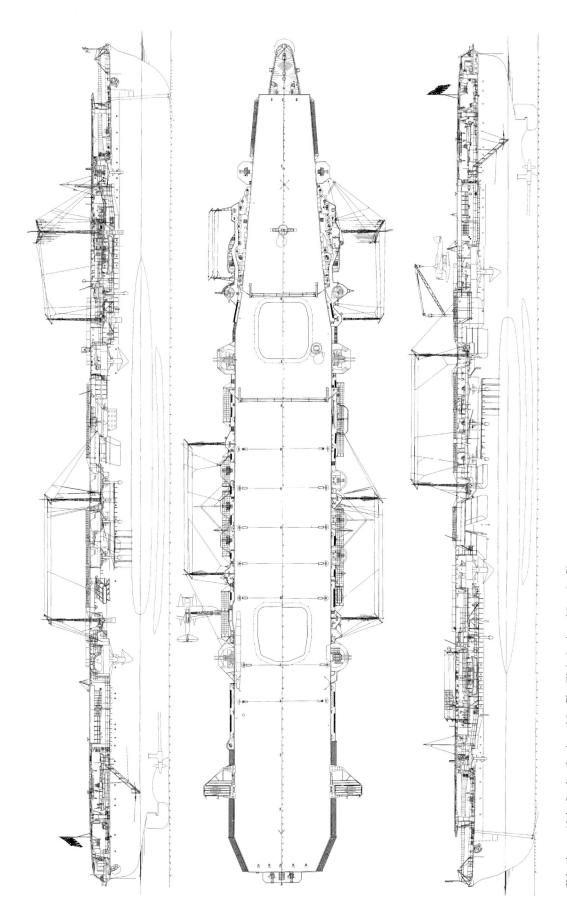

Chiyoda as a Light Carrier: Starboard Profile, Flight Deck and Port Profile
(The plans, which are adapted from original blueprints in the possession of Todoka Kazushige, are courtesy of Maeshima Hajime, by kind permission of Hajime Maeshima, Kokubunsha Publishing)

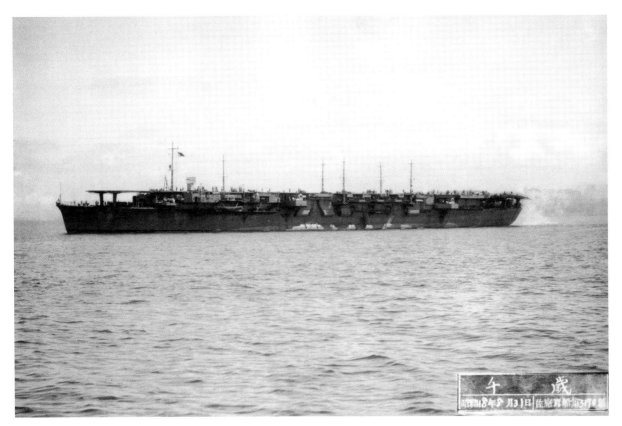

Chitose on 31 August 1943 following her conversion to a light fleet carrier. Note the extensive ventilation ducting on the side of the hull, the elevated mattress antenna of the No 21 air search radar, and the hinged signal and W/T masts in the upright position. (Author's collection)

subdivided, as its floor was only one metre above the waterline. If water flooded into this large area stability would deteriorate very quickly, resulting in the loss of the ship. There was a serious danger of flooding from the stern, so the after part of the hangar was converted into crew spaces – this precluded the stowage of more than two aircraft aft of the after lift – and these spaces were subdivided by three transverse bulkheads, arranged at different intervals. However. this measure has to be regarded as improvisational; it could never provide sufficient protection.
- The increase in displacement was compensated by the fitting of small bulges; this broadening of the beam effectively prevented an increase in draught and preserved stability. The bulges also made a small contribution to protection.
- The machinery installation was unchanged, but new funnels in the typical style of Japanese carriers, curved downwards, were fitted to starboard (see the photo p 178). The fore funnel, situated abaft No 1 12.7cm HA gun mounting, was for the boiler exhaust gases; the after funnel, which was located forward of No 3 12.7cm HA gun mounting, was for the diesel engines. On official trials a speed of 29 knots was attained on a displacement of 13,650 tonnes. It was calculated that the 2,687 tonnes of fuel carried was sufficient for an endurance of 11,810nm at 18 knots.

The HA outfit comprised four Type 89 twin 12.7cm 40-cal guns distributed symmetrically on both sides and mounted on newly-fitted sponsons that projected from the sides of the upper hangar below the flight deck. Type 94 HA FC directors were located on sponsons abreast the forward crash barrier, which was fitted directly in front of the forward aircraft lift. For close-range AA defence ten Type 96 25mm triple MG were mounted: four on sponsons on either side and two at the stern. The three mounts located abaft the funnels on the starboard side had enclosed gunhouses for protection against smoke and fumes. The number of Type 95 FC directors is not stated, but between three and five can be assumed.

After completion twelve single 25mm MG were added, and in the course of the general reinforcement of anti-aircraft weapons after the Marianas sea battle a further 18 single MG were mounted; this brought the total number of 25mm MG to 60. The ammunition hoists were located on both sides of the hull, and their arrangement is shown in the drawings of Frame 51.

Operational History

Chitose

Chitose was in Shortland from 12 October to 3 November 1942 and was assigned to the Combined Fleet when she departed there on the 3rd. After a stay at Truk

A starboard side view of *Chiyoda* as a light fleet carrier taken on 1 December 1943, immediately after the completion of her conversion. Note the smoke drifting aft from the downward-facing funnel vent and the numerous crew members beneath the after part of the flight deck. A considerable swell is visible at the stern. (Author's collection)

6–10 November she arrived at Sasebo on the 15th and was docked on the 28th.

In Sep 1943 she was assigned to the Training Force of the First Mobile Fleet (*Kidō Butai*) and was in the Inland Sea. In October she was at Yokosuka (5–9), Saeki (10–11), Sama (12–16 and 30–31). Between the stays in Sama she was at Singapore 19–26 October. From 5 to 26 November she was in the Inland Sea and was docked at Kure NY 19–21 November. Afterwards she moved to Truk and was there 1–7 December before returning to

Yokosuka, departing there on the 16th after a three-day stay in the Inland Sea where she remained until 11 January 1944. Since 1 January she had been attached to the First Mobile Fleet, and she was in Singapore from the 20th to the 25th. On 1 February she was assigned to Carrier Division 3 belonging to the Third Fleet, and was in Sasebo from the 4th to the 15th, then Kagoshima from the 16th to the 20th before moving to Saipan, where she stayed from the 26th to the 29th. Returning to Yokosuka on 4 March she was in that port until the 13th and again

A port side view of *Chitose* as a light fleet carrier, claimed to have been taken between January and March 1944. (Author's collection)

from the 17th to the 25th, during which period she was docked 19–24 March; in the meantime she had been in Kisarazu from the 13th to the 17th. After leaving Yokosuka she was in the Inland Sea from 27 March until 10 May. On 5 May she embarked aircraft belonging to Air Group 653, and arrived at Tawi Tawi on the 16th following a sortie on the 11th in preparation for operation *A gō sakusen*, together with her sister *Chiyoda* and many other ships.

After participating in the Marianas sea battle both carriers returned to Kure on 1 July, where *Chiyoda* was docked from the 2nd to the 9th and *Chitose* from the 20th to the 26th. After that both ships remained in the Inland Sea/Kyushū area during August and September before meeting their demise in the Battle for Leyte Gulf.

Chiyoda

Chiyoda was in Kure naval port until 6 January 1943 and then in Yokosuka from the 8th. She remained there until 29 January 1944 but her status as reserve ship (for the duration of the conversion) changed on 21 December 1943 when she was assigned to the Northeast Area 2nd Base Force (12AF, 51Afl). On 1 February 1944 she was assigned, like her sister, to Carrier Division 3, 3rd Fleet. She arrived at Truk on the 3rd and entered Kure on the 15th after departing Truk on the 10th. On 26 February she moved to Yokosuka, and departed there on 1 March for Saipan (5–6 and 8–10), Guam (6–8), Palao (12–15 and 24–27) and Balikpapan (19–21 and 30 March to 1 April). She sortied from Balikpapan on 1 April, was in Davao 3–5 April and then in the Inland Sea (Kure area) from the 10th· Like her sister she embarked aircraft belonging to Air Group 653 5–6 May and sortied on the

11th. After that her movements conformed to those of her sister.

Both ships were sunk on 25 October 1944 in the Battle off Cape Engaño following air attacks, although the seriously damaged *Chiyoda* was given the *coup de grâce* by US cruisers.

Endnotes:

1. A similar problem was experienced with the design of the contemporary cruisers of the *Mogami* class, which were to have their 15.5cm triple turrets replaced by 20.3cm twin turrets; all problems relating to structural strength had to be resolved beforehand, and particular dimensions had be standardised.
2. *Japanese Naval Vessels Illustrated, 1869–1945*; Vol 3 *Aircraft Carriers, Seaplane Tenders & Torpedo Boat & Submarine Tenders*, 336.
3. See the author's article in *Warship 2011*, 148–164.
4. The investigation of structural strength after the Fourth Fleet Incident in September 1935 resulted in the adoption of thicker plates for the upper deck, which was the strength deck – see the author's article in *Warship 2013*, 30–45.
5. The carriages comprised two-wheeled trolleys connected at their lower ends to form a single unit.
6. *Chiyoda* would be the first ship to embark all 24 floatplanes to test the time required for catapulting.
7. See the author's article in *Warship 2007*, 91–110.
8. This corresponded to a Vulkan gear reduction ratio of 0.841.
9. According to Makino Shigeru, *Kansen nōto*, the NGS requirement was 33 knots.
10. Only *Kitakami* and *Oi* underwent conversion.
11. For a full description of this programme see the author's *BB Kongō class & CV Unryū class*, Model Hobby, al Korfantego 8, (Katowice, Poland, 2010), 136–44.
12. Official dates; see also under Operational History.

T 53 DUPERRÉ

This short feature showcases the Editor's line drawings of the French T 53 fleet escort *Duperré* as completed in the late 1950s. The drawings are adapted from the official plans drawn up at Lorient and dated 18 November 1960.

The programme drawn up in the late 1940s for a new generation of 'fleet escorts' (*escorteurs d'escadre*) envisaged the construction of eighteen ships, to be authorised between 1949 and 1953. The lead ship, *Surcouf* (T 47 Type – see *Warship 2020*, 180–82), was duly authorised in 1949 and a sister *Kersaint* in 1950. The programme was then accelerated, with four ships authorised in 1951 and six in 1952.

The final batch of six was due to be authorised in 1953, but the *Marine Nationale* decided that they would be built to a revised design designated T 53 R. These later units were intended as air defence escorts for France's first purpose-built carriers *Clemenceau* and *Foch*, which were to be ordered under the 1953 and 1954 Estimates. The order for one of the ships was then held over to allow her to be redesigned as the prototype for a new type of anti-submarine escort, armed with a new generation of A/S weapons and sensors; she would become the T 56 *La Galissonnière*.

The five ships of the T 53 type were to have the latest generation of French air surveillance radars, which besides being more reliable than their predecessors would have an increased capability for tracking and controlling aircraft in the vicinity of a task force. Other key modifi-

cations to the design included a continuous deckhouse from the forecastle to the quarterdeck (albeit with a narrow passageway between the after torpedo tubes to allow them to train on the beam), thicker plating on the bridge, and the replacement of the 'short' KT 50 torpedo tubes by a six-barrelled Bofors 375mm rocket launcher at the forward end of the after deckhouse.

The weapon and sensor outfit of the original T 47 type was detailed in the short article on *Surcouf* published in *Warship 2020*, so only the changes which were features of the T 53 design are listed here:

550mm AT 47 torpedo tubes
It was envisaged from the outset that the two triple AT 47 tubes would be capable of launching either 'long' anti-surface torpedoes or 'short' anti-submarine torpedoes. There is, however, no evidence that the ships of the T 53 sub-group ever embarked the K2 high-speed A/S torpedo, which was effectively replaced by the Bofors 375mm rocket launcher (see below) and superseded from 1960–61 by the L3 A/S torpedo.

L3 anti-submarine torpedo
The L3 'short' A/S torpedo was the replacement for the

The name-ship of the T 53 series, *Duperré*, in February 1964. Note the continuous deckhouse running from the forecastle to the quarterdeck and the triple AT 47 torpedo mounting, which could fire both 'long' anti-surface and 'short' anti-submarine torpedoes. (Author's collection)

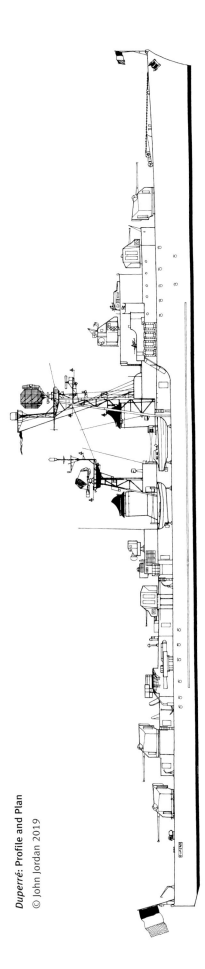

Duperré: **Profile and Plan**

© John Jordan 2019

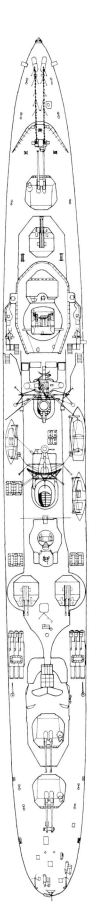

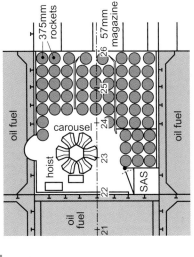

Duperré: **Bofors 375mm Rocket Magazine**

© John Jordan 2019

Characteristics

Displacement:	3,300 tonnes trial
	3,830 tonnes full load
Dimensions:	
Length	123.50m pp; 128.86m oa
Beam	12.71m wl
Draught	4.33m aft, 3.51m fwd
Propulsion:	
Boilers	4 Indret boilers
Engines	2-shaft Rateau geared turbines
Horsepower	63,000CV
Speed	34 knots (designed)
Endurance	5,000nm at 18 knots

Armament:	
Main guns	6 – 127/54 Mle 1948 (3 x II)
AA guns	6 – 57/60 Mle 1951 (3 x II)
	2 – 20/70 Mk IV
U/w weapons	Bofors 375mm R/L Mle 1954 (1 x VI)
	6 – 550mm AT 47 TT (2 x III)
	(12 torpedoes)
Electronics:	
Air/surf search	DRBV 22A
Height finding	DRBI 10B
Fire control	DRBC 11 (127mm)
	DRBC 30 (57mm)
Sonar	DUBV 1B, DUBA 1B
Complement:	350

Building Data

Hull No	Name	Builder	Laid down	Launched	In service
D 633	*Duperré*	Arsenal de Lorient	2 Nov 1954	23 Jun 1956	8 Oct 1957
D 634	*La Bourdonnais*	Arsenal de Brest	7 Dec 1954	15 Oct 1955	3 Mar 1958
D 635	*Forbin*	Arsenal de Brest	7 Dec 1954	15 Oct 1955	1 Feb 1958
D 636	*Tartu*	A C Bretagne	14 Dec 1954	2 Dec 1955	5 Feb 1958
D 637	*Jauréguiberry*	F C Gironde	Sep 1954	5 Nov 1955	15 Jul 1956

K2. Unlike the latter, which was a high-speed, straight-running model, the L3 had active acoustic homing and was designed to counter submarines capable of 20 knots underwater. The torpedo had a length of 4.32 metres and an all-up weight of 910kg which included a 200kg warhead. It was propelled by a 40kW electric motor, with power being provided by a nickel-cadmium battery; maximum range was 5,000m at a speed of 25 knots. Following launch the torpedo ran at a depth of 30 metres while executing a sinuous search pattern; acquisition range was 1,000m using a 32kHz active seeker, and targets could be engaged to a depth of 300m.

375mm rocket launcher Mle 1954

Developed by Creusot-Loire, the six-barrelled launcher used standard Bofors 375mm ammunition; it had remote power control for training (through 360°) and elevation (0°/+90°), and weighed approximately 16 tonnes. The rockets had proximity or time fuzing; one 250kg round (warhead: 100kg) could be launched every second, and the six tubes were reloaded vertically in 90 seconds. The rockets could be fired up to a 60-degree angle of elevation, and maximum range was 1,600m. The magazine was at hold level, directly beneath the launcher; the official plans of *Duperré* show a total capacity of 57 rounds, including nine exercise rounds (see drawing p 181). There were six rails on the sides of the launcher for illumination rockets, and a blast deflector directly behind the launcher prevented damage to the 127mm mounting on the same level.

DRBV 22A radar

Mounted atop the mainmast, this 'D'-band combined air/surface surveillance radar effectively replaced both the DRBV 20A and the DRBV 11 of the T 47. It featured a parabolic lattice antenna, and was similar in configuration and performance to the contemporary US Navy SPS-6; detection range was 120–220km, depending on the size and height of the target.

DRBI 10B radar

The DRBI 10 was an 'S'-band height finder evolved from the Picador land-based system (TRS 2200); it was also fitted in the new carriers and in the anti-aircraft cruisers *De Grasse* and *Colbert*. The unconventional design combined a US Robinson-type feed with a broad, square lattice antenna with cropped corners that produced a relatively narrow search beam. The feed moved up and down as the antenna rotated to produce a wavy scanned area, enabling the DRBI-10 to be employed as a 3-D radar. Range against a fighter aircraft was 100–140km.

DRBV 31

This short-range navigation radar replaced the earlier DRBV 30; it had a similar range (40km).

Sources:

Official plans of *Duperré* held by the Centre d'Archives de l'Armement.

Robert Dumas & Jean Moulin, *Les Escorteurs d'Escadre*, Marines Editions (Rennes, 2012).

The T 53-type fleet escort *Forbin* tied up inboard of the name-ship of the T 47 type, *Surcouf*, at Antwerp in 1964. Note the very different radar outfit which was the principal distinguishing feature of the two types: *Surcouf* has the distinctive concave mattress antenna of the DRBV 20A air search radar on the foremast and the parabolic antenna for the DRBV 11 combined air/surface surveillance on the mainmast, while *Forbin* combines the DRBI 10 heightfinder (foremast) with the DRBV 22A combined air/surface surveillance radar (mainmast). Note also the six-barrelled Bofors anti-submarine rocket launcher forward of the 127mm mountings on *Forbin*. (Leo van Ginderen collection)

WARSHIP NOTES

This section comprises a number of short articles and notes, generally highlighting little known aspects of warship history.

REFUGEE BATTLESHIP

Sergei Vinogradov describes the final years of *General Alekseev*, 1921–1936. His note has been translated and edited by **Stephen McLaughlin**

More than any other battleship of the Russian Imperial Navy, *General Alekseev* experienced the vicissitudes of revolution and civil war. She was laid down in 1912 at Nikolaev as *Imperator Aleksandr III*, but progress was slow due to wartime conditions. She was completed only in 1917, and by that time the first wave of revolution had swept away the tsarist regime. While she was still fitting out for the Black Sea Fleet in April 1917 she was renamed *Volia* ('Liberty'). The Bolshevik coup in November set the stage for the next act in the battleship's career. When negotiations with the Central Powers for a separate peace stalled, the Germans launched an offensive, and by the end of April they had reached the Crimea and were on the outskirts of Sevastopol, the main base of the Black Sea Fleet. The major ships of the fleet, including *Volia*, fled to Novorossiisk on 30 April, but under German pressure most were forced to return to Sevastopol in June. In October the ship was commandeered by the Germans and even made a short cruise under their control, but reports that she was renamed *Volga* or *Wolga* are incorrect.

After the collapse of Germany in November 1918, the ship was seized by the British and on 22 December 1918 a navigating party from HMS *Agamemnon* took her across the Black Sea to Ismid near Constantinople. She was interned there for almost a year, during which her triple turrets were closely examined by the Royal Navy, which was planning to use such turrets in its postwar battleships.[1]

In the autumn of 1919 the ship was transferred to the naval forces of General Anton Denikin, the leader of the White (anti-Bolshevik) movement. On 17 October, in anticipation of her return, she was renamed once again, this time in honour of the founder of the White Movement, General M V Alekseev, who had died a year earlier. On 29 October the ship, escorted by HMS *Iron Duke*, arrived in Sevastopol and on 1 November was officially transferred to Denikin's forces.[2] Although there was nothing to match her in the weak and scattered Red flotillas, which consisted primarily of armed river steamboats and barges, the giant ship was also something of a burden. *General Alekseev* required 40 tons of coal per day while at anchor, and far more if she put to sea. Coal was a scarce commodity, as were trained seamen; the battleship's full crew amounted to 1,250 officers and men, but she had to make do with no more than 400, fewer than half of whom had any experience of the sea. Nevertheless, by the spring of 1920 the ship was cleaned out, most of her guns were in service, and she was able to steam at 10 knots.[3]

However, by the time the ship was ready for service, Denikin's fortunes had taken a dramatic turn for the worse. Over the winter of 1919–1920 the Red Army decisively defeated the White forces in the northern Caucasus. Denikin, now discredited, turned over command to General Pëtr Wrangel, whose territory was limited to the Crimean Peninsula. Meanwhile the Reds were gathering in vastly superior numbers for the final destruction of the White movement. However, Wrangel's regime received a stay of execution when Poland attacked the Bolshevik state in April 1920. With Red Army units shifted eastward to deal with this new enemy, Wrangel made use of the opportunity by launching an attack into the southern Ukraine's rich agricultural region. *General Alekseev* was called upon to deal with the Red artillery positions that dominated the mouth of the Dnieper River, and on 31 July she put to sea, flying the flag of fleet commander Admiral M P Sablin. Escorted by minesweepers, on 3 August she anchored 19,400 yards off the enemy batteries and began firing single shots from her forward turret. The shelling continued on 4 and 6 August. There was no reply, and on 18 August the battleship returned to Sevastopol.[4]

Wrangel's offensive proved to be the last gasp of the White cause; on 12 October the Bolsheviks signed an armistice with Poland, allowing the Red Army to transfer overwhelming forces back to southern Russia. On 28 October they launched an offensive, and by 11 November

For centuries Russian nationalists had dreamed of Russian warships and armies at Constantinople, but when this finally happened the ships and soldiers were not conquerors but refugees from the failed White cause. Here is *General Alekseev* anchored off Constantinople in November 1920. (Author's collection)

the defences of Wrangel's Crimean stronghold were shattered. Wrangel, however, had foreseen the outcome and had made effective preparations. In these Admiral M A Kedrov, who had taken over command of the fleet upon the death of Admiral Sablin in September 1920, played a leading role. As Red forces approached Sevastopol a massive evacuation of White soldiers, officials and civilians began; by the time it ended on 16 November 146,000 people had been taken aboard 126 ships of every description. Amongst them was *General Alekseev*, which had some 3,000 people on board, including the ship's crew, cadets of the Naval Corps, and refugees. There were only five trained stokers on board, so speed was limited to 5 knots and stops were frequent; she arrived in the Bosphorus on 16 November, and by the evening of the next day she had anchored off Constantinople. Here the refugees disembarked; they would eventually disperse across Europe, forming Russian colonies in many cities.

However, a different fate lay in store for *General Alekseev* and the other warships of the refugee fleet. With her bunkers replenished and her crew reinforced, on 8 December the battleship, accompanied by the transport *Kronshtadt*, the collier *Foros* and the destroyer division, left Constantinople for the French naval base of Bizerta in Tunisia, arriving on 27 December. By the end of the year a total of 33 ships with 5,200 men had arrived. Once there they were led through the canal to the Lac de Bizerte, where they anchored. The mood on board the ships was optimistic: the difficult journey was behind them, they had escaped the vengeance of the victorious Reds, and most believed that the day was not far off when they could return to Russia – despite recent victories, the power of the Bolsheviks still seemed precarious, a belief no doubt reinforced by the Kronshtadt rebellion in March 1921, when even the ardently revolutionary sailors rose against the oppressive policies of the new regime.[5]

In the meantime, *General Alekseev* and the other ships of the Russian squadron off-loaded their ammunition to the French arsenal at Sidi Abdallah. The battleship was being operated by only about 350 men – one quarter of her full crew. Nevertheless, officers and volunteers, sustained by hopes of once again serving under the flag of St Andrew, kept the ships in order: watches were regularly maintained on the ships and necessary repairs were carried out.

In the spring of 1921 gunnery classes for young officers and cadets of the Naval Corps were held aboard *General Alekseev*, the first batch of eleven students embarking on 11 May. Cadet N N Aleksandrov recalled:

> ... it was dark when our boat approached *Alekseev*. It seemed to me somehow dead, gloomy. The lighting was poor, it was difficult to get one's bearings. ... They took us to a cabin. It was a mess – dirt, bugs. Somehow we settled in for the night, but we slept badly....
>
> [Gunnery] Classes were held until noon and resumed in the afternoon.... In our free time, individually and in groups, we wandered around the ship wherever we could. The ship was being placed in long-term conservation. The turbines and auxiliary machinery were cleaned and thickly lubricated, the boilers were gradually put in order ....[6]

On 19 July Admiral Kedrov visited *General Alekseev*, inspected the battleship and expressed his satisfaction with her condition. His visit ended with a gala dinner in the ship's wardroom. There was reason to celebrate: Kedrov had just returned from Paris, where he had carried out difficult negotiations with the French regarding the status of the squadron and its financing. Among the issues settled was the docking of the Russian ships, and on 31 August, *General Alekseev* entered one of the graving docks, which was barely large enough for

General Alekseev in dock at Bizerta, in August or September 1921. Note the framework on the bow; this had supported the 'fore-trawl', a minesweeping device. A boom pivoted from this fixture at the bow, and when lowered into the water it streamed a pair of paravane-like kites. The fore-trawl proved too delicate in service, and was removed from ships, but the bow fixture was often retained. (Author's collection)

the dreadnought. The last time she had been docked was July 1917 at Sevastopol, so when the water was pumped out, it revealed a hull covered with marine growth. As the shells were scraped off they were collected by enterprising Frenchmen and Arabs in buckets, bags, and baskets and taken to the market. The 'beard' of algae and the rust were also scraped away. Hull damage sustained in April 1918 when she fled Sevastopol ahead of the German advance was repaired. Meanwhile her Russian crew and the cadets worked on the ship's interior, cleaning the hold and magazines. On the morning of 22 September the dock was flooded and *Alekseev* was again afloat; no leaks were detected, and by 1000 the ship was once again anchored in the roadstead.[7]

In November the squadron received a shock: the crews were to be reduced. Only 50–60 men were left aboard *General Alekseev* to maintain and guard the ship. This situation persisted for almost two years until 29 October 1924 when, as a result of France's recognition of the USSR, the flag of St Andrew was lowered on all the ships of the squadron. The ships were turned over to the French authorities, and the few remaining Russian sailors were replaced by French security teams. The history of

the Russian squadron in Bizerta was over, and the ships, in particular *General Alekseev*, faced an uncertain future.

The Soviet naval command had been monitoring the squadron's ships all along, hoping that this remnant of the tsarist navy could be added to the USSR's meagre fleet. In April 1923 there was considerable anxiety in the Red armed forces over rumours that the greater part of the ships at Bizerta would be transferred to Rumania on 10 May. The deal, it was alleged, had been arranged due to British pressure in order to prevent the ships from returning to Russia.[8]

However, these fears proved unfounded, and in early December 1924 a Soviet delegation headed by the former tsarist naval officer E A Berens arrived in France to resolve the issue of the squadron's transfer. Berens' party included the famous naval constructor and scientist A N Krylov and four other technical specialists. During negotiations the French government seemed favourably disposed, and was even willing to waive compensation for the costs of the squadron's maintenance, provided that all repair work was done in French ports. Starting on 28 December 1924 the Soviet team began a thorough examination of all the ships. They found that *General Alekseev* was in fairly

A fine shot of *General Alekseev* soon after her period in dock. Note that the bow fixture for the fore-trawl has been removed. (Author's collection)

good condition: she had been painted inside and out, her turbines were well lubricated, and twelve of her twenty boilers had been cleaned. As a result, the battleship could steam under her own power at economical speed, although frequent stops would be necessary. It was esti-mated that the remaining repair work could be carried out in a shipyard in a period of about two months.[9] The only other vessels deemed of interest to the Red Navy were six *Novik*-type destroyers and four submarines; the other ships were good only for their scrap value.

A detail view looking forward from *General Alekseev*'s mainmast; turret no 3 with its four sighting hoods is in the immediate foreground, then comes the second funnel with the sides of turret no 2 visible to its left and right, then the narrow forefunnel, and just visible to the right is the starboard side of turret no 1. The peculiar Russian method of deck planking can just be made out, with athwartships strips holding the short longitudinal planks in place. (Author's collection)

To the Soviet naval authorities the prospects looked bright. Estimates were prepared for the repairs the Bizerta ships would need once they returned to Russia. On 17 January 1925 the head of the Naval Forces, V I Zof, presented a report to the government on the state of the ships. With regard to *Volia* (the 'White Guard' name *General Alekseev* was avoided), he noted that:

> ... in order to steer the battleship while under tow [it will be necessary] to bring into full service two boilers, one turbo-generator and the electric steering gear, as well as a capstan ... [R]epair of the remaining boilers, anchor gear and the entire steering gear ... cannot be performed either in Bizerta or in other facilities in Tunisia and she would have to be transferred to one of the large shipyards in southern France. All other repairs can be performed *in situ*.'[10]

In March 1925 the Technical Department of the Navy estimated the cost the battleship's overhaul and modernisation at 4,775,250 rubles. The fire control system would be upgraded with rangefinders in all the turrets and director control, six anti-aircraft guns would be installed, as would paravanes; catapults would be fitted on the second and third turrets. In view of the danger of poison gas shells (a concern of many navies at the time) the ventilation system would need gas filters, while hatches, manholes, gunports and scuttles would need to be made gas-tight. The estimated date of the battleship's entry into service was 1 May 1927.[11]

In the meantime, however, the outlook for the ships' return had dimmed. In April 1925 Zof reported that:

> ... the situation is now worse than it was about four months ago. In December, when negotiations began for the return of the Bizerta squadron, things went so well that we assumed it would be possible to recover a significant portion of our ships from Bizerta in the very near future .... However, it is not only the government of [Prime Minister] Herriot, a government of the French bourgeoisie, but obviously the Entente as a whole that has done everything to prevent the ships from being returned.[12]

In the French Senate there were speeches opposing the transfer of the ships to the USSR for fear it would weaken the position of France and its allies in Europe. Great Britain maintained an unfriendly silence, while the governments of the Black Sea states and Baltic republics were nervous about this potential addition to Soviet naval power. And the Soviet government – as opposed to the armed forces – expressed little interest in the recovery of the ships and undertook few diplomatic initiatives to support it.[13] As a result, the return of the ships foundered on the issue of the repayment of tsarist debts.

Nevertheless, efforts were made to maintain the Russian ships. In early November 1928 *General Alekseev* was docked to paint her underwater hull. The work was paid for out of the remaining funds the French government had allocated to the squadron.[14] With the transfer of the ships to the Soviet Union a dead issue, in 1933 the French government decided to sell them for scrap. The last to go was *General Alekseev* in 1934. Her guns and fire-control equipment were declared the property of France and deposited in the arsenal at Sidi Abdullah. The dreadnought herself was dismantled by a company founded by the engineer A P Kliagin, an *émigré* Russian who specialised in handling the leftover war materiel of his homeland.

By this time there was a small but well-established colony of Russian *émigrés* in Bizerta, and in the late 1920s they decided to build a small Orthodox church in memory of the ships of the squadron. It took several years to raise the necessary funds, and when construction finally began in 1937 *General Alekseev* contributed to its furnishing: Kliagin, who had purchased the battleship in its entirety for scrap, donated many items from the ship to the church, including a chandelier, anchors, anchor cables, two large marble slabs, searchlight reflectors, floor tiles, etc, plus direct financial contributions that eventually totalled 15,000 francs.[15] In 1938 the church was consecrated in honour of the Holy Blessed Grand Duke Alexander Nevskii, victor over the Teutonic Knights in 1242, and still retains relics of the Russian battleship.

Some more substantial items from *General Alekseev* also had an interesting fate: her 50-ton 12in guns. They not only survived the Second World War, but several guns exist to this day in museums in Moscow and St Petersburg, as well as in Finland. But that is another story entirely.

Endnotes:

1 Norman Friedman, *Naval Weapons of World War One*, Seaforth Publishing (Barnsley, 2011), 255.
2 David Snook, 'British Naval Operations in the Black Sea 1918–1920', *Warship International*, vol 26 (1989), no 1, 36–50, at 49.
3 Russian State Archives of the Navy (RGAVMF), F r-72, Op 1, D 34, L 74–76 [F = *fond*, record group; Op = *opis'*, inventory; d = *delo*, item; l = *list*, sheet].
4 P Varnek, 'Lineinyi korabl' "General Alekseev" i istoriia ego pushek', *Flot v Beloi bor'be*, Tsentrpoligraf (Moscow, 2002), 279.
5 *Ibid*, 281.
6 N N Aleksandrov, 'Na lineinom korable "General Admiral"', *Zhurnal kruzhka Morskogo uchilishcha vo Vladivostoke*, no 1, Litografiia Morskogo korpusa, RFK, Inv No 10, 148 [58] (Bizerta, 1922), 1–2.
7 *Ibid*, 7–8.
8 RGAVMF, F r-1 Op. 3, D 3161, L 16.
9 RGAVMF, F r-1, Op3, D 3161, L 115.
10 A Iu Tsar'kov, 'Nesostoiavsheesia vozvrashchenie Bizertskoi eskadry', *Gangut*, no 61 (2011), 61, citing Russian State Military Archives (RGVA), F 4, Op 1, D 97, L 37.
11 RGAVMF, F r-12, Op 2, D 87, L 232, 348, 399.
12 From a speech to the IX Party Conference of Baltic Fleet sailors, 16 April 1925, Zakliuchitel'noe slovo t. Zof po dokladu 'Mezhdunarodnoe polozhenie i zadachi morskoi oborony SSSR', *Morskoi Sbornik*, 1925, no 5, 18.
13 Tsar'kov, 'Nesostoiavsheesia vozvrashchenie Bizertskoi eskadry', 67.
14 *Zarubezhnyi Morskoi Sbornik*, 1930, no 2, 34.
15 *Morskoi zhurnal*, 1936, no 12 (108), IX god izdaniia, 6.

PARIS ON THE TAMAR

Aidan Dodson describes the Second World War career of the French battleship *Paris*.[1]

Paris was laid down in 1911, launched in 1912 and commissioned in 1914 as the third French all-big-gun battleship, armed with twelve 305mm (12in) guns in twin turrets, and twenty-two 138.6mm (5.4in) in single case-mate mountings. She was extensively refitted during 1922–23, when the elevation of her main guns was increased from 12 degrees to 23 degrees, and she received a modern HA armament of four 75mm Mle 1918 guns. At the same time she was fitted with a British-style control top, and in a further refit during 1927–29 she received a director control system of French design and manufacture. However, by 1939 *Paris* was obsolete, and with her sister *Courbet* had been operating primarily in a training role since 1931. The condition of the two ships was also now poor. On 17 March 1938, Flag Officer Training Division reported[2] to the Commander-in-Chief Mediterranean that:

> [o]f the two battleships PARIS and COURBET it is diffi-
> cult to tell which is the more worn out.
> The PARIS has a slightly superior armament, 305 MM
> guns changed in 1935 and a more complete watch organ-
> isation. She has an Admiral's Bridge and Signal Office, the
> COURBET has not. On the other hand the COURBET has
> eight oil-fired boilers and PARIS only four.[3] The
> COURBET's life of boilers is till 1944, the PARIS till 1941.
> Above all I am of opinion that the repairs which the
> COURBET is undergoing at this moment will make it pos-
> sible to keep her in commission with <u>less risk of surprises</u>
> than the PARIS.

On 23 September 1938, he further reported as follows:

> PARIS
> <u>Port H.P. Turbine.</u> This engine should have been over-
> hauled in 1937. The overhaul was put off till March 1938,
> and then, till the next refit of the ship.
>   This refit was fixed for the period 15 Sept. – Nov. 1938.
> It has been again postponed as a result of the diplomatic
> tension.
>   During the overhaul of this turbine, the "PARIS"
> should have a main test of the Steam-pipes.
>   <u>Starboard L.P. Turbine.</u> Overhaul due January 1939.
>   <u>Boilers 11, 12, 13, 14</u> (Heavy Oil Belleville). Should be
> partly re-tubed from June 1939 onward.
>   <u>Refit contemplated.</u> Oct. 1st – Nov. 15th according to
> the progress of diplomatic tension.
>   <u>Dry-docking.</u> Jan. 15th – Mar. 15th 1939.

The international situation never actually permitted any meaningful work, and on 10 June 1939 the Training Division was dissolved and the two old battleships became the 3rd Battle Division, initially as part of the 5th Squadron, then under the Prefect of the 2nd Maritime Region at Brest, serving as gunnery training ships.

Paris is brought alongside at Devonport, possibly to begin her September 1940 refit. She still sports the false bow wave that was painted on during her short period on offensive operations leading up to the French armistice. (Editor's collection)

They continued in this role until May, when the German offensive that began on 10 May led to their being given 96-hour refits at Brest and deployed to Cherbourg on the 28th with a view to operating in support of the French Army's rearguard actions. When *Paris* was taking up position off Le Havre on 11 June she was damaged by a German bomb on the port side of the forecastle and by near-misses, and had to withdraw to Cherbourg and then Brest for docking. She was still in dock when the French government requested an armistice on the night of the 17th, but was refloated just in time to take part in the exodus of 83 warships from Brest the following day as German troops approached the port. *Paris* limped to Devonport with another cripple, the cruiser-submarine *Surcouf*, which had only her electric motors operational; vessels unable to sail were scuttled, except for a few old hulks.[4] She arrived at Devonport flying the flag of Vice Admiral Lucien Cayol and carrying 1,600 naval cadets. Her sister *Courbet* arrived at Portsmouth on the 20th.

Between 0430 and 0445 on 3 July, the French ships in UK waters were boarded and taken over, part of a process that included the tragic British attack on the French fleet at Mers el-Kebir in Algeria and the intern-ment of the French warships at Alexandria. At Devonport, one French and three British lives were lost aboard *Surcouf*. While all vessels hoisted the *tricolore* at 0800, all were firmly under the control of the Royal Navy, their crews disembarked and interned.[5] The ships were then assessed by the British, it being noted on the 7th that *Paris* was still leaking from her bomb damage and required docking for a final fix. It was also noted that the Poles were interested in using *Paris* as a depot and training ship on the Tyne in place of the ex-passenger vessel *Gdynia* (ex-*Kościuszko*, 1915), and that C-in-C Western Approaches had no objection to such a transfer.

On the other hand, given the threat to Atlantic convoys from German raiders, it was also queried whether *Paris* and *Courbet* might be suitable as ocean escorts, alongside unmodernised British battleships. Information was thus requested on the 9th by Deputy Controller on the likely work needed to fit them for this, the complement required and whether the sisters possessed complete outfits of ammunition. However, three days later, C-in-C Western Approaches minuted that he considered such employment 'quite impracticable', quoting the 1938 French reports on their condition, copies of which had been found aboard *Paris*.

Accordingly, the initial plan was that *Paris* should go to the Poles, but this was cancelled in favour of employing her at Devonport with the triple roles of accommodation ship for the *Defiance* torpedo training establishment, base and repair ship for Captain Motor Launches (ML), his staff and the 1st ML Flotilla, and base and repair ship for the 1st MTB flotilla. Taken in hand on 2 September 1940 for the necessary repairs and alterations, the battleship commissioned on the 12th as 'French Ship *Paris*', for special service as a tender to HMS *Drake*. Work continued through October and November, a minute by C-in-C Western Approaches on 17 October noting the need to refit boilers and auxiliary machinery to provide essential services, as well as to provide power for base ship roles. The minute also commented on the priority of the work, as *Defiance* ratings were currently living under canvas. In addition, the current ML base-ship, the target ship (ex-battleship) *Centurion*, was being overwhelmed by the need to look after additional auxiliary patrol vessels, and had insufficient capacity to look after the MLs properly as well. The decision was now taken to fly both the White Ensign and the French ensign.

It was also noted that Vice Admiral Émile Muselier, appointed on 1 July 1940 by General de Gaulle as commander of the Free French Naval Forces (FNFL), had informally requested that eight 138.6mm guns be removed from each of *Paris* and *Courbet* for use elsewhere. This was regarded very much as a positive, as it would clear out two broadside batteries and greatly increase light and air on mess decks.

On 3 November 1940, it was decided that the scope of the refit should be reduced, as all electrical power would now be provided from shore. This meant that there was no need to undertake work on the boilers and main machinery, and on the auxiliary machinery only as necessary to use shore power for workshop requirements and ship's domestic purposes. On the other hand, two auxiliary boilers were to be fitted to meet hot water, mess deck heating and drying room requirements. All secondary guns were to be removed, and the main armament placed in a state of preservation; only the light AA battery was to be retained for defence purposes. Costs were estimated at around £50,000. Final work included a pair of deckhouses fore and aft of the ship's wing turrets.

Paris's seven 75mm HA guns were also removed and sent to Douala and Libreville in West Africa, along with nine 138.6mm guns removed from *Courbet*, in February 1941; they were transported by SS *Henri Gasper*, which was, however, mined and beached *en route*. Of the twenty-seven serviceable 138.6mm weapons that remained from the two battleships, twenty-six of them were made available for coastal defence purposes in the UK, seven batteries being so equipped from January 1942.[6] The guns were returned to France at Cherbourg in September 1945.

On 23 May 1941, C-in-C Plymouth made a request to the Admiralty that for 'local reasons connected with the efficient administration of trawlers and small craft based at Plymouth, it is urgently necessary to commission F.S.[7] PARIS as an independent command as Auxiliary Vessels Base Ship', also asking for a decision as to whether she should be called *Paris* or by some other name as a Royal Navy ship. The initial decision was that she should be 'His Majesty's French Ship *Paris*', commissioning as such on 9 June, but becoming simply FS *Paris* on 3 July.

Paris would continue to service her flock of small vessels until the summer of 1945. In February 1943 a report into

Paris at Devonport in 1941, repainted and riding higher following the disembarkation of equipment including the secondary guns, searchlights and 75mm HA guns (although one – presumably British – weapon has been fitted ahead of no.1 turret). Deckhouses have also been erected forward of the wing turrets. (Author's collection)

possible use in trials work noted that 'She cannot steam nor can she be towed.' In February 1944 *Paris* came under consideration as one of the obsolete warships to be employed as breakwaters during the prospective Normandy landings. However, while her sister *Courbet*, now laid up idle in the Clyde after a period as a target ship,[8] was selected for scuttling alongside *Centurion*, the Dutch cruiser *Sumatra* and the British cruiser *Durban*,[9] it was felt that the removal of *Paris* from her role at Plymouth would lead to too much disruption.

The important services provided by *Paris* once more came into question when, on 21 February 1945, a request came from the French authorities that she be returned to them for use as accommodation ship for the naval schools at Brest to replace shore accommodation that had been destroyed between 1940 and 1944. In response, on 25 March it was noted that:

– Current duties of *Paris* included being the base for 140 vessels, with requirement enduring for some time after defeat of Germany.
– She would need to be replaced by another ship or building new shore facilities (latter impracticable).
– 18 months work would be needed to get ship fit to steam at 12 knots, to include:
  new boilers
  complete overhaul of main engines: one shaft immovable on arrival in 1940; no astern power on another
  complete refit of all auxiliary machinery
  renewal of pipework
In connection with the need to do so much work, it was commented that '[i]t is considered that it was a remarkable engineering feat to get the ship here at all in her present condition, and normal experience is that miracles are not often repeated'.
– Even if another ship could be made available to replace *Paris*, the change-over would be extremely inconvenient. Accordingly, it was suggested that the battleship be retained, and that one or more of the cruisers due to reduce to Category C reserve be offered on loan to the French for use at Brest.

Director of Plans suggested a number of potential vessels for loan on 24 April 1945: the aircraft carrier *Furious* (noted as 'unable to steam – best offer') and the cruisers *Caledon*, *Colombo*, *Delhi* and *Hawkins*. The latter two were included on the proviso that agreement could be reached on the reduction to Category C reserve. On the other hand, it was noted that in spite of the inconvenience of handing back the French battleship, there was no real justification for withholding her from the French. It was therefore agreed four days later that there was indeed no British case for retention, although the ship's return should wait until the end of the war against German submarines. It was also noted that *if* a vessel were loaned instead, *Furious* was certainly big enough to accommodate the required 600 French trainees, but was in a poor condition and might

not be able to raise steam even for auxiliary purposes.

On 8 May it was decided that the cruiser *Colombo*, due to arrive at Plymouth about 12 May to reduce to Category C reserve, would be an appropriate choice to relieve *Paris*, a decision promulgated on the 22 May. *Paris* therefore paid off on 30 June and *Colombo* commissioned the next day; she would remain in service until March 1946.[10] The battleship was handed back to the French Navy on 14 July and towed to Brest on 21 August by the tugs *Champion*, *Attentif* and *Mammoth*. She then took up her duties as base-ship for the 2nd Maritime Region, and continued in this role until the end of 1955, when new shore facilities were ready. *Paris* was stricken on 21 December and given the designation Q64, pending sale. This was not long coming, the ship being condemned and sold for demolition early in 1956; she was towed to La Seyne during May/June to be broken up.

Endnotes:

1 This note is based primarily on UK National Archives file ADM 199/664, 'Relations with neutral countries and commissioning of French Ship PARIS as tender to HMS DRAKE', with some data from B Warlow, *Shore Establishments of the Royal Navy*, revised edition, Maritime Books (Liskeard, 2000), 106, with some corrections from ADM 199/664. For technical and broader career information on *Paris*, see J Jordan and P Caresse, *French Battleships of World War One*, Seaforth Publishing (Barnsley, 2017).
2 In documents found aboard by the British in 1940, and translated in ADM 199/664; the text published here reproduces the typography of the original.
3 Built with twenty-four coal-fired boilers, *Paris* had the forward four replaced by oil-burning units during 1922–23, but although the boiler outfit was overhauled during 1934–35, she remained primarily coal-fired, with all her coal-burning boilers dating from her original building. *Courbet* had received four oil-fired boilers during 1923–24, but during 1927–31 all her original boilers had been replaced by units originally made for cancelled *Normandie*-class vessels: four oil-fired from *Normandie* herself and sixteen coal-fired from *Flandre*. She had also received new cruise turbines.
4 On the evacuations surrounding the French armistice, see L Hellwinkel, *Hitler's Gateway to the Atantic: German Naval Bases in France 1940–1945*, Seaforth Publishing (Barnsley, 2014), 12–16; among the latter were a number of former armoured cruisers – see Dodson, *Before the Battlecruiser: the Big Cruiser in the World's Navies, 1865–1910*, Seaforth Publishing (Barnsley, 2018), 152.
5 In all, two battleships, two large destroyers, eight destroyers and torpedo boats, seven submarines, thirteen sloops, sixteen submarine chasers, three minelayers, one target ship, seven MTBs, 98 patrol craft/minesweepers, 42 tugs and harbour craft, and 20 trawlers were taken over.
6 J Dorman and J Guy, 'The French 13.8cm gun in British coast defence service', *Warship 2013*, 168–171.
7 French Ship.
8 For *Courbet*'s post-1940 career, see S Dent, 'The battleship *Courbet* and Operation "Substance"', *Warship 2016*, 152–160.
9 On these vessels' preparation, see UK National Archives ADM 1/17053.
10 *Colombo* then paid off; she was sold on 22 January 1948, and arrived at Newport to be broken up on 1 May.

THE ATTACK ON THE CRUISER *LEIPZIG* ON 26 MARCH 1945

Andrei Latkin and **Sergei Trubitsyn** recount a little-known incident from the last days of the *Kriegsmarine.*

The participation of the cruiser *Leipzig* in the Second World War is well documented in the literature, but there are gaps in the account. One of them is the passage from Gotenhafen to Danish waters on 26 March 1945, during which the cruiser was attacked by Soviet aircraft of the Red Banner Baltic Fleet.

During the Second World War, *Leipzig* repeatedly received combat damage. In December 1939 she was torpedoed by a Royal Navy submarine. The damage was significant, and after repairs were completed she did not return to the active fleet but was employed as a training ship. In the autumn of 1941 she participated in the operation to capture the Moonsund Islands in the Baltic. *Leipzig* almost became a victim of Hitler's order of January 1943 to scrap all major surface vessels. The cruiser spent some time in reserve and then again began to be employed as a training ship. Throughout 1944 the situation on the Soviet-German front deteriorated for the Third Reich, and the sector of the front adjacent to the Baltic Sea was no exception. In October 1944, *Leipzig* received an order to prepare for a minelaying operation. During this operation, on 15 October the cruiser collided with *Prinz Eugen*. The damage was severe. It was not possible to make a proper repair, so *Leipzig* was reclassified as a floating barracks. When the front approached Gotenhafen, she was used as a floating artillery battery. At the end of March 1945, the German Navy Command decided to transfer her to the west, and she went to sea as part of a convoy.

On the morning of 26 March 1945, reconnaissance aircraft of the 15th Separate Reconnaissance Tallinn Red Banner Aviation Regiment (hereinafter 15 ORAP) embarked on their customary sorties to search for German ships, which were carried out on a regular basis subject to weather conditions.

The attack on the German convoy Pi.274 by A-20 Boston torpedo bombers of 51 MTAP led by AA Bogachev, in which the tanker *Sassnitz* was sunk.

A port quarter view of *Leipzig* on 4 March 1943. (Leo van Ginderen collection)

The first major contact was made by the crew of Lieutenant I A Yaskin, who piloted A-20G[1] aircraft No 12. At 0714 a radio report was received: '0705, 56°10N 20°20E, 2 transports, 2 patrol boats, 1 minesweeper. Course 40°, 4 knots, line ahead.'

This was convoy Pi.274, sailing from Pilau to Libau and comprising the tanker *Sassnitz* (694 tonnes brt), the gunnery training ship *Delphin*, and the motor minesweepers *R.145*, *R.249* and *R.260*. As a result of a massive aerial assault launched by the 11th Assault Novorossiysk Red Banner Aviation Division (hereinafter 11 ShAD) and the 51st Tallinn Mine-Torpedo and Aviation Regiment (51 MTAP) of the 8th Mine Torpedo Gatchina Red Banner Aviation Division (8 MTAD) the convoy was eliminated: *Sassnitz* and the minesweepers *R.145* and *R.260* were sunk, and *Delphin* and the minesweeper *R.249* were damaged.

Prior to this attack, at 0645 a reconnaissance aircraft of 15 ORAP (Petlyakov Pe-2 No 61: pilot Lieutenant V I Shapochkin) was launched to reconnoitre the middle part of the Baltic as far as 16°00E and the Bay of Danzig. Shapochkin's aircraft transmitted three reports. At 0746 the following report was received at HQ: '0733, 54°45N 18°58E, 4 transports, 4 minesweepers, 1 torpedo boat. Course 180°. Air defence formation. Clear, visibility 1 km, haze.'

Half an hour later, at 0817, a report was received of a second convoy: '0804, 55°17N 17°38E, 3 transports, 1 patrol boat, 2 minesweepers; course 280°, 8 knots. Air defence formation, Clear, visibility 1 km, haze.'

At 0915 a report was received of a third convoy, which is the one which interests us: '0850, 55°12N 17°30E, 1 cruiser, 3 transports, 2 torpedo boats, 4 minesweepers. Course 280°, 8 knots. Air defence formation. Clear, visibility 1 km, smoke.' At 0922 reconnaissance aircraft No 61 landed safely.

The third convoy spotted by No 61 was designated Go.648 and comprised the cruiser *Leipzig* (under tow by a small uncompleted ship), the transports *Zephir* and *Ellen*, the auxiliary *Herkules*, tugs *A3* and *A6* towing MTBs, torpedo recovery vessels *T.123* (ex-*S.23*, ex-*Komet*) and *TF.18* with a barge in tow, the patrol boat

Pilot Junior Lieutenant V I Dolbin (51 MTAP), who died during the attack on the cruiser *Leipzig* March 26, 1945.

Navigator Junior Lieutenant P M Belyakov (51 MTAP), who also died during the attack.

Pilot Junior Lieutenant G S Khodorenko (51 MTAP), whose torpedo bomber took part in the attack.

V.313, the minesweeper *M.401*, the torpedo training ship *TS.4*, the floating battery *Orion*, and the motor minesweepers *R.242* and *R.245*.

This convoy was of particular interest to the Red Banner Baltic Fleet Air Force command because it included a cruiser, but in order to organise an attack it was necessary to assemble a large group of torpedo planes and bombers, and almost all aircraft available had already been committed to attack the convoys previously sighted, including Pi.274. At 0800 on 26 March the 1st Guards Mine-Torpedo Red Banner Aviation Regiment (hereinafter 1 GMTAP) had 19 aircraft: 11 A-20G, of which only seven were serviceable, and eight Ilyushin Il-4, of which seven were ready for combat. (The Il-4 was engaged only in laying mines and did not participate in torpedo attacks.) 51 MTAP had 23 A-20G, of which 16 were serviceable.

However, the command of the Red Banner Baltic Fleet Air Force could not ignore the presence of a convoy that included a cruiser. At 0956 Pe-2 No 21 of 15 ORAP (pilot: Lieutenant A P Arisov) took off, tasked with conducting reconnaissance of enemy ships in the south-western part of the Baltic Sea. Like the first scout, he found three convoys. His first report was received at 1102: '1044, 55°16N 17°40E, 2 patrol boats, 1 unidentified vessel. Course 70°, 8 knots. Clear, visibility 2 km, haze.'

The second report was received at 1130: '1100, 55°15N 17°00E, 1 cruiser, 2 torpedo boats, 3 patrol boats. Course 270°, 8 knots. Clear, visibility 4 km, haze.' Twenty minutes later, at 1150, a third report was received: '1105, 55°17N 16°54E, 1 transport, 2 minesweepers, 3 patrol boats, 1 floating dock. Course 270°, 5 knots.' At 1247 Pe-2 No 21 landed.

Based on the information received from this aircraft, it became clear that the weather in the area of the convoy that included the cruiser would permit an attack, and the torpedo planes and bombers due to return after the attacks on the other convoy could be directed against the cruiser.

While the headquarters of the Red Banner Baltic Fleet Air Force were planning an attack on the convoy, the ships of the latter rescued the crew of one of the 51 MTAP torpedo bombers which had come down during the first attacks. At 1256 the convoy picked up an inflatable boat with three Soviet aircrew. According to German documents these were the pilot, Lieutenant G G Enikeev, the navigator, Lieutenant N E Danilin, and the gunner/radio operator, Junior Sergeant D N Suleymanov. They never returned; Soviet documentation records them as missing in action (MIA).

Based on the data provided by Arisov's Pe-2, the following order was given: 'The commander of the Red Banner Baltic Fleet Air Force orders the commanding officer of 8 MTAD to organise a concentrated attack of torpedo bombers on the force comprising 1 cruiser, 2 torpedo boats, 1 minesweeper, 3 patrol boats, detected at 1100 at 55°15N 17°00E with the task of sinking the cruiser.'

These orders were promptly executed. The commanding officer of 8 MTAD assembled a force comprising eight A-20 torpedo bombers from 51 MTAP and three A-20 torpedo bombers from 1 GMTAP. Their primary mission was to sink the cruiser, their secondary mission to sink the transport in transit. The first group was to be escorted by six Yak-9 fighter aircraft from 21 IAKP, the second by two Yak-9s from the same formation.

The necessary orders, which are recorded in the war log of 8 MTAD, were transmitted between 1320 and 1323. The focus now was on keeping the convoy under constant surveillance so that its position could be monitored in real time. At 1300 Pe-2 No 10 of 15 ORAP (pilot: Lieutenant N S Vereshchagin) took off with orders to search the southwestern area of the Baltic as far as

15°30E. Then, at 1445, an A-20 from 8 MTAD (pilot: Lieutenant P N Koltashenko) took off, tasked with finding and shadowing the cruiser and directing the strike groups towards the target. Unfortunately, he was unsuccessful and the cruiser was not found. However, the Pe-2 managed to locate the convoy, and at 1515 the following report was received: '55°00N 16°00E, 1 cruiser, 2 torpedo boats, 3 transports, 2 minesweepers. Course 340°, 12 knots. Clear, visibility 2 km, haze.'

What happened next is best described by two combat reports, one from 1 GMTAP, the second from 51 MTAP.

Combat report No 64 of the headquarters of 1 GMTAP, Grabshtein Airfield, at 2000 26 March 1945:

Between 1534 and 1536 three A-20, one armed with a torpedo (pilot: Guards Lieutenant Skryabin) and two armed with bombs (pilots: Guards Junior Lieutenants Mozhakin and Masur), accompanied by aircraft from 51 MTAP, took off from Grabshtein airfield to attack the enemy cruiser reported at 55°00N 16°00E.

At 1658, position 55°24N 16°08E, an enemy force comprising a cruiser of the *Köln* class, three torpedo boats and three patrol boats was sighted, proceeding on a course 270° at 10 knots.

At 1700, with an angle of descent 20°, at a height of 25–50 metres, at a range of 1,000–1,400 metres, a speed of 300–480km/h and a bearing of 70° from the port side of the ships, the flight crews launched a simultaneous torpedo and bombing attack on an enemy cruiser of the *Köln* class. The crews of Mozhakin and Masur's aircraft observed their bombs fall short by 800–900 metres. Scryabin watched the wake of his torpedo but did not see it explode.

The cruiser attacked had two funnels and two masts,[2] with a bridge structure forward of the funnel. It had a light grey paint scheme, freeboard of about 3–3.5 metres, a length of 180 metres and a displacement of 6,000 tonnes.

When approaching the enemy and when turning away after the attack, the aircraft were subjected to heavy AA fire from medium- and large-calibre guns, automatic cannon and machine guns.

The aircraft landed at their home airfield between 1834

An A-20 Boston attacks a German convoy.

Attack by aircraft from 1 GMTAP on an enemy convoy on 26 March 1945, led by Guards Lieutenant AE Scriabin. [All photos are from the collection of the Central Archive of the Ministry of Defence of the Russian Federation (Archive of the Navy) at Gatchina.]

and 1836. ... There were no casualties, no accidents or encounters with enemy aircraft.

Operational summary No 170 drawn up at the HQ of the Red Banner Baltic Fleet Air Force, Palanga, at 2000 on 26 March recorded the results of the strike as follows: '... Due to poor visibility (thick haze, visibility 1.5–2km) the results of the strike were not observed.' Based on German documents, this first attack on this convoy did not yield results; however, the Germans recorded one downed Soviet aircraft, which was not the case.

Combat report No 53 of the HQ of 51 MTAP, Grabshtein Airfield, at 2000 26 March 1945:

Two groups took off between 1528 and 1535 to attack a light cruiser reported at 55°00N 16°00E. The first comprised five A-20 led by Captain Makarikhin: his own aircraft and those of Lieutenants Polyushkin and Piznik were armed with torpedoes, those of Junior Lieutenants Khodarenko and Dolbin with bombs. The second group comprised four A-20: the aircraft of Lieutenants Shklyarevsky and Gorbushkin and 2nd Lieutenant Petrov were armed with torpedoes, that of Junior Lieutenant Balakin with bombs The two groups had an escort of Yak-9 fighters from 21 IAKP (leader: Chistyakov), and landed back at the airfield at Memel between 1830 and 1847.

At 1707 the group of Captain Makarikhin discovered a light cruiser at 55°03N 16°00E accompanied by three torpedo boats, four patrol boats, one minesweeper, course 240°, 12 knots. At 1712 the five A-20s of his group attacked. The three aircraft armed with torpedoes focused on the cruiser, while the two armed with bombs attacked the torpedo boat stationed off her bow. An explosion was observed in the after part of the cruiser, presumably from a torpedo. A thick haze and low visibility (up to 1 km) prevented detailed observation of the results. A fire broke out on the torpedo boat attacked by the two bombers.

At 1715, Lieutenant Shklyarevsky's group came in for the attack. The three aircraft armed with torpedoes again concentrated on the cruiser, while the single bomber attacked the torpedo boat off her port bow. An explosion was observed on the cruiser, but the prevailing weather conditions precluded a detailed analysis of the results.

All the attacking aircraft were met by an intense fire from the HA guns, automatic cannon and machine guns. Aircraft No 13 of Captain Makarikhin's group (pilot 2nd Lieutenant V I Dolbin; navigator Junior Lieutenant P M Belyakov; gunner Sergeant Mescherinov) was shot down while attacking the cruiser. The aircraft was set on fire as the pilot turned away from the attack and went down in flames; after that there was no sign of the crew.

In operational summary No 170 the attack by 51 MTAP again recorded: ' ... Due to poor visibility (thick haze, visibility 1.5-2 km) and strong opposition, the results of the strike were not observed.'

The Germans treated the second and third attacks as one, and they again reported one Soviet aircraft shot down; on this occasion the report was correct. The cruiser *Leipzig* was under tow by a small ship. The latter was struck by a torpedo; the cruiser was undamaged.

The ship set on fire was the floating battery *Orion*, which also suffered flooding in her engine room from four near-misses. The engines and their auxiliary machinery were damaged. As a result, the ship lost way and had to be towed into Swinemünde, which it reached on 29 March. It is not clear whether *Orion* was repaired before the end of the war. Despite the best efforts of the Soviet pilots, the cruiser *Leipzig* was undamaged.

Endnotes:
1   The Douglas A-20G Boston bomber was supplied to the Soviets under Lend-Lease.
2   *Leipzig*, unlike the earlier German cruisers of the *Köln* class, in fact had only a single broad funnel, and like the *Köln*s had only a single tubular foremast.

SMS *LEITHA*

Aidan Dodson describes the former Austro-Hungarian river monitor preserved in Budapest.

On the river Danube in the centre of the Hungarian capital Budapest, and dwarfed by the country's huge Parliament building, lies the 300-tonne river monitor *Leitha*. She was launched at Budapest in 1871 and commissioned the following year, serving in the Austro-Hungarian Navy's Danube flotilla until its demise at the end of the First World War. As completed, she and her sister *Maros* were armed with a pair of 15cm/21 Wahrendorf breech loaders in a single turret, with a pair of 25.4mm machine guns abreast the funnel.

She was rebuilt in 1894, including the complete replacement of her armament, with a single shielded 12cm/35 Krupp gun forward, plus two 47mm/33 and an 8mm. The secondary armament was changed to one 6.6cm/45, one 6.6cm/18 and three 8mm weapons in 1915, mounted on an enlarged superstructure. Paid off in April 1918, *Leitha* was reactivated in early 1919, playing a role in events surrounding short-lived Communist regime in Hungary.

While other ex-Austro-Hungarian monitors were distributed between Yugoslavia and Romania in December 1920, the old *Leitha* and *Maros*, plus the newer *Szamos* (1894), were sold in 1921. *Maros* was broken up, but the other two were stripped down to their bare hulls and rebuilt as the unpowered dredgers *József Lajos* and *Tivadar*. They served in this role under various ownerships, from 1948 known as *FK-201* and *FK-202*, until the end of the 1980s. *FK-202*, long laid up, was then sold for scrap, but the ex-*Leitha* was still in service in 1992, when she was sold to the Swiss firm Kies und Beton.

Various proposals had been made since the 1970s to preserve one or other of the ex-monitors, which finally

Broadside view of the restored *Leitha* at her moorings in front of the Hungarian Parliament building. She is part of the Military History Museum.

Stern view of *Leitha* with Budapest's Margaret Bridge, built in 1876 and thus almost the same age as the monitor, in the background.

A forward view of *Leitha* showing her replica turret and pilot house. When preserved, she was a bare hull, much of the final reconstruction being carried out at Komarno, in Slovakia, during 2009–10. (All photographs by Aidan Dodson)

came to fruition when the hull of the former *Leitha* was donated by her owner to the Hungarian Military History Museum. Following initial conservation, she was displayed for five days in central Budapest in May 1996, before beginning a full restoration to her original appearance in 2005: she was inaugurated in her new permanent

berth next to the Parliament building on 20 August 2010.

Leitha is actually not the only survivor of the old Austro-Hungarian Navy on the Danube. The hull of the 1904 monitor *Bodrog*, later Yugoslav *Sava,* used from 1962 as a barge, was formally preserved in 2015, completing restoration in 2019, for display at Belgrade.

A's & A's

THE MOBILE NAVAL BASE (*WARSHIP 2020*)

Aidan Dodson provides some additional notes on the fate of HMS *Agincourt.*

Regarding David Murfin's excellent note on the abortive conversion of the battleship HMS *Agincourt*, it may be noted that preparations for her reconstruction got farther than is implied. *Agincourt* – which had been in reserve since March 1919 and in care & maintenance since March 1920 – actually recommissioned on 2 November 1921 to facilitate preparations for her refit. However, the latter was cancelled on 23 February 1922 (17 days after the signature of the Washington Treaty) and, having completed de-storing by the middle of March, the ship paid off for disposal on 7 April 1922. Her care & maintenance crew was then retained at Rosyth to aid in de-storing *Lion* and *Princess Royal* (information from TNA ADM 1/8620/35, minute dated 7 March 1922).

Agincourt was sold to J&W Purves on 22 January 1923 for £25,000 (bid accepted 16 December 1922) as a 'job lot' with *New Zealand* (£21,000) and *Princess Royal* (£25,000), all ships having been laid up at Rosyth. They were handed over on 25 January 1923 to the Rosyth Shipbreaking Co, to whom the contract had been transferred, to be broken up by '18 months from date of ratification of [Washington] Treaty' (*ie* January 1925; data from Admiralty Sales Ledger, Naval Historical Branch, Portsmouth. The ships' demolition, in a leased location within Rosyth Dockyard, is documented in the album R1836075 in the collections of Archives New Zealand – two images from the album have previously appeared in Ian Buxton's *Metal Industries* [1992]), and while *New Zealand* and *Princess Royal* had been adequately dealt with by the deadline, the lower part of *Agincourt*'s hull was deemed to be insufficiently mutilated to meet the terms of the contract. Accordingly, it was agreed with the Admiralty that the hulk, which still watertight, should be cut in two, and so deemed contractually 'destroyed'.

It was originally intended to do this at low tide on the breakers' beaching ground, outside the Dockyard basin, but weather conditions on the day chosen for the transfer (the highest Spring Tide) were such that it was judged too hazardous for the hulk to pass out of the Dockyard entrance locks. This created a problem, as the next appropriate Spring Tide would be after the 'destruction' deadline had passed. A solution was found in carrying out the work in one of the Dockyard's dry docks, the cutting between two midships transverse bulkheads being complete even before the dock was fully empty. Within 48 hours of entering the dock, both halves had been

floated out and later taken separately to the beaching ground for final breaking up.

A similar 'bisection' had been carried out on the cut-down hulk of HMS *Lion*, on the Tyne at Hebburn, in November of the previous year, 1924. Its purpose had been to permit the hull to enter the port of Blyth, where final demolition was to be carried out (see White, *Battleship Wharf* [1961]), but one wonders whether the feat might have provided Rosyth Shipbreaking with inspiration for their solution to the '*Agincourt* Problem'.

THE BEGINNINGS OF SOVIET NAVAL POWER (*WARSHIP 2020*)

Author Przemysław Budzbon has written in to point out an error in his Table 2 published on page 91. The Date of Specification for the Series I submarine should read 17 Feb 19<u>2</u>7 [not 1937].

WARSHIP GALLERY (*WARSHIP 2020*)

Readers have contacted us to point out errors in the captions for last year's Warship Gallery. In the caption for the upper photograph on page 220 there is a reference to an article by Peter Marland 'in this year's annual'. Peter's article was indeed due to be published in 2020, but because of lack of space had to be held over until 2021 (see pp 150–164); this decision was only taken relatively late on in the production process, and the cross-reference in the caption was missed.

Also in the 2020 Gallery, there were two references to an ESM array referred to as a 'Dunce's Cap': to clarify, this term, as well as being archaic (and arguably offensive), is strictly unofficial as used here.

LCT 7074 (*WARSHIP 2020*)

We had intended to be able to provide a detailed update to the story of *LCT 7074* in *Warship* 2021, covering the move to her final berth on Southsea Common, planned for April 2020. It was noted in Stephen Fisher's article that this was dependant on 'unpredictable environmental factors'; what no one could have predicted at the time was the appearance of the coronavirus pandemic. The result was that the move did not eventually take place until the night of 23–24 August – the unusual timing being to reduce the likelihood of crowds of onlookers. Even then unseasonally high winds meant that it took two attempts to get the LCT loaded onto the barge that transported her from Portsmouth's Historic Dockyard to the beach at nearby Southsea. After that came a final short journey by road to where she now rests, on public display.

NAVAL BOOKS OF THE YEAR

Aidan Dodson & Serena Cant
Spoils of War: The Fate of Enemy
Fleets after the Two World Wars
Seaforth Publishing, Barnsley 2020; 328 pages,
illustrated with almost 250 B&W photographs
and numerous drawings and maps; price £35.00.
ISBN 9781526741981

In *Spoils of War*, Aidan Dodson and Serena Cant have
tackled a subject that has, until now, received relatively
little attention in maritime history. Cox's recovery of 35
ships of the Imperial German High Seas Fleet from the
cold waters of Scapa Flow is well known, and the fate of
many other vessels formerly in service with the Central
Powers or the Axis has been covered in articles and
online, but this is the first book to bring all of these
stories together in such impressive detail.

Aside from a brief introduction at the start of the book
and an equally brief conclusion at the end, the book is
divided into two independent parts of broadly equal
length. Part One opens with a summary of the disposition
of the German, Austro-Hungarian, Bulgarian and
Turkish fleets at the end of the First World War, and their
immediate fates after the armistice. In most cases this
included their surrender and subsequent arrival at Allied
ports, but in some instances (especially in the Balkans)
the political situation dictated different outcomes. The
following chapter details how the ships were intended to
be – and how they ultimately were – divided amongst the
victors. A third chapter covers the stories of individual
ships under their new owners and concludes with a 50-
page table covering all the Central Powers' ships in
commission at the end of the war, of which there were
more than a thousand, and their ultimate fates.
Understandably, not every individual story can be told
and those that are recounted here are necessarily
restricted by space. Major vessels are generally dealt with
individually, with smaller ships (ranging from destroyers
to coastal forces) covered by group or class. As might be
expected, most were scuttled or expended as targets, but
some were commissioned into Allied navies. Often
forgotten are those that remained in their original navies
and those that were converted for peaceful maritime
purposes. Three appendices include a 1921 report
detailing the dismemberment of the High Seas Fleet, a
section on the reuse of U-boat engines in the UK, and an
impressive eight pages on the surviving archaeology of
German warships in UK waters.

Part Two, which looks at the fleets of Germany, Japan,
Italy, Romania, Bulgaria and Finland after the Second
World War, follows much the same structure, closing
with a 70-page table, although the appendices are
arguably less informative, being primarily reproductions
of postwar documents.

All the well-known aspects of both stories feature,
including the Grand Scuttle in 1919 (and the changes in
the allocation of vessels that resulted), Operation
'Deadlight' and the Bikini Atoll trials. None are covered
in great detail, but more context is provided than in
previous published works.

There are some small criticisms that relate principally
to the layout of information. For example, the sub-
section on submarines in Chapter 1 flows from the
U-boats' surrender in UK ports into the individual fates
of some U-boats that were used in a 'publicity' tour
around Britain. This feels out of place, not least because
the same subject opens the second chapter. Similarly,
many of the photographs are separated from their
coverage in the text. However, these are minor quibbles
and certainly do not detract from the impressive detail
and accuracy on display throughout. This book will be
invaluable for maritime researchers; to have so much
material combined into one comprehensive volume is a
Godsend for historians and archaeologists alike.

Stephen Fisher

Conrad Waters
British 'Town' Class Cruisers:
Design, Development &
Performance, *Southampton* and
Belfast Classes
Seaforth Publishing, Barnsley 2019; large format hardback,
320 pages, illustrated with numerous black & white photographs
and drawings, one four-page colour foldout and eight colour
plates; price £40.00.
ISBN 978-1-5267-1885-3

This book is intended as a technical history of this
famous class, though it also provides sufficient operational history to evaluate performance in action. It has
used much additional information now in the public
domain and it also covers the postwar period. It is generously illustrated in black-and-white (many photographs
are from the author's collection) and in colour.

The first four chapters cover, respectively, the origins of
the class, the design process, construction, and the
detailed design of the ten ships. That on the design
process, from staff requirements to final specification,
successfully sets the scene for the next two chapters.
Waters suggests that 'the class's powerful anti-aircraft
armament was the envy of many contemporary navies'.
Rather than to the 4in, this must refer to the close-range
armament of two quadruple, and in *Belfast* and
Edinburgh two octuple 2pdr (40mm) pom-poms.
However, there should be a caveat that initially these
mountings were controlled with the Mark II pom-pom
director that relied on 'eye-shooting' with a simple ring
sight. During the war, this was superseded by the Mark

IV director with Gyro Rate Unit (GRU) and later Type 282 radar; unfortunately, their dates of introduction remain uncertain due to gaps in the wartime record. The same problem is found for the evolution of the much-criticised HACS system controlling the 4in guns. Originally this could deal only with aircraft flying the steady level courses necessary for high-altitude bombing (and perhaps also with torpedo bombers making their approach). It was enhanced during the War by the addition of the GRU and associated GRUB to give it the tachymetric capability needed to counter diving attacks, but again the dates of introduction are, it seems, uncertain. At the beginning of the War, HACS had little success at shooting down aircraft, though it was able to deter and disrupt their attacks; *Gloucester* was sunk in 1941 only after she had exhausted her 4in ammunition. However, there is insufficient information to judge how far the improvements to HACS (only the majority of the dates for the installation of Type 285 radar are known) made up for the improved methods of attack developed by the Axis air forces. (It is interesting that three attacks by torpedo aircraft were beaten off with barrage fire from the main armament.)

Chapter 5 describes the wartime improvements, not least the plethora of radars (including gunnery radars) that these cruisers proved capable of accommodating. Chapter 6 analyses performance in wartime operations. Four of the class were sunk: *Southampton* and *Gloucester* by bombs, *Edinburgh* and (in controversial circumstances) *Manchester* by flooding following torpedo hits; *Manchester* had earlier survived similar flooding. Following DK Brown, Waters is critical of the risk of rapid capsize inherent in the wing compartments flanking 'B' boiler room in British cruisers with the unit system of propulsion. However, as Waters acknowledges, despite the instances of serious flooding in the 'Towns' none experienced the special conditions necessary for this type of capsize – they seem to have arisen only in smaller cruisers.

Chapters 7 and 8 complete the story of the 'Towns'. *Glasgow* and *Liverpool* remained as they had been at the end of the war, though the other four received 'large repairs' that fitted them for nuclear-biological-chemical warfare and for tachymetric control of their 4in and (by that time) Bofors 40mm guns. Fortunately, *Belfast* remained in the reserve ship organisation until she could be preserved in 1971. This excellent book will surely be of interest to many of her visitors and to *Warship* readers seeking a detailed study of her and her nine sister ships.

John Brooks

Peter Hore
Henry Harwood: Hero of the River Plate

Seaforth Publishing, Barnsley 2018; hardback, 256 pages, 63 plans & photographs; price £25.00.
ISBN 978-1-52672-529-5

Written by former RN officer and Head of Defence Studies Captain Peter Hore, *Henry Harwood* is the first published biography of the victor of the famous December 1939 Battle of the River Plate. Later promoted to Commander-in-Chief, Mediterranean, Harwood's subsequent career was marked by a series of setbacks from which his reputation never recovered. Supplementing material drawn from an extensive range of archival and published sources with exclusive access to Harwood's private family papers, Peter Hore looks to use this biography to rehabilitate the admiral's legacy.

Following a foreword from former First Sea Lord Admiral Sir Jock Slater and a brief Introduction, the book comprises 20 chapters chronicling Harwood's career. The initial five cover the period from his entry into Stubbington House – a preparatory school for the naval college entry examinations – up to his appointment as commodore in command of the Royal Navy's South American division in 1936. This period saw him establish relationships that were to have a major influence on his wartime service, perhaps most significantly with future Admiral of the Fleet Sir Dudley Pound. An interesting aspect of his career was his Catholicism during a period when 'to be Roman Catholic in the Service gave you another kind of oddness'.

Immediately before taking up his South American command, Harwood had lectured on the attack and defence of trade at the Royal Naval College's senior officers' war course. Hore demonstrates how this and the knowledge of South American waters gained from Harwood's subsequent posting were to stand him in good stead during the River Plate engagement. The battle and its aftermath inevitably receive considerable discussion, and while this adds little new information to an action that has already been extensively analysed, it does demonstrate the wide acclaim which Harwood's conduct earned at the time.

Harwood's subsequent tenure as Assistant Chief of Naval Staff (Foreign) at the Admiralty, covered rather more succinctly, served as prelude to his surprise appointment as C-in-C Mediterranean in succession to Andrew Cunningham in April 1942, a double-jump promotion from rear admiral to acting admiral that Cunningham opposed. Much of the rest of the book – some 70 pages – is focused on his performance in this role.

It is widely acknowledged that Harwood assumed his command at a time when the Mediterranean Fleet's fortunes were at a low ebb, with cumulative losses leaving few major ships ready for action. This situation was soon exacerbated by Rommel's capture of Tobruk and drive into Egypt. Hore assesses the consequences on the unsuccessful 'Vigorous' convoy and the temporary evacuation of the base at Alexandria. A constant underlying theme is lack of support from other service heads in the theatre, initially Air Marshal Sir Arthur Tedder.

Matters did not improve when the tide began to turn, most notably manifested in a difficult relationship with Eighth Army's commander, Bernard Montgomery. Montgomery's complaints about the Navy's tardiness in reopening a captured Tripoli to shipping – considered unjustified by Hore – were a major factor in a decision to reassign Harwood to the Eastern Fleet as second-in-command. Hore speculates that antipathy towards Harwood's Catholicism from the Protestant Ulsterman might have impacted the relationship. In the event, a heart attack prevented Harwood assuming his new position and he never enjoyed another fleet role.

Hore makes strong arguments, backed by in-depth research, that many of Harwood's misfortunes during his Mediterranean tenure were not of his making. However, doubts remain. Harwood was undoubtedly the right person in the right place at the time of the Battle of the River Plate; the result was a resounding victory. Appointment to the Mediterranean command brought new challenges at a difficult time and required different skills. Importantly, Harwood was unequal to the challenge of establishing effective inter-service relationships with his counterparts, and this played a major part in the reverses that followed.

Conrad Waters

Andrew Choong
Armoured Cruiser *Cressy*: detailed in the original builders' plans
Seaforth Publishing in association with the National Maritime Museum, 2020; hardback, 128 pages, complete sets of full colour plans; price £30.00. ISBN 978-1-5267-6637-3

The latest in the series by Seaforth drawing on the collection of original plans held by the National Maritime Museum, this volume focuses on HMS *Cressy*, name-ship of a class of six armoured cruisers designed by Sir William White to match the new 'fleet' cruisers being built for Italy and France. Intended not only to police the trade routes but to operate as a fast wing of the battle fleet, they had a deep belt of 6in Krupp cemented armour amidships, with plating of similar quality and thickness on the turrets and casemates, to keep out the latest steel high-capacity shell filled with Lyddite or Melinite. As the author points out, the protection of these cruisers was almost on a par with that of the contemporary battleships of the *Canopus* class – also designed by White – but they were 3–4 knots faster, and therefore anticipated the 'battle cruiser' concept.

Andrew Choong is Curator of Historic Photos and Ships' Plans at the NMM's outstation at the Brass Foundry in Woolwich, and is therefore ideally qualified to comment on the draughts published in this book. After a clear and concise introduction to the *Cressy* design, he has opted to group the first batch of plans under the headings Layout and Structure, Propulsion Machinery, Armour, Armament, and Accommodation and Habitability. There are gatefold plans of the Arrangement of the Belt Armour, the Profile as Fitted and a Sketch of Rig. There is a section on fittings to cover the arrangements of the fore and aft bridges, deck coverings, ventilation and magazine cooling. The final section of the book has enlarged profile and section plans of the ship, as in the earlier books of the series, and enlargements of the deck plans, all of which are accompanied by a

detailed commentary. Where the plans of *Cressy* are lacking, the author has selected plans of her sisters *Sutlej* and *Euryalus* to fill the gaps.

The plans themselves are stunning, well-chosen and printed on thick, high-quality paper, and Choong's commentary shows an in-depth appreciation both of the principal considerations that provided the context for the design and of the technology of the day. The attention given to oft-neglected aspects such as the ventilation of the lower decks, the heating of accommodation spaces, cooling and refrigeration, fresh water supply, and the fireproofing of wooden fittings is admirable.

The book is not without its faults. The text on page 13 ends halfway through a sentence and is followed by several pages of plans and a new section. On page 20 there is a reference to a structural drawing of *Euryalus* rather than *Sutlej*. The author's description of the ammunition supply arrangements for the main 9.2in guns is initially confusing, and the plans of the shell hoists on page 41 are clearly front and side elevation views (not 'plan and elevation'). Many of the profile and deck plans in the first half of the book are reproduced across a double page and significant details are lost in the gutter – in one case an entire funnel.

Despite these (relatively minor) blemishes, this is one of the best books in the series and is particularly impressive in its scope and ambition. Having just spent two years researching the French counterparts of the *Cressy*s, the reviewer was struck by how many of the primary considerations apparent in the design, how much of the technology, and how many of the practices were common to both navies.

John Jordan

Kevin Brown
Fittest of the Fit: Health and Morale in the Royal Navy, 1939–1945
Seaforth Publishing, Barnsley 2019; hardback, 276 pages, 33 B&W illustrations; price £25.00. ISBN 978-1-5267-3427-3

The late David K Brown, naval architect and regular *Warship* contributor, was a great believer in the dictum that a truly successful warship was more than just a satisfactory balance of firepower, protection and propulsion, and that the human element – a happy, healthy, efficient crew – was just as vital as these material factors. He would surely approve of Kevin Brown's book. Having already written two volumes for Seaforth covering the sailing navy era, in *Fittest of the Fit* Brown turns his attention to the Second World War, when the RN went through a period of enormous change (much of which reflected the wider transformation in British society as a whole) and expansion.

Brown's style of writing, while clear and lucid, can be helter-skelter, bounding around across the subject, peppering the reader with a fusillade of facts and figures, dates and details; scarcely alighting on one area before moving on to the next; the result is that it is sometimes difficult to divine a full picture. There are a few typos, some repetition, and the occasional error (*Juno* was a destroyer, not a cruiser). The breadth and depth of the research is clear, however, and although Brown sometimes assumes a greater degree of medical knowledge on the part of the reader than might actually be the case, this may simply be the product of writing a book potentially aimed at two distinct, specialist readerships, and those with a medical background may be just as unsure about what an AMC or HF/DF was as this reviewer was about sulphonamide or neoarsphenamine.

The story is one of often impressive advances in techniques not always matched by their implementation. The pioneering work of Gillies and McIndoe in reconstructive surgery is given due credit; in contrast it is clear that in another relatively new field, blood transfusion, the Navy was rather slower than the Army in adopting a comprehensive system. Penicillin was adopted and manufactured successfully (albeit initially in a somewhat extemporised manner), tuberculosis dealt with effectively, malaria rather less so. Combat stress does not seem to have received anything like the attention it needed, while there was at least one other area which was initially neglected: at one point in 1943 there were more servicemen out of action from sexually transmitted diseases than from combat wounds. One underlying trend that does become evident was the reluctance – and even active resistance – on the part of those in authority to adopt new practices until forced to do so by overwhelming evidence or the pressure of the situation. Similarly the pernicious influence of the British class system appears regularly, from the senior officers who believed that it gave the country an advantage over its enemies (and its allies ...), to the complaints about country houses being requisitioned for use as auxiliary hospitals and so depriving the well-to-do of their use – not to mention the RNVR officer who gained twelve years seniority over the course of a single gin-lubricated lunch in a club in Ceylon.

There are numerous comparisons with the efforts of other combatant navies. In general the RN comes out of this pretty well, certainly doing better than the *Kriegsmarine* in just about everything except the welfare of submariners, whom the Germans looked after very well, and the British rather less so. The United States, while not particularly innovative, took an approach that was consistently pragmatic, results-based and highly effective. The Germans were particularly keen on appearance, the Americans on education. The Japanese seem to have been chaotic and sometimes callous, while the Soviets were under-resourced, prone to political interference, and generally a mess. All seem to have suffered from at least one problem in common: widespread poor diet meant that sailors everywhere suffered from terrible teeth.

Although *Fittest of the Fit* is essentially 'social history', the ships themselves are not ignored. An intriguing section covers the conversion of merchant vessels into hospital ships and the problems of their subsequent operation. Many, for example, were coal-fired, and coal is a 'dirty' fuel, while hospitals need to be clean... On a more general note, facilities which were perfectly satisfactory

in northern European waters were not just inadequate but quite inappropriate in, for example, the Middle East. A notable change in the layout of major warships came as a result of the realisation that the peacetime standard of a single sickbay, usually on the upper deck amidships, was not only inviting the possibility of the entire medical complement (and their patients) being wiped out by a single unlucky hit, but could also make getting attention to wounded men while in action far too slow. This arrangement was therefore replaced with the far more practical one of medical stations distributed throughout the ship. In terms of overall health, an analysis by ship type suggested that aircraft carriers were the healthiest ships to serve on, depot ships the least.

Much more is covered in this fascinating book, which leaves the reader having simultaneously learned a lot and curious to learn more.

Stephen Dent

Hilary Rubinstein
Catastrophe at Spithead: The Sinking of the *Royal George*
Seaforth Publishing, Barnsley 2020; hardback, 288 pages, illustrated with 8 pages of colour plates plus B&W line drawings and maps; price £25.00.
ISBN 978-1-5267-6499-7

On 29 August 1782 the first rate three-decker HMS *Royal George* capsized while at anchor at Spithead, the great fleet anchorage outside Portsmouth Harbour, amidst a great fleet assembling for the relief of Gibraltar. The loss of an iconic vessel, flagship at the battle of Quiberon in 1759, named for the monarch and, by extension, the ruling Hanoverian dynasty of Georges, prompted much soul-searching and speculation. The rest of the fleet, led by Admiral Lord Howe, with his flag in HMS *Victory*, sailed. Gibraltar was relieved, Britain survived the War of American Independence and built another *Royal George*. Hilary Rubinstein's previous book was a biography of Admiral Sir Philip Durham, a survivor of this disaster, a prominent officer in the next round of French Wars, and a significant contributor to the history and mythology of the *Royal George*, a disaster that continues to fascinate.

This book opens with a fine study of Admiral Richard Kempenfelt, who was lost with the ship. After a slow start to his career Kempenfelt had emerged as a rising officer in the Channel Fleet from 1779, advising a succession of elderly admirals before demonstrating his talent for command in 1781, cutting up a major French convoy. Greatly valued by key figures ashore and afloat, the Admiral was below deck when the ship sank; his body was never recovered.

The bulk of the book examines the disaster, the court martial, the attempts at salvage, the demolition of the wreck by underwater explosions in the late 1830s, and finally the fate of the survivors, many of whom sailed with the fleet to Gibraltar. The big question remains: why did the ship sink? At the time various explanations were offered. The *Royal George* was being heeled over to expose the underwater body on the starboard side so that a defective water pipe could be replaced. This pipe was used to draw seawater into the ship to clean the lower decks. It was not an essential repair; it could have been left for the next time the ship went into dry dock. The ship's carpenter persuaded the Admiral and Flag Captain to have the work completed at Spithead before sailing. With the ship already heavily loaded for the voyage she had to be heeled over further than expected, allowing water to enter the lower deck gun ports, which had not been caulked. Last-minute attempts to right her failed. The ship was lost through a failure of basic seamanlike precaution. Claims that the fabric of the ship gave way, that the timbers were too weak to stand the strain, lack credibility. She had been inspected in dry dock within the past 12 months, and while defective timbers were found, no one expressed any concern about the ship's seaworthiness for the forthcoming 1782 campaign. Kempenfelt was far too good a seaman to sail in a defective ship.

The responsible warrant officer, Ship's Master Richard Searle, only reached the *Royal George* as she was about to sink, having spent the night ashore with his family. He was lost, along with several junior officers who were involved in the sinking. The survivors, including Captain Martin Waghorn, were acquitted. In simple terms those responsible lost control of the process amid the press of other business, and no one thought there to be any danger until the ship began to sink. This exemplary study surpasses all existing texts.

Andrew Lambert

Aaron Steven Hamilton
German Submarine *U-1105*, Black Panther: The Naval Archaeology of a U-boat

Osprey Publishing, Oxford 2019; hardback, 134 pages, 76 illustrations in colour and B&W; price £25.00.
ISBN 978-1-4728-3581-9

This is a slightly unusual book both in format and content, being not quite a 'coffee-table' book but still of a size well-suited to the reproduction of photographs. In terms of the content, the book takes us through the development of the three late-war German innovations fitted to this Type VIIC/41 U-boat, and provides an account of her brief wartime career, extensive postwar testing by the Royal Navy and US Navy, and then final disposal by the latter. This is followed by a detailed account of diving on the vessel, which now lies at the bottom of the Potomac River. This makes the book unusual as technical histories go, and is more common with ships such as the *Titanic*.

There are nine chapters, plus three appendices covering technical specifications, crew and a chronology, with a foreword by Dr Innes McCartney of Bournemouth University. *U-1105*, nicknamed the 'Black Panther', was fitted on completion with a black, anechoic coating of rubber sheets code-named 'Alberich' (after the dwarf in Wagner's Ring), which was designed to reduce sound

wave reflections, rendering her almost invisible to a ship's active sonar. She also received a *Schnorchel*, which allowed a boat to remain submerged at periscope depth when running on diesels; using this equipment, the *Kriegsmarine* set several endurance records for submerged passage that were not surpassed until many years later. The third innovation was a much-improved passive sonar, code-named GHG *Balkon* (see Warship Gallery).

After a relatively successful first and only wartime patrol in which she made good use of all three of these features, *U-1105* was surrendered to the RN, which recognised her significance. Trials were carried out over several months before it was agreed to hand her over to the US Navy which, it seems, was less impressed and simply used her as a submarine target. The author hints that the USN demanded her simply to prevent her falling into Soviet hands; however, the Soviets acquired access to these technologies via shared reports, and applied anechoic coatings to their submarines well before the West.

U-1105 was towed across the Atlantic through a winter storm while surfaced, such was the USN's determination. She was initially assumed lost as the communication aerials had all been washed away, along with the 3.7cm deck gun. Following her recovery she was laid up, then used in trials for recovering sunken submarines, ending her career as a target.

The wreck lies off Piney Point, Maryland, and is accessible to divers; diving on the wreck was the author's inspiration for the research that led to this book. All in all, an interesting story, well told, despite the odd typo and a careless mix of US and UK spellings, and a useful book for the library.

W B Davies

Les Brown
Black Swan Class Sloops: detailed in the original builders' plans

Seaforth Publishing, Barnsley 2020; hardback, 128 pages, illustrated throughout with full colour plans; price £30.00.
ISBN 978-1-5267-6596-3

The sixth in Seaforth Publishing's expanding series reproducing original builders' drawings held in the National Maritime Museum's archives, *Black Swan Class Sloops* focuses on the distinguished wartime escort vessels that played a leading role in winning the Battle of the Atlantic. 'High end' ships utilising steam turbine propulsion and built in relatively limited numbers, the *Black Swans* were widely regarded as the 'Rolls-Royces' of their type in Royal Navy service.

The book broadly follows the successful format established by its predecessors. The essence of this approach is the large scale, full colour depiction of a comprehensive selection of 'as fitted' general arrangement plans showing all aspects of the subject's design in an exquisite level of detail. However, one significant departure compared with previous volumes is the decision to cover an entire class of vessels rather than a single ship. This decision – undoubtedly aided by the relatively diminutive nature of the 300ft-long design – opens up a number of interesting opportunities. Not only do the four sets of plans selected – of *Black Swan*, *Flamingo*, *Starling* and *Amethyst* – serve to show the development of the *Black Swan* and Modified *Black Swan* designs at various stages of their service careers, but they also evidence evolving practices by draughtsmen at different yards and at different times. The colourful and elaborate drawings produced by Yarrows at the beginning of the war make an interesting contrast with the more austere plans produced by Fairfield and Alexander Stephen in the tougher, mid-war years.

The book benefits from the succinct introductions provided by author Les Brown to both the original and modified class variants, including an overview of the evolution of previous Royal Navy sloop designs. His captions to the various plans are extensive and are particularly strong on describing how the ships were modified in the light of practical experience. The plans themselves are superb and provide a clear insight into the class's internal workings. The reviewer was fortunate enough to visit *Whimbrel*, last surviving member of the class, in Alexandria and can well remember the ease of access provided by the broad passageways to port and starboard at upper deck level that are clearly depicted here. A less positive memory was recalled by the profile drawings illustrating the depth of the boiler rooms, a fact brought home vividly when he was stranded in pitch darkness halfway down an access ladder after an Egyptian rating snagged the temporary lighting cable being used to illuminate the space.

There are a few minor criticisms. Notably, the splitting of plan views between pages has sometimes been calculated too finely, leading to small slices of the drawings being omitted. In this regard, it is a pity that the publisher's largesse has not extended to replicating the excellent gatefold arrangement used for the *Black Swan* profile and plans to illustrate some of the other ships, thereby eliminating this potential problem. This would also have provided some compensation for the fact that the book is somewhat shorter than its predecessors. However, these are mere quibbles. Overall *Black Swan Class Sloops* is a well-produced, high-quality book that maintains the high standards being set by an innovative and highly informative series.

Conrad Waters

Robert C Stern
The Modern Cruiser: The Evolution of the Ships that Fought the Second World War

Seaforth Publishing, Barnsley 2020; hardback, 288 pages, numerous photographs and diagrams; price £35.00.
ISBN 978-1-5267-3791-5

In *The Modern Cruiser*, distinguished author Robert C Stern sets out to describe the various factors that influenced the design of the ships of this type that served during the Second World War. The book's scope is ambitious. It examines the evolution of the cruiser type from its origins in the mid-nineteenth century, encompasses the ships built by all the major naval powers, and extends to assessing strategic, political, economic and technological considerations.

Structurally, the book is focused on twelve chapters arranged in broadly chronological order. The first three essentially set the scene, describing the evolution and performance of cruiser design from the iron-hulled cruising ships of the 1860s to the immediate aftermath of the First World War. The following six chapters – encompassing around half the book – are focused on the influence of the naval limitation treaties that dominated the interwar period and the cruisers that they spawned. Two chapters examine the cruisers built during the Second World War and their performance in battle. A short, concluding chapter gives an overview of post-war cruiser construction. The text is supported by an impressive selection of photographs and numerous tables summarising class characteristics. While there are some helpful diagrams, it would have been useful to see the inclusion of more drawings illustrating the general arrangements of the ships described.

The Modern Cruiser is an impressively comprehensive book written by an author who clearly has considerable knowledge of the subject. Indeed, it is the book's broad scope that is its most significant strength, enabling the reader to place particular cruiser designs into a wider context, not only in terms of foreign designs but also with respect to other relevant developments. Stern is good on the political machinations that surrounded the various treaty negotiations, with a particular emphasis on the US perspective. Less strong is his analysis of cruiser performance, which – in contrast to the book's overall approach – is restricted to describing a handful of engagements in some detail. The lessons that can be drawn from these are inevitably limited and exclude some important areas of discussion, such as the comparative performance of 'heavy' cruisers armed with 8in guns and 'light' cruisers armed with 6in guns, and the increasingly important influence of technological innovations such as radar.

For a British reviewer with a particular interest in Royal Navy cruisers, the book contains additional frustrations. Perhaps inevitably, Stern tends to see treaty discussions through an American prism, according insufficient weight to how calculations with respect to Japanese naval strength rather than Anglo-US parity invariably dominated British thinking. This results in some rather unbalanced assertions as to the British negotiating position. Additionally, there are some misunderstandings as to how the Admiralty's warship design process worked and a not-insignificant number of detailed errors when describing the evolution of specific Royal Navy designs.

While not without its flaws in areas of both interpretation and detail, *The Modern Cruiser* is an important and informative book that should be required reading for all those with an interest in 20th century cruiser design.

Conrad Waters

Kure Maritime Museum and Kazushige Todaka (eds), translated by Robert D Eldridge

The Battleships *Yamato* and *Musashi*

144 pages; price $75.00/£74.50.
ISBN 978-1-68247-385-6

Submarines and Submarine Depot Ships

240 pages; price $75.00/£71.95.
ISBN 978-1-59114-337-6

Destroyers

232 pages; price $75.00/£71.95.
ISBN 978-1-59114-630-8

All published by the US Naval Institute Press, Annapolis 2019/2020 (original Japanese editions 2005); landscape-format hardbacks, illustrated throughout with B&W photographs (plus plans in *Yamato* volume).

Given the limited scope of the published photographs of warships of the Imperial Japanese Navy in western secondary sources, these three early volumes of a new series (overall title: 'Selected Photos from the Archives of the Kure Maritime Museum') are particularly welcome. The majority of the photographs reproduced here are from the collection of Shizuo Fukui; these are complemented by images from lesser-known collections, notably those of Koyishi Nagamura and Yoshiyuki Amari. All are of considerable historical value because of the widespread destruction of IJN documents at the end of the Second World War. Despite this depredation, the Museum (of which the author is Director) has over the years amassed a collection of some 20,000 photographs, of which these volumes present us with but a small selection.

Of the three books, that covering *Yamato* and *Musashi* will undoubtedly attract the most interest. All the well-known images are present, but pride of place must go to a series of 19 photos taken on board *Musashi* during her trials; although the quality is variable, these show aspects of the ship in considerable detail. A number of the best-known (and highest-quality) images are reproduced both over a full page and also as a series of detailed enlargements. While this could be interpreted as an exercise in padding, it also serves to highlight features which might otherwise be missed, and is particularly useful from the model-maker's perspective. (It should be noted, however, that a few of the less well-known and/or poorer-quality images are given the same treatment, arguably to less effect.) The author has taken the conscious decision not to interfere in any way with the original images, meaning that all dust marks and scratches on the original negatives remain. This can be irritating when a major blemish disrupts the composition, thereby diminishing the photograph's impact; given what modern digital re-touching can achieve if skilfully done, it does seem a genuine pity. This caveat apart, the reproduction is first-rate, with good quality printing on heavyweight coated paper, resulting in the best presentation of these images that this reviewer has seen.

The other two volumes cover a much larger number of less famous ships in correspondingly less detail (though both have considerably more pages), but with the same emphasis on quality of reproduction. Both feature more types of warship than implied by their titles, including minelayers, minesweepers, torpedo boats and escort vessels. There is a preponderance of standard ship portraits, but there are also numerous unusual close-ups, onboard views and dockyard shots. The latter are a particular favourite of this reviewer, often with unidentified vessels visible in the background along with details of the IJN's shore-based infrastructure – a mixture of the latest technology and things that had barely changed in centuries. Throughout, the level of detail is outstanding; these are pictures that really reward close study. In contrast the captions are in most cases minimal, giving little more than name, date and location (if known); there is the very occasional error – in one caption the battleship *Satsuma* is described as a destroyer – and the translation is sometimes less than idiomatic.

It should be noted that the books are very expensive. Nevertheless, for the dedicated IJN enthusiast they surely constitute a must-have, and likewise for anyone considering modelling any of these perennially fascinating but under-documented warships. Hopefully further volumes will maintain the same high standard.

Stephen Dent

Richard Endsor
The Master Shipwright's Secrets

Osprey Publishing, Oxford 2020; large format hardback, 304 pages, with more than 230 images, mainly in full colour; price £65.00/ $85.00.
ISBN 978-1-4728-3838-4

A massive tome with a price to match, this seems at first glance to be the archetypal coffee table book, certainly best read flat on a horizontal service. However, initial impressions can be deceptive. The book is clearly a labour of some scholarship, translating for the lay reader much of the Restoration era naval shipwrights' skills, with extensive use of documents collected by the redoubtable Samuel Pepys, all gathered around King Charles II's close interest in <u>his</u> navy. The many drawings, both from the author's own hand and from the Van der Veldes, father and son, are veritable works of art, and there are several gatefolds with plans drawn to a scale of 1/72.

After the expected introduction, the book is set out in ten chapters with two appendices, endnotes and an index. It revolves around the life of the *Tyger* in her various manifestations, as an example of the processes employed by various shipwrights and the Royal Dockyard at Depford, and each chapter relates directly either to a specific version of the *Tyger* or the process of planning and building a wooden warship. Like many ships at that time, and for quite a few years subsequently, *Tyger*'s last manifestation was effectively a mythical rebuild of the remains of a barely surviving predecessor – an accountancy myth used on many occasions by the practical shipyard officers to overcome a treasury refusal to fund new ships down the years both in Britain and abroad.

The final chapter sets out in some detail contemporary shipyard contracts, and made interesting reading for this reviewer, who spent several years in the early 1980s interpreting 20th century government shipyard contracts. It seems that not much has changed, although I was allowed a few more overseers than seemed to have been employed in Pepys' day.

This is a gloriously illustrated description of the art and science of the Restoration shipwright, reflecting well on the forethought of Pepys, the skills of the artists, both past and present, and the scholarship of the author in bringing it all together. However, it is unclear at whom the book is it aimed given the high price. The plans are a modeller's delight, but I suspect the average modeller would rather spend £65 on models! Hopefully, in due course it will become available at a slightly more modest price, either discounted or second-hand.

W B Davies

John Jordan & Philippe Caresse
French Armoured Cruisers 1887–1932
Seaforth Publishing, Barnsley 2019; hardback, 272 pages, illustrated with numerous B&W photographs and drawings; price £40.
ISBN 978-1-5267-4118-9

In 1888 the French Navy laid down *Dupuy-de-Lôme*, the world's first modern armoured cruiser with both a protective deck and side armour. She and her successors were very influential, yet this is the first book to give a full history of these ships. Like the authors' previous *French Battleships of World War One*, this work is illustrated by John Jordan's excellent drawings and by superb photographs from the Caresse collection, all clearly laid out by Stephen Dent.

In both *Dupuy-de-Lôme* and the following smaller ships of the *Amiral Charner* class, the medium as well as the main guns were mounted in turrets, but the hulls were still of the preferred French form with prominent plough bows and tumblehome; in a seaway, they were very wet forward and rolled heavily. But the next true armoured cruiser, *Jeanne d'Arc*, had more modern lines. She was designed by Louis-Emile Bertin and introduced many features that remained characteristic of all subsequent French armoured cruisers. The engines were placed amidships with boiler rooms ahead and astern; thus the funnels, which had Bertin's distinctive 'pagoda' profile, were in two well-separated groups. *Jeanne d'Arc* and her successors also had a forecastle deck that extended a long way aft; hence they were high-sided with two unarmoured decks above the belt. *Jeanne d'Arc's* medium QF guns were in shielded mounts but, beginning with the *Duplex* class, at least half the medium armament was in single or twin turrets. This gave them wide firing arcs and an unusually high command compared with guns in casemates.

In the years 1898–1901, France laid down no fewer than thirteen armoured cruisers. The smaller *Dupleix* class, with a uniform armament of eight 164.7mm guns, were for service on foreign stations, but the remaining ten, as well as a further five laid down between 1903 and 1906, were fleet cruisers. Of these five, the final two of the *Edgar Quinet* class also had a uniform armament of fourteen 194mm guns – the main guns of their predecessors – principally, it seems, to increase the number of guns able to penetrate 6in armour.

Britain laid down no armoured cruisers until 1898 but by the end of 1901 she was already one ahead in numbers building; the margin then grew rapidly. Britain also built these ships more quickly; by the end of 1902, both countries had six in commission but thereafter British numbers powered ahead, and by 1908 the Royal Navy had commissioned 35 armoured cruisers while France did not reach her final total of 24 until 1911. As to the quality of the French designs, John Jordan suggests that they would have held their own against their British counterparts. Certainly the high command of their turret guns gave them an advantage. But the high and largely unprotected sides of the French ships increased the danger space, making them easier as well as more vulnerable targets. Also, the French turrets and working chambers were designed to hold considerable amounts of ready-use ammunition, including bagged charges, albeit in flash-proof lockers. However, in the First World War, these ships were never tested in a gun action.

This book is particularly good on the many French technical innovations, not least in boilers. *Dupuy-de-Lôme* had cylindrical boilers (though, while these were fire-tube boilers, they were not of the locomotive type); thereafter water-tube boilers, usually the large-tube Belleville or Niclausse, were fitted. Bertin favoured the small-tube Guyot–Du Temple design, though it proved to be less reliable and performed no better in full-power trials. In fact, the long-established Belleville type was consistently most economical and for two classes gave the highest speed.

Now, in place of a previous dearth of information, Jordan and Caresse have given us a comprehensive work of reference, covering all manner of interesting topics, about these fascinating ships.

John Brooks

Stefan Draminski
The Battleship USS *Iowa*
Osprey Publishing, Oxford 2020; hardback, 352 pages illustrated with over 1000 images, many in full colour; price £40.00.
ISBN 978-1-4728-2729-6

This is the latest in the 'Anatomy of the Ship' series (nearly forty titles to date). It follows the long-established format; however, as was the case with its two immediate predecessors, the author has produced the 1000-odd illustrations and drawings detailing the anatomy of the vessel using a series of seven ultra-detailed 3D computer-generated models of the ship to reflect the changes made over her service life, each with a phenomenal level of detail.

Section One has the customary technical description, together with a lengthy operational history of *Iowa*, illustrated in the main with official US Navy photographs. Section Two comprises nearly 40 pages of what are here termed 'primary views'. These cover the overall appearance of the vessel at the various stages of her career, with port and starboard profiles and a plan view for each stage, together with additional perspective views where appropriate; all are in full colour, being generated from the 3D computer models. Section Three has some 260 pages of detail in nine separate sections covering topics from hull structure to the ship's boats, using a mix of coloured illustrations and 2D line drawings.

The result is an absolute dream for the ship modeller, although the more casual reader may find the book a little overpowering in comparison with the original format. One can only wonder at the computing skills and the time invested by the author in generating the basic models, which are superb.

W B Davies

Graham T Clews
Churchill's Phoney War: A Study in Folly and Frustration

Naval Institute Press, Annapolis 2019; hardback, 339 pages, 11 B&W photographs; price £28.78/$34.95.
ISBN 978-1-68247-279-8

Winston Churchill was such a commanding figure in 20th century British history that it is unsurprising that well over fifty years since his death he remains in many respects an enigma whose life historians continue to pore over and debate. In this instance the focus is on the brief period between September 1939 when he returned to the Admiralty as First Lord and May 1940 when he was appointed Prime Minister. It is an interlude when Churchill's personal decision-making has been much criticised and about which the man himself was uncharacteristically defensive, even humble, in his memoirs. Regarding the Norwegian campaign, he confessed 'Considering the prominent part I played in these events … it was a marvel I survived and maintained my position in public esteem and parliamentary confidence'.

In this book, Graham Clews examines in minute detail Churchill's actions and, in particular, his strategic decision-making at the time. This includes the offensive anti-submarine patrolling policy advocated at the outbreak of war, his vigorous and lasting promotion of Operation 'Catherine' (wresting control of the Baltic Sea from the enemy), his plans for creating an 'Inshore' or 'Close Action' Squadron and, of course, the disastrous Norwegian Campaign. The above were issues directly within his brief as First Lord of the Admiralty, but Churchill was not content to restrict himself to purely naval matters. *Churchill's Phoney War* also encompasses his involvement in Britain's aerial bombing campaign, France's strategic role pre-invasion and the devolvement

of the war budget in which he surprisingly sought to deny funds for new naval construction in favour of the Army in particular.

The author's previous career as a teacher is evident in his approach: in some senses the book reads like a school report. Each subject is examined dispassionately; a conclusion is then reached. Along the way, Clews' overarching aim is to separate truth from conjecture in relation to how much the decision-making process was Churchill's alone and how much it was part of a corporate responsibility, as the Admiralty struggled with myriad and often novel challenges at the start of hostilities.

Belief in the strategy of taking the battle to the enemy was one of Churchill's instinctive traits. Offensive anti-submarine patrolling, which required the Navy to deploy some of its precious escort ships away from convoy duties, seemed to fit the First Lord's *modus operandi*, yet Clews concludes that Churchill's personal involvement was less pronounced than thought hitherto. His advocacy of the audaciously offensive Operation 'Catherine' was strongly, and ultimately successfully, opposed by the Admiralty and by his First Sea Lord, Dudley Pound, in particular. He did not always get his own way.

The climax of the book concerns the German invasion of Norway in April 1940. As before, Clews seeks to tease out how much Churchill was personally responsible for the succession of strategic errors which led to ultimate defeat. His conclusions are somewhat surprising in view of what we know about Churchill, although the latter's inclination to interfere unnecessarily is highlighted.

Churchill's Phoney War is quite demanding on the reader, as Clews delves into both the political machinations of the day and the minutiae of the debates within governmental departments. These require considerable prior knowledge in order fully to appreciate their significance in the wider context of the war. However, the author is a good guide, and the systematic approach he adopts is helpful in gaining a better understanding of this unique period of the war and the part played by this famous leader.

Jon Wise

Evan Mawdsley
The War For The Seas: A Maritime History of World War Two

Yale University Press, New Haven & London 2019; hardback, 557 pages illustrated with 59 B&W plates and 9 maps; price £25.00.
ISBN 978-0-3001-9019-9

Evan Mawdsley's new book gives a clear and detailed account of the Second World War at sea. It is less an operational narrative than a thorough-going analysis. Echoing Admiral Sir Herbert Richmond, Mawdsley asserts that while Allied sea power did not *win* the war, it *enabled* the war to be won. Without Allied command of the seas, Britain could not have been supplied from across the Atlantic, nor could the USSR via the Arctic convoys;

Italy could not have been defeated, nor Europe invaded.

Mawdsley has organised his book into five parts covering the maritime efforts of all of the major powers; his text embraces national strategies, capabilities, inter-service rivalries and cooperation, advances in technology and the role of intelligence and logistics.

Part I covers Northern Europe from the outbreak of war until 1940, ending with the actions against the French at Mers el-Kebir and Dakar, judged necessary at the time both to remove a latent threat to British sea power and to convince the United States that Britain was determined to continue the fight. Part II takes us from July 1940 to April 1942, with the British Empire under threat from German forces operating from bases stretching from Norway to the French Atlantic coast as well as from the Italians in the Mediterranean. Despite this pressure, Britain was able to use the global shipping networks relatively unhindered. Soviet naval activity in both the Baltic and Black Seas, a topic often neglected by historians, is not ignored. Nor is America's undeclared war in the Atlantic, where US warships escorted convoys to the Mid Ocean Meeting Point, meaning that even before Pearl Harbor the United States was essentially at war. Part II concludes with Japan's attacks throughout the Far East, turning the war into a truly global one.

Part III covers the pivotal period of the war at sea April–December 1942, when Axis expansion reached its peak. American forces increasingly undertook the bulk of the fighting in the Pacific theatre and the US Navy was able to inflict decisive defeats at the battles of the Coral Sea and Midway. Meanwhile, the British in the Atlantic were able to defend trans-Atlantic shipping in the face of strong U-Boat attacks. Tensions between the proponents of an early return to Europe and those who advocated fighting in North Africa and the Mediterranean were resolved. Operation 'Torch' and the invasions of Sicily and Italy all put pressure on the Axis forces and these, together with the continued supply of materiel to Russia and the build-up of forces in Britain for the eventual return to Europe, were all made possible by Allied command of the seas.

January 1943–June 1944, the period covered by Part IV, saw the UK and USA winning the maritime war. Italy was defeated and the United States continued its advance westwards in the Pacific, decisively crushing Japanese naval power at the battle of the Philippine Sea. Part V, the final section of the book, covers the remainder of the war, when complete command of the sea contributed in no small part to the defeat of both Axis powers, Germany and Japan.

This book has been the subject of favourable reviews elsewhere, and this reviewer was not disappointed. The author's straightforward approach and accessible writing style is to be commended. There are a few small editing errors, but these in no way detract from the quality of the narrative. Mawdsley's book gives a thorough grounding in the global war at sea that experts and amateurs alike will enjoy, and represents good value for money.

Andrew Field

Steve R Dunn
Southern Thunder: The Royal Navy & The Scandinavian Trade in World War One

Seaforth Publishing, Barnsley, 2019; 304 pages, illustrated with 39 B&W photographs; price £25.00.
ISBN 978-1-5267-2663-6

Southern Thunder is Steve Dunn's seventh book about the Royal Navy during the First World War. He notes that during the conflict the RN never lost command of the sea. This enabled Britain, albeit only just, to ensure that enough foodstuffs arrived in its ports for the survival of the population, while maintaining a blockade on goods reaching Germany that significantly contributed to the latter's capitulation in 1918. It is this theme, in relation to trade with the neutral countries of Denmark, Norway and Sweden, which is the subject of *Southern Thunder*.

Britain's links with the Scandinavian countries consti-tuted a 'double-edged sword'; while it was necessary to try to prevent their exports reaching the enemy, it was also vital to maintain a flow of goods between these neutrals and Britain. Naturally the German Navy sought to interdict and to disrupt as well, particularly with its U-boats. *Southern Thunder* reveals a clear change in the tactics used by the Royal Navy following the introduction of the convoy system in the spring of 1917. Initially, the emphasis had been on actively enforcing the naval blockade and, through economic and political measures, disrupting the Scandinavian trade with Germany. After the convoy decision was taken, it became a logistically demanding task for the RN to manage and protect an unending flow of traffic both ways across the North Sea. The bold and aggres-sive tactics used by the Germans tested the anti-subma-rine capabilities of the Navy to the limit.

Many of the individual engagements, which would otherwise have been lost in the mists of time, are brought to life with the aid of some meticulous research. There was much individual heroism in these life or death strug-gles alongside glimpses of the true horrors of 'total war' when ships were sunk without warning in the icy waters of the North Sea. Central to the narrative is the long and painful process whereby the Royal Navy finally accepted that the convoy system was necessary. Dunn argues that the social and educational upbringing of the officer class in the years leading up to the war flew in the face of the essentially defensive posture associated with the convoy principle. The public school/Britannia College philos-ophy, steeped in the traditions of Nelson, was directed towards seeking out and destroying the enemy, not sitting back and waiting for him to come to you.

This was an unglamorous, tedious war fought in small, often inadequate ships, in an inhospitable and dangerous environment. To the writer's credit, he does not shy away from discussing in detail the political, diplomatic and economic framework underpinning the narrative, a feature too often left by other works to more academic studies. In this respect, the author's grasp of often complex issues is impressive. This is an erudite book, which makes a significant contribution to the naval literature of the Great War.

Jon Wise

John Lambert & Al Ross
Allied Coastal Forces of World War II
Volume I: Fairmile Designs and US Submarine Chasers

Seaforth Publishing, Barnsley 2018; 262 pages, illustrated with more than 200 B&W photographs and numerous line drawings; price £40.00.
ISBN 978-1-5267-4449-4

First published by Conway in 1990, this book and Volume II (Vosper MTBs and US Elcos) set the bar for well-researched, well-illustrated books on the design of a long-underrepresented type of wartime vessel. John Lambert died in 2015 and his vast collection of highly detailed technical drawings was acquired by Seaforth, who are now republishing the series, with the much-anticipated addition of a third volume to follow. The original books have been difficult to obtain for some time, so this reprint is very welcome.

The heart of this book is the story of Fairmile Marine, an innovative wartime company who produced wooden launches in kit form that could be assembled in even the smallest boatyards. These vessels went on to serve as motor launches, landing craft and as the famous motor torpedo boats and motor gun boats of Coastal Forces. Seven chapters detail the various designs, along with one each covering the Admiralty's 72ft Harbour Defence Motor Launch and the US 110ft subchaser. Additional chapters detail equipment, machinery, weapons and camouflage, while twelve appendices provide figures on wartime construction and testing, plus a few sample service histories.

The book is primarily a reference work, with more than 120 pages given over to drawings alone and almost 50 to tables of construction. Each chapter provides a detailed background and class evolution, followed by a comprehensive collection of detailed drawings. Depending on the boat, these show everything from lines and profiles of the hull design and its construction, right down to the arrangement of the voice pipes and guard rails. Lengthy tables record the location and date of each vessel's construction – no mean feat given that some 700 Fairmile Bs alone were ordered during the war.

So, aside from bringing this book to a new generation of researchers, what does this reprint bring us? The only new addition is eight pages of colour artwork showing the camouflage schemes of 36 boats. The colours themselves are based on recent research using period fleet orders to create the correct mix formulae, and have been applied to vessel profiles according to the patterns visible in wartime photographs and Admiralty instructions. As the original book was entirely black and white this is a welcome addition; however these illustrations lack the high quality of the existing drawings, and the work on

which they are based is not listed in the references. Additionally, the new pages have been somewhat clumsily inserted approximately a quarter of a way through the book, while the chapter on camouflage is 100 pages away.

Apart from these illustrations and a new front cover, nothing else appears to have changed in this reprint. This is perhaps unsurprising given the high quality of the original, but one might have expected a new foreword or author's note, which is sadly absent. Another section that could have been updated and where there has been considerable change since 1990 is the final chapter: the survivors. At the time of the first imprint there were a number of HDMLs and Fairmiles still in UK waters, including two operating as ferries; by 2018, this had been reduced to a mere six Fairmiles in the UK (in 2020 that has slipped to three), although conversely several more boats, both Fairmiles and HDMLs, have been identified abroad. It is a shame that the opportunity wasn't taken to bring this chapter up to date.

This doesn't detract from an excellent and authoritative work, and the primary benefit of the new edition is in making it more widely available to historians and modellers alike. No one else has attempted the level of technical history in *Allied Coastal Forces*, and Volume I remains an important source for anyone researching these boats.

Stephen Fisher

Lawrence Paterson

 ### Eagles Over the Sea 1935–42: A History of Luftwaffe Maritime Operations

Seaforth Publishing, Barnsley 2019; 464 pages, 154 photographs; price £30.00.
ISBN 978-1-5267-4002-1

 ### Eagles Over the Sea 1943–45: A History of Luftwaffe Maritime Operations

Seaforth Publishing, Barnsley 2020; 432 pages, 110 photographs; price £30.00.
ISBN 978-1-5267-7765-2

These two volumes tell the story of German maritime air operations from the official re-establishment of military aviation to the end of the Second World War. An underlying theme of the first is the German Navy's struggle to maintain an independent naval aviation capability in the face of Hermann Göring's insistence that all 'air' should answer solely to him; then, after that battle had been lost, to obtain adequate support for maritime operations. Thus, while at the beginning maritime aircraft operated largely under *Kriegsmarine* direction, soon only catapult-launched aircraft on board warships were left under the Navy's control. All other units became absorbed into the wider *Luftwaffe*, although a few, in particular KG (*Kampfgeschwadern* = Bomber Group) 26 and 40, retained a primary maritime function – albeit often with little real naval input into their deployment and resourcing. This was particularly the case while Erich Raeder remained as naval C-in-C, although Karl Dönitz was able to achieve more influence via his better personal relationship with Adolf Hitler.

Each volume begins with a Glossary that includes translations of German organisational terms and an equivalent rank table. (The latter has, however, a glaring error in both volumes: a reversal of the terms *Konteradmiral* and *Vizeadmiral*.) This section also sets out and explains the complex organisational and unit designations of the *Luftwaffe* – an essential adjunct to the main text.

Chapter 1 of the first volume deals with German naval aviation during the First World War although, curiously, all but two of its illustrations are of between-the-wars subjects. Chapter 2 looks at the clandestine rebirth of German airpower up to 1935, after which the Navy's attempts to build up a true air arm of its own were undermined by Göring. Chapter 3 covers the participation of these aircraft in the Spanish Civil War, together with the ordering of two aircraft carriers (*Graf Zeppelin* and 'B') and their prospective air-groups. The outbreak of war, which led to the suspension of their construction (followed by the cancellation of 'B'), is the subject of Chapter 4, which covers events to the end of 1939. Chapter 5 covers the Norwegian campaign and the occupation of western Europe, with the further erosion of maritime air power by the conversion of many specialist units to a general bombing role, especially during the Battle of Britain. Chapter 6 deals with the advent of the Fw 200 Condor force of long-range maritime patrol aircraft – albeit in sub-optimal numbers, and accompanied by a further transfer of naval-affiliate units to the *Luftwaffe*. Chapter 7 covers operations in the Mediterranean and Baltic theatres, which were rewarded with a wide range of successes, especially during the British evacuation of Greece and Crete, while Chapter 8 continues the Mediterranean story with the Malta convoys, together with the assault on the Arctic convoys and operations in the Black Sea. This brings us up to the end of 1942, leaving us on the eve of the Operation 'Torch' landings in French North Africa.

The second volume takes us through the remainder of the war, and covers operations off Norway and in the Bay of Biscay, the Mediterranean and the Black Sea, and in the North Atlantic in support of the U-boats. Chapter 5 details the events surrounding the invasion and surrender of Italy, during which the *Luftwaffe* first employed its Fritz-X (SD 1400) and Hs 293 guided bombs, most spectacularly in sinking the Italian battleship *Roma* while en route to internment in September 1943, while Chapter 6 covers the German successes in the Aegean and in the bombing of Bari. However, the later chapters tell the tale of a steady decline in the effectiveness of German maritime air power as losses and shortages combined with increasing Allied strength and vacillating priorities to make it more and more difficult for the units to perform their roles. Each of the two

volumes concludes with an Appendix that covers the characteristics of the principal aircraft used by the *Luftwaffe* in maritime roles.

Eagles over the Sea provides a comprehensive treatment of its topic, with a good balance between aircraft and people, the narrative being enhanced by eyewitness accounts from airmen who took part in some of the operations covered. The choice of illustrations is generally good, with a number of images of individuals alongside those of aircraft and ships. There are some negative points: occasional unnecessary repetitions, misspellings and factual errors, including some that betray the author's occasionally-flawed understanding of aircraft types and designations. Nevertheless, *Eagles* is to be recommended as likely to be the definitive work on the subject for a long time, of interest to both those concerned with the Second World War in the air, and those whose interests lie in its conduct at sea.

Aidan Dodson

Bill Clements
Britain's Island Fortresses: Defence of the Empire 1756–1956

Pen & Sword, Barnsley, 2019; hardback, 304 pages, illustrated with photographs, plans and maps; price £25.00.
ISBN 978-1-52674-030-4

The British Empire, primarily concerned with maritime trade, was necessarily protected by the Royal Navy, the world's most powerful fleet for all but the last fifteen years covered by this book. Many of the naval bases that repaired, refuelled and re-armed the fleet were located on islands, large and small, which were more easily protected than continental positions. These bases needed defensive works equipped with artillery against hostile naval or amphibious forces intent on destroying the base or using it to raid British shipping. In essence the fortifications served a larger, maritime purpose. Steam propulsion greatly reduced the cruising radius of warships, requiring additional coal depots and sources of fresh water for inefficient boilers, while the new submarine telegraph cable which wired up the Empire in the second half of the 19th century needed secure shore relay stations on British sovereign territory, thereby creating an 'all-red-route'.

Fortress expert Bill Clements focuses on ten of these insular bases, from Bermuda to Hong Kong, placing the evolution of the defences in the wider naval and strategic context. His introduction provides the broad context of fortress and artillery development, relations between the Admiralty, the War Office, the Ordnance Board, the Colonial Office and the Treasury, which as ever was anxious to control costs. He also provides a generic analysis of the evolution of coast defence systems from muzzle-loading smooth-bores and stone walls to 15in gun mountings capable of 45-degree elevation with a range of 35,000 yards. While the largest gun used in defences in Britain after 1890 was the 9.2in, 15in weapons were installed at Singapore to counter Japanese battleships. The evolving size of local garrisons and the installation of minefields are also addressed. This wider perspective makes the book more useful for naval historians than older fortress histories that scarcely looked over the parapet, let alone addressed the strategic context in which they existed. The operational history of each fortress provides a useful foil for the essentially theoretical calculation of threat.

The coverage is selective, Clements arguing that Malta is too big and familiar a subject. Bermuda, Jamaica, St Helena, Antigua and St Lucia share an entry. Ceylon, Mauritius, Ascension Island, Singapore and Hong Kong were all significant naval bases, but their histories varied greatly. Bermuda was developed as tensions with the United States rose after 1803, was a vital position in the War of 1812, while further development between 1815 and 1890 was prompted by steam shipping, new artillery and armoured warships, in a climate of Anglo-American hostility. When the political climate changed around 1900 the defences were no longer a priority and American forces used the island in the Second World War. By contrast Hong Kong and Singapore were developed in the interwar era against Japan; both fell to Japanese armies, being too close to the mainland for effective naval defence. Ascension was garrisoned to prevent it being used as a base for an attempt to free Napoleon from St Helena, and became successively a coal depot, the base for the Navy's West African Anti-Slave Trade Patrol, and finally a cable relay. The author notes that the volcanic island was terraformed by the Royal Navy after eminent Victorian scientists Charles Darwin and Joseph Hooker, who landed from RN ships, recommended introducing soil and plants onto the upper slopes of the extinct volcano. The Navy took their advice, creating the current lush forest and improving the food and water supply of the base. The Americans installed an air strip in the 1940s, and the island was a vital relay point for the 1982 Falklands Campaign. Island bases still matter, but by 1982 they were not fortified. Today the Falklands are protected by land, air and sea forces.

Island Fortresses is well-written and well-researched, and is highly recommended.

Andrew Lambert

John Henshaw
Liberty's Provenance: The Evolution of the Liberty Ship from its Sunderland Origins

Seaforth Publishing, Barnsley 2019; landscape format hardback, 128 pages, many B&W photographs and ship drawings; price £25.00.
ISBN 978-1-5267-5063-1

John Henshaw is an Australian chartered surveyor with an enduring interest in ships and the sea. He has owned several yachts and is both an accomplished watercolour artist and ship model maker. The drawings in this book are all his own work, and his very readable text explains how the design of the iconic Liberty ship had its origin in the Sunderland shipyard of Joseph Thompson & Sons.

By 1935 Robert Thompson had produced a design for a 10,000-ton break-bulk tramp steamer that would be economical to build and operate. The SS *Embassade* launched by his yard in July 1935 was the first example, and he refined the design to produce the slightly larger SS *Dorington Court* in May 1939. The Admiralty assumed responsibility for merchant ship construction on the outbreak of war and quickly became aware of Thompson's abilities as a designer and manager. In September 1940 he was asked to lead an Admiralty merchant shipbuilding delegation to the USA with instructions to place contracts for the construction and delivery of 60 break-bulk tramp vessels as quickly as possible. He took his own designs with him to illustrate what he wanted, including his design for *Empire Liberty*, evolved from *Dorington Court* to facilitate series production in Sunderland. Most US shipbuilders had full order books by then and showed little interest, but the industrialist Henry Kaiser was the exception. The US Maritime Commission was also astute enough to realise that the UK would have to cover the cost of building two new shipyards in order to get its ships and that these could be turned over to US production when they were complete. Contracts were duly signed with Kaiser on 1 December 1940 for the two yards, one on the east and one on the west coast of the USA, and for the ships to be built for Britain in them. They were built by Todd as part of the organisation known as the Six Companies under Kaiser's leadership. On his journey back to the UK, Thompson's ship was torpedoed but he kept his briefcase, containing the contracts, with him when he abandoned ship. After spending nine hours in a lifeboat in freezing conditions he was rescued and took the wet but preserved contracts to the Admiralty.

Henshaw explains how the *Embassade* and *Dorington Court* designs led to the *Empire Wind* and *Empire Liberty* classes of standardised vessel built by Thompsons during the war. The plans of *Empire Liberty* shown to the American naval architects were, therefore, the origin of the 2,710 Liberty ships built in Kaiser's yards between 1941 and 1944. Further evolutions of the design became the Fort, Park and Victory ships built in Canada as well as the USA. While the Sunderland origins of the Liberty ship design have actually been well documented in the past, Henshaw believes that a myth has grown that the Liberty ships were an 'all-American' design with no outside involvement, and he wrote this book to ensure that Thompson's shipyard is given the credit it deserves for originating the design.

Both during and after the war a number of Liberty ships and their derivatives saw both naval and commercial service in roles other than the carriage of break-bulk cargoes and one of this book's strengths is the inclusion of chapters in which specific modifications are described. These include the British wartime CAM ship *Empire Moon*, the merchant aircraft carrier *Empire MacAlpine*, the repair ships HMS *Assistance* and HMS *Hartland Point*, and the guided missile trials ship HMS *Girdle*

Ness. American derivatives described include colliers, tank transports, aircraft transports, hospital ships, missile range monitoring ships and many others. All of them are illustrated with photographs and line drawings.

This is a book for readers with a wide interest in ships as well as the Battle of the Atlantic and the war of logistics that underpinned the land campaigns of the Second World War. The author is clearly passionate about his subject and his argument accords with earlier historiography on the subject. The price is very reasonable for a 128 page book of this quality, and the sections on Liberty ship derivatives and conversions elevate it above earlier basic descriptions of these vessels.

David Hobbs

Thomas J Cutler (ed)
The Battle of Leyte Gulf at 75:
A Retrospective

US Naval Institute Press, Annapolis 2019; hardback, 336 pages, 7 maps, B&W plate section; price $34.95/£31.50.
ISBN 978-1-68247-461-7

Thomas J Cutler's seminal book *The Battle of Leyte Gulf, 23–26 October 1944* was first published by HarperCollins in 1994. The author was recently commissioned by the US Naval Institute to edit a volume of essays to mark the 75th anniversary of the battle, arguably the greatest of modern times in terms of its scope and the number of ships involved. He has opted for a book in two parts, the first of which comprises a series of specially-commissioned papers by prominent naval historians, while the second is a compendium of articles published in journals such as USNI *Proceedings* and *Naval History* over the years.

There are problems inherent in this approach. It is difficult for an editor to establish tight parameters for the distinctive aspects of the battle which have been identified and devolved to the various authors, and this has resulted in a significant degree of overlap and duplication. The modern contributions are also uneven in terms of style and, perhaps more importantly, of quality. The second part of the book shares some of the same problems and, although these articles are important to the retrospective aspect of the work, for some of the issues discussed the argument has moved on since they were written.

Of the modern contributions the most striking are those of Trent Hone, who places Halsey's fateful decision to take the Third Fleet north in pursuit of Ozawa's decoy carrier force in the context of prewar US Navy wargaming and doctrine, and Vincent P O'Hara, who has provided a thought-provoking analysis of the aftermath of battle, when the US Navy failed to exploit its victory by remaining east of Leyte rather than employing its surface forces to cut Japanese supply lines through the Camotes Sea.

Highlights of Part 2 include Halsey's justification of his actions, originally published in the April 1952 issue of *Proceedings*, Oldendorff's discussion of the operational and tactical issues he faced when confronting the

Japanese two-pronged surface attack on Leyte Gulf, and a riveting, atmospheric account of the night destroyer action against the southern prong in the Surigao Strait by Admiral James L Holloway III, then gunnery officer on the destroyer Bennion and subsequently a Chief of Naval Operations.

There is a lot of interesting material here, and yet somehow the book still feels 'bitty' and incomplete – there is a reference on page 52 to a non-existent map. Operational issues such as Halsey's decision to take his battleships north with the carrier task force receive a huge amount of coverage while other, more technical aspects of the battle tend to be neglected. Why did *Yamato*'s captain opt to use AP shell against carriers, even if he did mistake Sprague's CVEs for fleet units? Would the same issue of unsuitable ammunition have arisen in Leyte Gulf against the transports? Did Oldendorff, whose elderly battleships had been equipped primarily to support the landings and had expended most of their AP shell against *Fuso* and *Yamashiro* at Surigao Strait, suffer from the reverse problem? These issues are touched on in the book, but are not explored in any great depth. Cutler's compendium is a valuable contribution to the literature on Leyte Gulf, but it is unlikely to be the last word on the subject.

John Jordan

John Roberts
Battlecruiser *Repulse*

Seaforth Publishing, Barnsley 2019; hardback, illustrated with full-colour plans throughout, 160 pages; price £30.00.
ISBN 978-1-5267-5728-9

Author John Roberts holds a pivotal place in chronicling the story of British capital ships, having written a number of excellent books including *British Battleships of World War Two*, co-authored with Alan Raven. In the early 1980s this book started the ball rolling in the analysis of British warships, setting a high standard in the process which almost 40 years later shows little sign of faltering.

Battlecruiser Repulse is one of the new series from Seaforth in which a study of a ship is based on the plans held by the National Maritime Museum. These plans are stunning on many levels, not least of which is their great visual appeal in addition to the detailed information they contain. The ability to depict something approaching 800 feet in length in ink on linen and at a human scale, the standard of draughtsmanship, and even the corrections and additions appended in pen or pencil, all add up to make them a delight to behold. These precious documents also show the patina of age through discoloration and fraying edges, which adds to their fascination.

In the introductory section to the plans, Roberts writes about the origin of the battlecruiser type and how it had fallen out of favour in the period immediately before the start of the First World War. Events conspired to resurrect the concept, resulting in the elegant *Repulse*, *Renown* and the last of the type, *Hood*. This is an outstanding essay in its own right, contributing insight into this enduringly interesting class of warship.

The author then presents a wealth of informed detail on the ship before concentrating on the plans. This follows the established pattern of the series, in which the plans have been split into sections to allow enlargement and render the annotations legible. The general arrangement drawings, profiles, decks, bridges, rig etc, have been reproduced together with a gate-fold section devoted to three of the more significant drawings: the profile as built in 1916, the rig following the major 1936 refit and, across the four pages of the centre section, the 1936 profile. The plans have been overlaid with extended captions and it is here that the author's detailed knowledge is most apparent.

The result is a comprehensive and definitive description of a handsome and important British warship by an expert on the subject. It must also be said that producing a book like this is no mean feat given the complexity involved in organising the material, and the final result is a tribute to all involved. There was only one error that I came across: the battlecruiser *Queen Mary* was built by Palmers and not John Brown, although the ship's machinery was constructed by the latter.

The book is fully up to the high production standards that Seaforth has set. For those interested in the battlecruiser as a concept and the graphical documentation of these complex machines, this title is a must.

Ian Johnston

James J Bloom
Rome Rules the Waves: A Naval Staff Appreciation of Ancient Rome's Maritime Supremacy 300 BC – 500 CE

Pen & Sword Military, Barnsley 2019; hardback, 316 pages, two B/W illustrations and seven maps; price £25.00.
ISBN 978-1-7815-9024-9

Despite his title, James Bloom admits in his foreword that Rome's eventual command of the Mediterranean hardly resembled the era when Britannia ruled the waves. Nor does he seem to be aware that only the notoriously partisan Admiralty monograph on the Battle of Jutland was entitled a 'Staff Appreciation'. He informs his readers that the present work began as a straightforward operational history of the Roman navy from the early Roman republic to the late Western empire; but when he found that his work had been anticipated by Michael Pitassi, he changed his approach to apply modern strategic theory, especially that propounded by Alfred Thayer Mahan and Sir Julian Corbett, to the classical period. Ancient sources say almost nothing about Roman strategic intentions while, of the two modern authors, only Mahan wrote briefly on just one of Rome's Carthaginian wars. Thus Bloom declares his intention to extrapolate what the ancients *might* have intended strategically and what the moderns *might* have written about naval warfare in Roman times.

However, after the introductory chapters the rest of the book seems to be much as originally intended. It shows that it was only during the First Punic War with Carthage that Rome decided she must have her own navy: and that, by the time of the Second Punic War, she exercised almost complete command of the Western Mediterranean. Yet subsequently her naval forces dwindled and, during the conquest of her Eastern empire, she relied largely on the navies of her allies, notably Rhodes. Rome herself did not again build large fleets until Pompey the Great's suppression of piracy and Octavian overcame the fleets of Sextus Pompeius and of Antony and Cleopatra. Only then did Octavian as the Emperor Augustus create a permanent naval establishment, with two main fleets based in Italy and smaller fleets near the boundaries, notably on the Rhine and Danube rivers. Most of the actual fighting was by both land troops and riverine fleets in these frontier provinces.

Thus it seems that Rome had no settled naval strategy in the Republican era. During the Empire's first two centuries, she commanded the Mediterranean because she had no significant naval opponents, while her actual campaigns were combined operations, many along rivers. This was all very different from the oceanic strategies anatomised by Mahan and Corbett.

Bloom's operational history is often difficult to follow, particularly where he skitters about in time and space. His text is repetitive, and the maps (which retain many of their original Italian captions) do little to assist the reader; better editing might have helped. Bloom provides little information on ancient warships and battle tactics, and is inconsistent even on the extent to which the Roman *corvus* demanded skilful shiphandling.

Contrary to Bloom's hopes that his book will appeal to a general readership, it will probably be of interest mainly to students of naval strategic theory. *Warship* readers who are interested in ancient naval warfare may well prefer to consult other works.

John Brooks

Zvonimir Freivogel
Warships of the Royal Yugoslav Navy 1918–1945: Yugoslav Warships and their Fates, Volume I

Despot Infinitus, Zagreb 2020; 363 pages, 382 photographs; price €49.90.
ISBN 978-953-8218-72-9

Over the past few years, the author has produced a significant number of books and articles on Adriatic naval topics, and this new book is the first of two volumes that will catalogue the ships of what began life as the Navy of the Kingdom of the Serbs, Croats and Slovenes in the wake of the First World War, formally becoming the Royal Yugoslav Navy (RYN) in 1929. This fills an important gap in the literature, as information on this force has been difficult to find outside contemporary yearbooks and compilations such as *Conway's All the World's Fighting Ships*.

The Introduction gives a brief overview of the history

of the RYN up to April 1941. It would have benefited from a little more general background information, as it alludes to matters that are likely to be obscure to anyone without some existing knowledge of Balkan history; this also applies to certain other parts of the book. The fighting ships are then described, beginning with the ex-German cruiser *Dalmacija*, purchased in 1925, and continuing with destroyers, torpedo boats, submarines, mine-warfare vessels and the seaplane tender *Zmaj*. A curious omission is any entry for the battleship *Kumbor* (ex-*Erzherzog Kronprinz Rudolf*), which is only mentioned and depicted in the Introduction. Although she was scrapped only a year after being handed over, one would have expected an entry on a par with the other vessels, if only for reasons of completeness.

A section then covers auxiliaries, ranging from submarine depot ships and the sail training ship *Jadran* to 'other' vessels, including harbour hulks inherited from the Austro-Hungarian Navy and customs launches. The following section deals with vessels serving on the Danube, most importantly four ex-Austro-Hungarian river monitors. It is followed by an overview of Yugoslav maritime aviation, then a section on the RYN between 1941 and 1945, when a corvette and some MGBs were added to the handful of vessels that had escaped from the Adriatic. The text is supported by an extensive selection of photographs, supplemented by drawings, giving an excellent coverage of the history of the RYN. A final section entitled 'Conclusions' is followed by 'Atachments' [*sic*] and a bibliography. The former comprise a set of tables providing data on the RYN's guns, torpedoes, seaplanes and ranks, and a list of Italian place names and their Croatian equivalents.

Unfortunately, as has been the case with other books by the same author and publisher – and highlighted by numerous reviewers – the text is in desperate need of editing by a native English speaker. Generally, this does not result in incomprehensibility, but there are points which come close, and enjoyment is constantly undermined by an inappropriate turn of phrase, exacerbated by frequent misprints. Similarly, the book also lacks an index, something that is wholly unacceptable in a work of this kind. Both issues materially detract from what is otherwise an admirable volume.

Aidan Dodson

Emily Malcom & Michael R Harrison
Glasgow Museums – The Ship Models: A History & Complete Illustrated Catalogue

Seaforth Publishing, Barnsley 2020; hardback, 384 pages, illustrated with 700 colour photographs; price £35.00.
ISBN 978-1-5267-5752-4

The first thing that should be said about this book is that it is very large and heavy, representing as it does the shipbuilding output of the River Clyde, which produced more than 25,000 ships from the mid-1800s onwards –

and indeed still does, albeit at a greatly reduced pace.

The first section, which accounts for about one third of the page extent, provides historical context, examining why and how the models were made, and for whom. The origins of the present collection are described against the backdrop of a rapidly expanding City of Glasgow that would become a major centre for shipbuilding, marine engineering, locomotive manufacture and structural steelwork of all kinds. The remainder of the book is devoted to the ship models themselves, which besides the Clyde-built vessels also include ships built elsewhere, the earliest of which pre-date shipbuilding on the Clyde.

The treatment of individual models is excellent, with each photographed and presented with basic dimensional data and a brief biography. In addition to models of full-hulled vessels there are various half-hull and some plating models. The latter are large-scale half-hulls on which the strakes of plating and individual plates, overlaps and butt straps have been marked. Plating models exist for *Aquitania*, *Queen Mary* and *Queen Elizabeth* and are among the largest in the collection.

Clyde river vessels are well represented, from harbour craft to the many steamers that once plied the river and waters of the Firth of Clyde, taking the citizens of Glasgow to Clyde coastal holiday destinations – a class of vessel represented today by the sole survivor, the paddle steamer *Waverley*. Dredgers, hoppers, and the river ferries and vehicular ferries that crossed the Clyde at Glasgow are included, representing the varied output of yards which, at the height of the industry, numbered more than thirty on the Clyde itself and its tributaries, the Rivers Cart and Leven.

Merchant vessels of all types are represented, from the crack record-breaking Atlantic steamers of the late-1800s, through the classic passenger cargo vessels in which many Clyde yards specialised, to the great iconic liners of last century, most notably the Cunarders *Queen Mary*, *Queen Elizabeth* and *QE2*, each of the three models being more than 5 metres long. Warships are equally well represented, from early torpedo boat destroyers through to the recent Type 23 frigates. Battleships and battlecruisers include *Indomitable*, *Colossus*, *Hood* and *Howe*, the latter two models more than 4 metres long. While most of the models are professionally built, there are some superb amateur models, including the cruiser *Dido*.

For those interested in ships and ship models this hugely impressive book offers the prospect of many hours spent happily dipping into what must be one of the finest collections of ship models in the UK. The book also brings the entire collection together, as in reality it is split between the Riverside Museum itself on the banks of the Clyde and the museum's resource centre located elsewhere in the city. The authors and Seaforth are to be congratulated for such a carefully thought-out and well-produced book. It is lavishly illustrated and well designed, with the high production values we generally associate with this publisher.

Ian Johnston

Admiral S N Timirev
The Russian Baltic Fleet: In time of War and Revolution, 1914–1918
Seaforth Publishing, Barnsley 2020; hardback, 238 pages, three B&W photographs, one map; price £25.00.
ISBN 978-1-5267-7702-7

For anyone who has read any of the recent books about the Royal Navy's role in assisting the Baltic States to achieve independence in the immediate aftermath of the First World War, this contemporary account will be particularly interesting. The book is based on an original memoir entitled 'Notes of a Naval Officer' written by Rear Admiral Sergei Timirev, a former officer in the Tsarist Navy, and covers the time he spent serving in the Baltic during 1914–18 when his country fought to resist the eastward advances of the Germans while at the same time experiencing the internal revolution that eventually brought the Soviet Union into being. It was written when Timirev was living and working in exile in China, probably in 1922. Some time later the manuscript was apparently smuggled to the USA, where it was translated into English and published in 1961 by a group of Russian *émigrés*.

In the confined waters of the Baltic Sea the threats of groundings, of undersea obstacles and of enemy mining were ever present; essentially, this was littoral warfare. In several places, Timirev refers to the Baltic Fleet protecting the 'right flank' of the Russian Army. Indeed, terms such as 'Chief of the Brigade' to describe Timirev's appointment to command what would be more familiarly be called a 'squadron' (of cruisers, in this instance), is indicative of the subservient role the Navy was expected to play in support of the Army.

More than half of the account deals with the period leading up to the abdication of Tsar Nicholas II in March 1917. This section is dotted with little pen-portraits of the higher echelons of the Russian naval fraternity, of which Timirev was a part; they illustrate the sort of close brotherhood and petty rivalries among the naval officer class that are probably present in most major navies in peacetime. The shock comes when Timirev witnesses the revolutionary turmoil in Petrograd while on leave. Thereafter, events follow in quick succession. Meanwhile Timirev plays a decisive and heroic role in the Battle of Moön Sound, the ultimately hopeless defence of Riga. The Russian Baltic Fleet becomes increasingly paralysed as the revolutionaries take over, and command and action planning are taken away from the Tsarist officers and placed in the hands of those who have pledged allegiance to the Bolshevik cause.

In the course of these tumultuous upheavals, Timirev reveals his true personality as a 'dyed in the wool' upper-class royalist. Revolutionary pronouncements by other ranks are dismissed as 'rabble-rousing balderdash', while native Estonians attempting to seize control of their own country are dubbed 'anarchists'. His brother officers become either conniving traitors or admirable loyalists.

This is a fascinating account of a little-known period in the history of this historic navy. While the excellent commentary by the translator puts Timirev's background and associates into context, one cannot help but feel some sympathy for the author, living in exile in Shanghai, deprived of his wife, his son and his pension, and trying to write a factual memoir of a period when he lost just about everything he cherished.

Jon Wise

Vincent P O'Hara
Six Victories: North Africa, Malta and the Mediterranean Convoy War, November 1941–March 1942
US Naval Institute Press, Annapolis 2019; hardback, 336 pages, many tables, plans & photographs; price $34.95/£33.95.
ISBN 978-1-68247-460-0

The author has selected the period November 1941 to March 1942 as the focus of his study of the naval war in the Mediterranean. This was a period marked by early British successes, followed by a turnaround in favour of the Axis as the Italians – with some initial reluctance – decided to utilise their superiority in big ships to boost the security of their convoy traffic between Italy and North Africa, and the Royal Navy came under increasing pressure with the sinking of the battleship *Barham* by *U-331* in November, the disabling of the battleships *Queen Elizabeth* and *Valiant* by Italian special forces at Alexandria during late December, the entry of Japan into the war, and the commitment of large numbers of German bomber aircraft to the central Mediterranean. It is perhaps a stretch to describe some of these actions as 'victories' – historians have long argued about who prevailed in the first and second naval battles of Sirte, the second of which was fought in appalling weather conditions – but there can be little doubt that the Axis forces achieved their overall objectives, or that the British suffered crippling losses of warships and mercantile vessels in the three months to March, or that the fate of Malta during this period hung by a thread.

One of the great strengths of Vincent P O'Hara's book is the tables, which are the result of extensive research. Each account of a naval action is preceded by a listing of force strengths: for convoys there is a detailed breakdown of the composition of each element, including the size/capacity of the merchant vessels and the organisation of the escorting and covering warships, and these are complemented by summative tables which show the percentage of cargo delivered. Appendix 1 consists of tables showing significant damage to, and losses of British and Axis ships (mercantile and naval), while Appendix 2 lists the convoys to and from Alexandria during the period. The accounts of the actions are illustrated by the author's own specially-drawn maps. Unfortunately, the small page format means that these are not always easy to read, particularly under artificial light, due to the comparatively dark grey employed for the sea – a measure adopted to enable the tracks of British forces to be rendered in white and those of Axis forces in black.

The role of intelligence, and particularly 'Ultra', is central to the narrative. As the author makes clear, it was less decisive than might have been anticipated: information regarding enemy sailings was not always timely and was often incomplete, and forces often failed to appear when and where expected due to delayed sailings and temporary reversals of course. The British had the advantage of search aircraft fitted with radar, but the poor weather conditions prevailing over the winter of 1941–42 meant that they frequently failed to locate Italian surface forces. British superiority in night fighting, due in part to training but also to the increasingly extensive fitting of radar, led the Italians to operate their major surface units with extreme caution, but from late December 1941 the Royal Navy was hamstrung by Cunningham's inability to support his light cruisers and destroyers with battleships in the Eastern Mediterranean. The elimination of the Malta-based Force K as an effective unit able to operate against Italian sea lines of communication in the central Mediterranean after it ran into a minefield off Tripoli further contributed to the reversal of fortunes.

The author, in his conclusion, highlights the extent to which the Italian Navy learned from its earlier experiences and was able to secure a level of sea control that enabled large quantities of fuel, weaponry and supplies to be safely transported to support the Axis land forces in North Africa. However, he tends to underplay the significance of the German contribution. Although the arrival of convoy MW.10 in Malta was delayed by the aggressive actions of the Italian battleship *Littorio* against an escort of light cruisers and destroyers, the cargo ships were sunk in harbour by the German aerial assault which began the following day. The handful of German U-boats deployed to the Mediterranean were far more successful than their Italian counterparts, which were available in much greater numbers and could have been more effectively handled. According to the author's own table A1.1, during the months November 1940 to March 1941 the Italians sank five Allied merchant ships of 12,547GRT, the Germans 18 ships of 104,239GRT.

John Jordan

M Ernest Marshall
Rear Admiral Herbert V Wiley: A Career in Airships and Battleships
US Naval Institute Press, Annapolis 2019; 400 pages, hardback, illustrated with 27 B&W photographs; price $38.95/£42.95.
ISBN 978-1-68247-317-7

Rear Admiral Herbert Wiley's name is synonymous with the history of lighter-than-air aviation – a lesser-known part of interwar naval history. In this painstakingly researched biography, the author has crafted an excellent book that details the story of Wiley's life as a well-respected advocate of airships for maritime patrol, intertwining his personal and professional life with a detailed account of the history, personalities and significant episodes of the American rigid airship program.

Born in a small Midwest town at the end of the 19th century, Wiley joined the US Navy and, ever keen to invest in his career at a time when rapid progress was being made in manned flight and the future balance between lighter- and heavier-than-air flight was unknown, he volunteered for lighter-than-air service, holding a deep conviction for it from the start. At the time airships had a much greater range than aeroplanes and Wiley witnessed the making of history in manned flight, helping to forge some of it himself. The United States Navy commissioned four rigid airships: USS *Shenandoah*, *Los Angeles*, *Akron* and *Macon*. Wiley served on all four, and the history of each is covered through the lens of his career.

Wiley built a reputation as a consummate airship officer in a field that was not without risk: airships were enormous and fragile, sensitive to the weather, and had to be handled carefully it they were not to be damaged. Three of the airships met with disaster, Wiley surviving the crashes of two of them; some of his friends did not. In the process he earned the recognition and respect of the major personalities in aviation. He had expected to see rapid advances both in aeroplanes and airships; he had not expected to see the end of the airship era. However, after the loss of USS *Macon* the US Navy moved away from rigid airships.

As war clouds loomed, Wiley pragmatically changed tack, returning to the surface fleet, and at the outbreak of the Second World War he commanded a squadron of destroyers in the Pacific; he was subsequently commanding officer of the battleship USS *West Virginia* at the Battle of Surigao Strait, the last battleship-on-battleship action in history. Along the way he received many accolades for his leadership and achievements. He suffered the personal tragedy of losing his wife, but remained a doting father who stoically balanced the needs of his family with service to his country.

The author is to be congratulated for a meticulously researched biography that is an invaluable reference for the US Navy's rigid airship program and, at the same time, provides an absorbing insight into the career and family life of a United States Naval officer at the cutting edge of technical development between the wars.

Phil Russell

Angus Konstam
British Escort Carriers 1941–45
Osprey Publishing, Oxford 2019; softback, 48 pages, illustrated with 40 photographs and 8 pages of full colour illustrations; price £11.99.
ISBN 978-1-4728-3625-0

For those who are familiar with Osprey's publications, this book for the most part matches expectations. It follows a familiar layout: text interspersed with photographs and colour illustrations specially prepared for the particular subject. It is here that the reviewer experienced a slight sense of disappointment. The graphic artist, Paul Wright, is well known for his extensive portfolio of ship paintings; however the aircraft profiles are not quite up to the same standard.

The book focuses on the development of what became known as the escort carrier from its inception to the ultimate *Ruler* class. The initial idea was simply to add a flight deck to a merchant ship, usually a tanker or grain ship, although it evolved in the US Navy to a purpose-built vessel, many of which were supplied under Lend/Lease to the Royal Navy. Each major class is described in some detail and illustrated with a full-colour profile and, in some cases, a deck view. There are selected operational notes on individual ships, together with two full-page reproductions of paintings and a colour cutaway of HMS *Attacker*. The six main chapters are complemented by a biography and an index.

This is a short but useful book for the personal library and a good source for modellers.

W B Davies

Steve R Dunn
Battle In The Baltic: The Royal Navy and the Fight to Save Estonia & Latvia 1918–1920
Seaforth Publishing, Barnsley 2020; hardback, 304 pages, illustrated with maps and B&W photographs; price £25.00.
ISBN 978-1-5267-4273-5

The presence of British naval forces in the Baltic months after the end of the First World War is, on the face of it, surprising. Steve Dunn explains that with Russian Communists bent on recovering the territory lost at Brest-Litovsk, Nationalists in the Baltic States pressing for the independence promised, a German-backed vagabond army bent on looting and destruction, and White Russian forces attempting to push back against the Reds, the British government, lacking the will to commit land forces after four years of war, decided instead on warships. Two squadrons were sent, the first only a fortnight after the signing of the Armistice. On 26 November 1918 Rear Admiral Alexander-Sinclair sailed with five cruisers, nine destroyers and a set of vague orders including the instruction that if he encountered Bolshevik warships off Estonia and Latvia, he was to presume they were hostile and they should be 'treated accordingly'.

Things did not begin well. One of his cruisers, HMS *Cassandra*, hit a mine and sank. Despite such setbacks, his remaining ships were active, bombarding Soviet troops outside Riga and capturing two destroyers. Dunn leaves us in little doubt as to the difficulties the squadron faced with extremes of weather, badly charted minefields and a complex political situation. The squadron withdrew in January 1919, and in March a new squadron under Rear Admiral Sir Walter Cowan arrived to face the same problems. Cowan's orders were to prevent the overrunning of Estonia and Latvia, and to resist the Russian fleet at sea. As Dunn points out, '… For a man of Cowan's aggressive character, such orders were virtually a licence to wage war …'. He took the offensive, and the story then details the many bombardments, the exchanges of gunfire with Russian battleships and

destroyers, the successful torpedo attack against the cruiser *Oleg* using coastal motor boats (CMBs), and the combined air and CMB attack against Kronstadt harbour that sank two battleships and a depot ship. There was a cost: by the time Cowan's squadron was withdrawn in December 1919, twenty British officers and 108 men had lost their lives in the Baltic; a light cruiser, two destroyers, two sloops, a submarine and smaller craft had been sunk.

The intervention was not popular in Britain and naval morale was suffering. Many of the seamen, veterans of the recent war, had anticipated demobilisation but were being asked to risk their lives in an undeclared war. There were small acts of mutiny on ships in the Baltic and at Port Edgar (Firth of Forth), where men of the 1st Destroyer Flotilla refused to board their ships to return to the Baltic. The government finally decided that the squadron should withdraw and not return, although in May 1920 HMS *Tiger* and the newly commissioned *Hood* visited the Baltic, 'showing the flag' and offering a potent reminder to the Russians of Britain's naval strength. In the following year Britain and the Russian Soviet Republic signed a trade agreement; the undeclared war was over.

This is an accessible book, and the reader gains a better understanding of the political turmoil in the Baltic and the indecision at home. Some of the author's conclusions regarding the success or otherwise of British intervention are debatable – he compares that in the Baltic to more recent interventions in Afghanistan and Iraq. However, the book raises some interesting questions and is recommended.

Andrew Field

Gerald Toghill
Dreadnoughts: An illustrated History
Amberley Publishing, Stroud (UK) 2019; paperback, 126 pages, illustrated; price £15.99.
ISBN 978-1-4456-8635-6

Gerald Toghill is ex-RN, having served for 25 years. Described as having 'a passion for naval history', he has already produced two volumes on trawlers in Royal Navy service and has now turned his hand to the ships at the other end of the scale: dreadnoughts. The result is this short volume, which could have been an enjoyable introduction to the subject but fails to do the topic justice.

The author adopts a 'broad brush' approach: the text lacks detail, as well as references and supporting notes. Information often seems broadly scattered, and because the author has chosen to adopt a 'country by country' approach the chronology is all over the place. The photographs are undated and are often wrongly captioned. The photo of HMAS *Australia* on page 40 is of the 'County' class cruiser, not the battle cruiser, and that of USS *Idaho* on page 75 shows the elderly battleship sold to Greece in 1914, not 'the last of the *New Mexico* class to be commissioned in 1919'. The name of the renowned Italian naval constructor is misspelled 'Caniberti', and HMS *Lion* and HMS *Tiger* are typed 'Beautiful [as opposed to 'Splendid'] Cats'.

Hopefully people with only a rudimentary knowledge of battleships will buy it and be motivated to find out more, perhaps from one of the six books listed in the bibliography. However, other than as a gift for a young relative, the *Warship* reader should look elsewhere.

Andrew Field

Anthony Sullivan
Britain's War Against the Slave Trade: The Operations of the Royal Navy's West African Squadron 1807–1867

Frontline Books, Barnsley 2020; 372 pages, 4 maps,
5 B&W drawings; price £25.00.
ISBN 978-1-52671-793-1

On the morning of 6 July 1841 the 12-gun brig *Acorn*, commanded by Lieutenant John Adams, was sailing 400 miles south-west of Freetown when she spotted a strange sail and gave chase. After a twelve-hour pursuit, *Acorn* closed the range and opened fire on the slave-ship *Gabriel*, which had cut away her anchors and boats to evade capture. *Gabriel* returned fire and a running battle was maintained for two hours until *Acorn* came alongside *Gabriel*, firing canister shot to clear the slaver's deck. Having lost most of her sails, *Gabriel* finally struck her colours and hove to. An examination revealed *Gabriel* to be equipped for 800 slaves, and she was dispatched to St Helena with a prize crew.

With encounters like this the Royal Navy fought almost single-handed to suppress the slave trade. Most slavers sought to evade capture, and the reception was often hostile. If that was not enough, sickness also took its toll and did not discriminate, with dysentery and fever affecting naval personnel, slavers and slaves alike. The plight of the enslaved Africans continued even after a slaver was captured, with sickness ever-present as prize crews contended with long journeys in unfamiliar ships, with pirate attack also a possibility. Only on arrival at Freetown or St Helena, for the case to be heard at mixed commission courts, could freedom be assured.

The trade in human lives has been an uncomfortable but enduring trait of history. At the time, Britain's trade in slaves was not unusual; what was unusual was the unilateral decision, in 1807, to ban the slave trade throughout the British Empire. Thus the Royal Navy went to war against the slavers, to end a centuries-old trade which, by the early 19th century, saw slaves being exported from every creek and estuary along 3,300 miles of west African coast. The scale of the undertaking grew until one fifth of the Royal Navy was employed against the trade. Between 1808 and 1860, the West Africa Squadron captured 1,600 slave ships and freed 150,000 Africans. However, the cost to the Navy was high: between 1839 and 1844 alone, 385 officers and men were killed in action or through disease and a further 495 were invalided out of the service.

Anthony Sullivan has produced a deep and comprehensive operational history of this little-known campaign. It is not a light read, but the author works hard to ensure the full story is told within the unfurling political

context, and in the process has produced an interesting, informative, and comprehensive account of the British Empire's war on the slave trade in the 19th Century.

Sullivan contends that, facing opposition at almost every turn, Britain did much to atone for its earlier slave trading operations. This was thanks to the professionalism and dedication of the Royal Navy, which undertook a protracted and costly operation to bring to an end an evil business which had existed for over 350 years. Given the recent upsurge in awareness, this is a book that anyone interested in the subject should read.

Philip Russell

Mark Stille
British Battleship vs Italian Battleship: The Mediterranean 1940–41

Osprey Publishing, Oxford 2020; softback, 80 pages,
data tables, photographs and colour artwork; price £13.99.
ISBN 978-1-4728-3228-9

The latest in a series of 'comparative' books published by Osprey, this little volume looks at the composition of, and interactions between the British and Italian battle fleets in the Mediterranean during 1940–41. The basic premise is undermined by the author's confession that there was only a single occasion when British and Italian battleships engaged one another in combat. Nevertheless, Stille gamely provides a technical, strategic and operational overview of the war in the Mediterranean with a particular focus on the battleships. The respective battle fleets are described in detail, and there are accounts of each of the actions in which battleships were present, illustrated with coloured situation maps.

Stille is more comfortable with the strategic and historical side, less so with less so with technical aspects. The horizontal protection of the *Littorio* class is significantly overstated, *Warspite* as reconstructed in the mid-1930s is credited with only two 8-barrel 2pdr pom-poms, and the author has a poor understanding of the key principles involved in underwater protection systems: the water-filled compartment in the *Nelson*s is described as a 'buoyancy chamber'(!), while he states that *Littorio* had an 'angled belt covering the bulged lower hull'.

The book is well illustrated with data tables, plans and photographs, although the colour artwork of the ships themselves is disappointingly indistinct. There are some excellent photographs, many from the collection of Maurizio Brescia. However, some are reproduced too small to show the details highlighted in the caption, and there are two significant errors: in particular, the photo of *Malaya* purporting to date from 1935 'after modernization' was clearly taken before the modernisation which began in 1934.

This is a useful introduction to war in the Mediterranean between the British and Italian fleets, but readers looking for a greater in-depth analysis of the conflict and more detailed and accurate information about the ships would be wise to look elsewhere.

John Jordan

WARSHIP GALLERY

The Trials of *U-889*

John Jordan presents a series of photographs of one of the last Type IXC U-boats to be completed taken during postwar trials at Halifax, Nova Scotia, in mid-1945.

The German submarine *U-889* was ordered from Deschimag, Bremen, on 2 April 1942, her yard number being W-1097. She was the last boat of the IXC/40 series to be built at that yard before work began on the revolutionary Type XXI 'electro-boat'. She was commissioned on 4 August 1944 under Kapitänleutnant Friedrich Braeucker and attached to the 4th Training Flotilla at Stettin. She then graduated to the 5th Flotilla to be fitted out for action, and was subsequently transferred to the 33rd Front Flotilla. Testing and work-up in the

Baltic followed, and *U-889* ran her final acceptance trials off Stettin on 10 December 1944. Training for active service was conducted at Pillau, Hela, Gdynia and Bornholm Island. She returned to Kiel in early March and her final degaussing (DG) ranging took place on the 22nd.

U-889 left Kiel on 28 March, arriving at Horten, Norway, on the 30th. After a pep talk by Kapitän zur See Hans-Rudolf Rösing, *Führer der U-boote West*, she left for Kristiansand on 2 April, arriving the following day. After taking on fuel and water, together with stores suffi-

U-889 running on the surface at her maximum speed of 17 knots in relatively calm seas and (right) schnorchelling at 4.5 knots. The scratch British/Canadian crew experienced considerable difficulty in maintaining trim when the boat was running close to the surface: first the screws came out of the water, then the bow. For the purpose of the initial trials the bow had to be kept beneath the surface in order to test the directional capabilities of the *Zweibel* array, but the control tower needed to be above water and open in order to maintain communication. On one occasion the crew lost control of the trim and *U-889* bottomed in Bedford Basin, but she managed to resurface – shortly afterwards, a message from Flag Officer Submarines was received stating that under no circumstances were these submarines [*U-889* and the other boat surrendered to the RCN, *U-190*] to be dived!

cient for 27 days (a 20-day patrol was envisaged) she sailed at 2001 on 5 April in company with *U-516* (Type IXC), *U-1226* (Type IXC/40), *U-2511* (Type XXI) and a surface escort.

The U-boats followed a course 30nm offshore parallel to the Norwegian coast as far as 62° North, then headed west, rounding the Faeroes (64°N) on 14 April, and proceeding to a point 100nm south of Iceland 62°N 16°W to strike soundings off Flemish Cap, an area of shallow water centred roughly on 47°N 45°W some 560km east of St John's, Newfoundland and Labrador, which *U-889* reached on 6 May. The passage was made at average speeds of 6.5 knots (605nm/93 hours) on the surface and 5.5 knots (1078nm/196 hours) using the *Schnorchel*, and at a speed of 1 knot dived using the main motors (516nm/515 hours).

Emissions from the boat were picked up on 7 and 8 April, when she was off Norway, and when north of the Shetlands. On 4 and 5 May she was radar-located when off Flemish Cap, but dived without being spotted. On 8 May *U-889* surfaced to refresh the air in the boat and an aircraft was sighted, but she dived without being spotted. Neither the radar detections nor the aircraft sighting on the 8th resulted in an attack, but *U-889* failed to sight any potential targets. She was heading to attack shipping in the approaches to Halifax when the order to surrender was received on 9 May. Having surfaced on the 10th and raised a black flag of surrender in accordance with instructions, she was sighted by a Liberator aircraft of No 10 Squadron RCAF, and was subsequently taken in charge by a Canadian escort group comprising the corvettes *Dunvegan* and *Saskatoon* and the minesweepers *Oshawa* and *Rockcliffe*. The boat was originally to have been directed to Bay Bulls in Newfoundland, but the orders was subsequently changed and *U-889*, now escorted by two Canadian frigates of the 28th Escort Group, was redirected on Shelburne, Nova Scotia, where she was formally surrendered on 13 May at the Whistle Buoy, off the entrance to the harbour.

As one of the last U-boats to be completed before the 'electro-boats' of the Type XXI series entered service, *U-889* acted as a trials ship for the latest German electronic equipment intended for installation on board the latter, and had been fitted with a *Schnorchel* in January 1945; she also had on board six of the latest T-5 GNAT acoustic torpedoes. She was therefore of considerable interest to the Royal Canadian Navy, which was in advance of both the Royal Navy and the US Navy in developing countermeasures to the latter model.

Initial trials of the submarine and her equipment were conducted off Halifax, Nova Scotia, under the aegis of the Naval Research Establishment (NRE), Halifax, between 1 June and 21 July 1945. The boat was under the command of Lieutenant Bruce Collins, RN, and had a mixed crew of RN and RCN personnel, many of whom were reservists with no experience of submarines. The Canadians were particularly interested in the performance of the *Schnorchel* and of the *Zweibel* passive array, which was unique to *U-889* and was intended to detect enemy surface units when the boat was running just below the surface. The results of the trials were recorded in a comprehensive typewritten report PHx 59, which was

A view of the starboard side of the conning tower, showing the tiered arrangement of the twin 3.7cm and 2cm AA mountings. The large-diameter pipe is the air inlet for the *Schnorchel*, which in this view is in the lowered position.

illustrated by German plans of the boat and by photographs of the submarine and her equipment taken during the trials. Many of these photographs are reproduced here with explanatory captions. More trials, this time with the now fully-instrumented GNAT acoustic torpedoes, subsequently took place between 23 September and 30 October, but these were interrupted by unfavourable weather conditions which prevented their completion.

Following the RCN trials *U-889* was allocated to the US Navy, arriving at Portsmouth Navy Yard on 10 January 1946; she was subsequently sunk as a target on 20 November 1947 by a torpedo from the US submarine *Flying Fish* off Cape Cod.

Acknowledgements:

The author wishes to thank Air Commodore Derek Waller, RAF (Rtd), who checked through my draft copy and suggested some small improvements. For a more complete account of the post-war history of *U-889* and *U-190*, see Derek's article for the journal *Argonauta*, Vol XXXV No 4, Autumn 2018, which is accessible online at: http://www.forposterityssake.ca/RCN-DOCS/U-Boats-in-the-RCN-by-Derek-Waller.pdf

Below: The bridge looking aft. In the foreground are the two twin 2cm C/30 light AA guns, and beyond them, at the lower level, can be seen the 3.7cm *Flakzwilling* M43U in the comparatively rare DLM42 twin mount.

Above: The starboard quarter of the hull, showing the three-bladed propellers, the after diving planes and the twin rudders.

Above: The upper deck looking forward. The *Schnorchel* can be seen in the stowed position in the recess to starboard. The broad, flat deck with its wooden planking serves to emphasise that the Type IXC boats, unlike their Type XXI successors, were designed to operate primarily on the surface. Lieutenant Bruce Collins, RN, commanding officer for the trials, expressed the view that this class was inferior to British submarines in diving qualities, and that maintaining trim at periscope depth or when schnorchelling required much concentration on the part of the planesman and the trimming officer due to the large planing area of the upper deck.

Left: The bridge, showing the masts for the detectors of enemy radar and radio emissions. Antennae for the *Borkum* German Search Receiver (GSR) of the *Rund Dipol* type were fitted permanently on the bridge (left of photo) and on the head of the *Schnorchel*. The device mounted on the bridge was for the detection of longer-wave transmissions (20–300cm), that on the schnorchel for transmissions in the 120–180cm range. The *Borkum* was crystal-based and unlike the earlier *Metox* emitted no radiation, so its use could not be detected by enemy aircraft; however, use of a single dipole meant that the device was unable to determine the direction of hostile aircraft, and it was ineffectual against the late-war centimetric Allied radars.

The FuMB-26 *Tunis* (centre picture) combined FuMB-24 *Fliege* (8–23cm) and FuMB-25 *Mücke* (2-4cm) antennae, and was designed to detect the transmissions of the latest Allied S- and X-Band radars; it entered service in mid-1944. The antennae were mounted on a wooden jackstaff, which was shipped into a holder on the fore side of the periscope standard on the bridge and rotated by hand. The device was portable: cables were plugged into the conning tower via an open hatch.

On the right of the photo is the standard MF direction finder to monitor enemy radio transmissions.

Left: The Balkon GHG (*Gruppenhorchgerät*) hydrophone array, which was mounted at the fore end of the ballast keel. It housed 48 Rochelle salt hydrophones (diameter: 75mm) set in a horseshoe array at the forward end of the ballast keel. It was capable of frequency selection, and used a combination of sector and phasing for directional determination. It proved ineffectual when running on diesels due to self-noise.

Right: The *Pillenwerfer* (lit 'pill thrower') was an ejector system for sonar decoys that entered service in U-boats in 1942. It launched a metal canister some 10cm in diameter filled with calcium hydride. When mixed with seawater the calcium hydride produced large quantities of hydrogen which bubbled out of the container, creating a false sonar target. A valve opened and closed, holding the device at a depth of about 30 metres. The bubble of hydrogen lasted for about 20–25 minutes. The Royal Navy referred to the device as a Submarine Bubble Target (SBT).

Right: The upper deck looking aft. The after escape hatch is in the raised position.

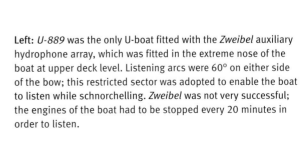

Left: *U-889* was the only U-boat fitted with the *Zweibel* auxiliary hydrophone array, which was fitted in the extreme nose of the boat at upper deck level. Listening arcs were 60° on either side of the bow; this restricted sector was adopted to enable the boat to listen while schnorchelling. *Zweibel* was not very successful; the engines of the boat had to be stopped every 20 minutes in order to listen.

The motor room looking aft. When submerged, *U-889* was powered by two electric motors each with a nominal rating of 500hp. Maximum speed when dived was just over 7 knots.

The forward torpedo room, showing the hinged doors for the four 533mm torpedo tubes with the handling gear above. When surrendered *U-889* had the full complement of twelve torpedoes on board. Six were T3 electric 30-knot torpedoes, of which five were fitted with the LuT and one with the FaT pattern-running device; the remaining six torpedoes were of the T5 *Zaukönig* type (GNAT), five flat-nosed and one round-nosed. The latter were of great interest to the RCN, which conducted trials with them in 1945 and 1946.

Above: The diesel room looking aft, with the two rev counters in the foreground. When running on the surface *U-889* was powered by two 9-cylinder MAN 4-cycle diesels, fitted with Buchi superchargers driven by exhaust gas turbines. Each of the diesels was rated at 2,200bhp, and shaft revolutions at normal speed were 460rpm which, according to the engineer's notebook found on board, could be increased to 480rpm for half an hour for a maximum 2,500bhp.

Left: Close-up of the head of the *Schnorchel*, showing the anti-radar covering and the *Rund Dipol* GSR, which was allied to a *Wanze* receiver. The head of the *Schnorchel* had a 3mm rubber coating with a raised square honeycomb pattern that contained fine iron particles. Early versions of the *Schnorchel* were found to reduce the U-boat's radar signature by 70% at minimum exposure, but by only 30–40% when 3–4 metres above the surface. When fitted with a rubber coating the radar signature was reduced to only 10% against surface ships, but the gain was less pronounced against aircraft flying above 100ft.

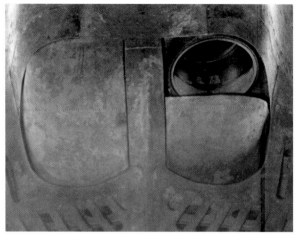

Above: The external shutters for the after pair of torpedo tubes.

Left: Side view of the conning tower with the *Schnorchel* in the raised position.

The *Schnorchel* in the lowered position; the recess in which it was stowed was in the starboard side of the outer hull casing just forward of the conning tower. The *Schnorchel* was raised by a ram on its heel, driven by the telemotor system, the pumps and controls being in the control room.